W0246034

STRUCTURAL AND FUNCTIONAL ASPECTS OF TRANSPORT IN ROOTS

Developments in Plant and Soil Sciences

VOLUME 36

Structural and Functional Aspects of Transport in Roots

Third International Symposium on
'Structure and Function of Roots'
Nitra, Czechoslovakia, 3–7 August 1987

Edited by
B.C. LOUGHMAN
Department of Plant Sciences
University of Oxford, UK

and

O. GAŠPARÍKOVÁ and J. KOLEK
Institute of Experimental Biology and Ecology
Slovak Academy of Sciences
Bratislava, Czechoslovakia

Kluwer Academic Publishers
DORDRECHT / BOSTON / LONDON

Library of Congress Cataloging in Publication Data

International Symposium on Structure and Function of
    Roots (3rd : 1987 : Nitra, Czechoslovakia)
    Structural and functional aspects of transport in
roots.

    (Developments in plant and soil sciences ;
v.   )
    1. Roots (Botany)--Anatomy--Congresses.
2. Roots (Botany)--Physiology--Congresses.
3. Plant translocation--Congresses. I. Loughman,
B. C. II. Gašparíková, O. III. Kolek, Jozef.
IV. Title. V. Series.
QK644.I58 1987   581.1'0428   88-27294

ISBN-13: 978-94-010-6889-5   e-ISBN-13: 978-94-009-0891-8
DOI: 10.1007/978-94-009-0891-8

Published by Kluwer Academic Publishers,
P.O. Box 17, 3300 AA Dordrecht, The Netherlands.

Kluwer Academic Publishers incorporates
the publishing programmes of Martinus Nijhoff,
Dr W. Junk, D. Reidel, and MTP Press.

Sold and distributed in the U.S.A. and Canada
by Kluwer Academic Publishers,
101 Philip Drive, Norwell, MA 02061, U.S.A.

In all other countries, sold and distributed
by Kluwer Academic Publishers Group,
P.O. Box 322, 3300 AH Dordrecht, The Netherlands.

All rights reserved
© 1989 by Kluwer Academic Publishers

No part of the material protected by this copyright notice may be
reproduced or utilized in any form or by any means, electronic or
mechanical, including photocopying, recording or by any
information storage and retrieval system, without written
permission from the copyright owners.

Contents

*Chapters indicated with an asterisk were first published in *Plant and Soil*, Volume 111 (1988).

Session 4: Root-shoot relationships with respect to transport processes

Session 5: Effects of stress on function of roots

Preface

The Third Root Symposium organized by the Institute of Experimental Biology and Ecology of the Slovak Academy of Sciences took place at Nitra in Slovakia in early August 1987. The theme of the earlier Symposia at Tatranska Lomnica and Bratislava was continued in the hope that recent discoveries at the anatomical level might be correlated with new approaches to physiological and biochemical aspects of roots. The organizers were very pleased with the response to the announcement of the meeting and benefitted from the presence of leading plant scientists who came to Nitra after the Botanical Congress in Berlin. Those participants particularly appreciated the contrast between the small specialist meeting where everyone was accessible to everyone else and the multi-faceted and somewhat bewildering Congress in Berlin.

One welcome feature of the Symposium was the manner in which discussion developed after lecture and poster sessions in which those working in the field of structural aspects were able to contribute to detailed arguments on mechanisms of ion transport and water flow while the biochemists were intrigued by the new ideas on the location of barriers to movement across the root. As always, the local organizers had provided excellent facilities with generous entertainment during the evenings, culminating in a weekend spent walking in the High Tatra for a small group of the more energetic members. The excellence of all the arrangements made on our behalf was much appreciated by everyone and it is to be hoped that an equally successful Fourth Symposium will be planned after a suitable interval. The Editors receive no payment for their services from Kluwer Academic Publishers but a generous arrangement has been made with the publishers whereby all participants receive a free copy of the published volume, a particularly important point for those from countries with currency restrictions. The editors have been assisted in their task by the very willing cooperation of H. Lambers, M. G. T. Shone, P. J. C. Kuiper and P. W. Barkow and we are very grateful for their help.

Structure and function of the root: Some reflections and current problems

O. GAŠPARÍKOVÁ

Institute of Experimental Biology and Ecology, CBES, Slovak Academy of Sciences, CS-814 34 Bratislava, Czechoslovakia

In his now classic studies devoted to root development in field crops Weaver in 1926 regretted the marked lack of knowledge on roots as one of the most neglected plants organs. In spite of the considerable efforts of experimental botanists to gain a clearer understanding of root function in the life of plants we still have not yet attained an ideal state of knowledge. Since our last meeting in 1980 other scientific symposia on the root and its functions were held in Austria (1982), in Anaheim (1982), in Utrecht (1983) and in Blacksburg (1985).

In recent years great attention has been paid to the structural and functional characteristics of particular regions of the root. The meristematic root cells form a cytologically and physiologically heterogenous complex and their structural characteristics reflect the principal features of their metabolism. They are marked by high rates of DNA and RNA synthesis, but have a relatively low respiration rate, low levels of substrates, lower rates of protein synthesis and a low activity of the majority of enzymes. This is the result of the weakly developed membrane system of juvenile mitochondria, the presence of many free ribosomes, only rare occurence of polysomes and endoplasmic reticulum characterised by short cisternae. Meristematic cells absorb a relatively low amount of mineral nutrients usually utilized to provide for their own synthetic processes concerned with protoplast formation. This is confirmed also by our results of $^{35}SO_4$ absorption which after short exposure is maximally accumulated in the growing part of the root and incorporated into proteins.

The metabolic activity of cells increases with cell extension and transition to cell elongation. The level of low and particularly of high molecular cell components such as DNA, RNA, proteins and polysaccharides increases. The profile of enzyme systems of basal metabolism is completed due to the increasing root autotrophy. This state is struc-turally manifested by changes in the size of the nucleus and nucleolus and in the chromatin structure; the rough ER attains maximal development. Ribosomes aggregate into polysomes and associate with ER as a structural assembly for enzyme protein synthesis. The well developed structure of the Golgi apparatus indicates active synthesis and transport of the cell wall polysaccharides. The quantitative proportions of single chemical components of the cell wall and their space distribution changes and the development of the apoplast occurs. The dimensions of intercellular and inter-fibrilar spaces available for diffusion and increased fixed electric charges of ionic groups of the matrix are obviously important. With the increasing vacuolation and the development of a large central vacuole, physical conditions develop for the storage and reutilization of ions as well as of organic substances to provide the turgor pressure of elongated cells. Selective ion uptake becomes active by cell membrane differentiation conditioning the active transport by the synthesis of the relevant ATP-ases and carrier molecules. Thus physical conditions arise for the development of electrochemical gradients generated by proton pumping as a driving force of solute transport. With the advancing development of the mitochondrial membrane system and the completion of the mitochondrial matrix the bulk of the whole mitochondrial respiration system increases leading to the activation of electrogenic proton pumps responsible for the maintenance or increase of cell wall extensibility. Thus the elongated cells have already produced conditions for increased ion uptake to maintain the osmotic equilibrium between newly formed elongated cells and their surrounding tissue. Another important aspect of cell growth is the capacity for transport of water into the cell.

The cells from which particular root tissues are

formed already differ morphologically and metabolically in the meristem. This is confirmed by the patterns of DNA and rRNA synthesis determined by the combined methods of cytophotometry and autoradiography in our laboratory together with changes in the chromatin complex structure. After completion of the non-dividing growth stage with its specific pattern for particular cell types another stage follows accomplishing the process of differentiation which is specific for certain types of cells.

The mature root cell represents three principal compartments clearly differing in their ionic composition. These are the cell wall, cytoplasm and vacuole. The techniques presently available for quantitative study of ion compartmentation in root cells do not provide sufficient resolution for separate analysis of the cytosol, ER and mitochondria of the cytoplasm without cell fractionation; hence, the cytoplasm is considered as a single compartment. The use of x-ray microanalysis but especially NMR spectroscopy is likely to contribute substantially to our knowledge of the quantitative ion compartmentation in root cells.

Differentation into particular tissue types allows for the principal root functions — uptake and transport of water and ions thus forming a precisely programmed harmony of absorption, transport and metabolic function.

Currently the movement of water and solutes in the root is the subject of numerous experiments, considerations and discussions. Ion transport in roots is often described as a centripetal flux of solutes in the apoplasm and symplasm of cortical cells to the Casparian strips of the endodermis and from there *via* the symplasm to the xylem. While the participation of epidermis and cortex in this pathway is relatively well known, the participation of the central cylinder tissues is a matter of dispute. The focus of this interest is the regulation of ion trasport to the xylem which is ascribed to the xylem parenchyma cells and the possible operation of a proton — translocating ATPase located in the membrane of these cells.

Our knowledge of mechanisms responsible for the flux of inorganic ions through cell membranes, particularly the nature of the driving forces is more complete. Considerable success has been obtained in understanding the proton translocating ATPases of the lasmalemma and tonoplast. In addition to the ATPase another possible source of driving force of the ion transport has been suggested. It seems that part of the energy linked ion transport in roots may be driven by an electron transport system at the plasmalemma. The kinetic processes of the transport of several anions as well as neutral molecules were elucidated. It is generally agreed that elucidation of the kinetic parameters of active transport do not lead to its full understanding. It is essential to understand its molecular mechanisms which are, however, associated with great methodological problems.

It is believed that the plant cell is able to maintain its ion status at an almost constant level so that there is a negative correlation between the absorption and the ion content in the cell. The adaptation of ion efflux with regard to their availability is then performed by changes in the activity of transport systems, *e.g.* by the regulation of the transport by induction or repression of carrier synthesis, as well as by efflux control. It seems that the steady state ion concentration in the cell is relatively stable due to the kinetic characteristics of carrier systems and their ability to adjust themselves to given conditions. It is also clear that a major role is played by growth.

Water flow in the root is mostly considered to be driven by the forces which can be described by the equations of irreversible thermodynamics. Professor Dainty at the 2nd symposium in Bratislava made clear the problems of such an interpretation when he said . . . "We need to know what the hydraulic conductivity and the reflection coefficients are for the individual barriers, . . . What the solute concentrations and water potential profiles are, where the pumping of solutes occurs, and other information."

The participation of the apoplastic cellular and symplastic pathways in the radial transport of water along with their capacitances and resistances has received considerable attention. Great differences between the hydraulic conductivity of whole roots and of individual membranes may indicate the importance of cell to cell transport of water in the root. When discussing the longitudinal flow of water and solutes in the root the complexity of the vascular system is often neglected. The studies in this field call for very active cooperation between physiologists and anatomists. Doubts about the low resistance of paths for axial flow are growing.

In our department we are engaged in the study of the causes of the differences between measured values of the relative conductivity and calculated values of the conductive capacity. On the basis of experiments with barley roots Luxova (1986) recorded the occurence of a hydraulic safety zone in which characteristics conditioning the hydraulic constriction and consequent hydraulic protection are compensated by a greater development of the vascular system.

The partial root structures and functions referred to earlier form the structure and functions of the whole organ. At preceding symposia several authors have documented the root integrity within the whole plant. Root and shoot functions are a steady-state equilibrium, but the mechanisms controlling this state are as yet unknown.

By experimental interference with these correlations it is possible to verify quantitative growth changes of the whole plant. Partial derooting leads to an enhanced absorption rate of solutes by the remaining roots and an increased relative growth rate of individual root members. However, such experimental interferences produce a certain risk in attempting to extrapolate from results obtained in normal conditions. In natural conditions in maize plants gradual formation of the primary seminal root, the seminal adventitious roots, mesocotyl roots and nodal roots growing out in crown roots may be observed.

According to our observations on maize and sorghum plants unable to form seminal adventitous roots because they only have the primary seminal root, the efficiency of the latter began to decline before the first crown roots appeared. It was manifested in the shoot by the decrease in the net photosynthetic rate obviously due to the transient lack of water in the leaves. The decrease of photosynthesis was not observed when the seminal adventitious roots emerged between the emergence of the primary seminal and crown roots. From this it may be suggested that seminal adventitious roots compensated for the growing disproportions between the efficiency of the seminal root and the increasing demands of the developing shoot.

The shoot may affect the transport in the root in two ways: directly *via* passive driving forces and indirectly *via* metabolism-dependent driving forces. The shoot supplies the root cells with metabolic products which are essential for active ion uptake and for the maintenance of root cell membrane potentials. Root-shoot interrelationships need more attention namely with respect to the flow of water as well as ions.

Correlations between growing roots and shoots are conditioned by the regulation of the photosynthate sources and sinks. This concerns also the regulation of the uptake and transport of ions probably by the effect of growth substances as was postulated by Pitman. The active loading and unloading of the phloem forms an osmotic gradient which generates the hydrostatic pressure gradient. This may lead to the Poisell flow along the phloem. It is suggested that source and sink activities control the several flow mechanisms of the assimilates.

The root-shoot relationship is a typical example of the problem of interpretation of experimental data obtained on a lower level of the biological organization of the plant. It is obvious that such problems become more complicated as we are faced with new data obtained at increasing levels of organization. Someone with intuition will build up a theory which will put all the pieces together into a unified model. There is a continuous interplay and contradiction between the accumulated knowledge and the meaning of it. Since the whole is more than the sum of the parts, it can be explained only when all structures and functions in their totality and in their interactions can be taken into consideration. We may therefore agree with Danilova, who in her important book "Structural Principles of Ion Uptake by Roots" postulates the inevitability of a need for close cooperation between plant physiologists and anatomists in studying the problems of solute transport in the root.

Session 1

Structural characteristics of roots with respect to absorption and transport of solutes and water

B. C. Loughman et al. (Eds.), Structural and functional aspects of transport in roots, 3–14.
© 1989 by Kluwer Academic Publishers.

Pathways and processes of water and nutrient movement in roots

M.E. McCULLY and M.J. CANNY
Department of Biology, Carleton University, Ottawa, Ontario, Canada K1S 5B6

Key words: air embolism, branch root vascular connections, hydraulic conductance, nutrient uptake, rhizosphere, root surface, xylem maturation

Abstract

Recent work in our laboratory provides evidence for a revised view of the functioning of roots of maize, and probably of all the grasses. The development of coherent soil sheaths on the distal 30-cm of these roots, and the loss of the sheaths further back, led us to investigate the differences in surface structure, anatomy, carbon exudation and microflora of the sheathed and bare zones. The significant differences are summarized. But the fact which underlies all these differences is the maturation of the late metaxylem (LMX). In the sheathed zones the LMX elements are still alive and non-conducting; only the early metaxylem (EMX) and protoxylem are open. In the bare zones they are open vessels. This leads directly to the dryness of bare zones and the wetness of sheathed zones, and indirectly to the other differences noted. Branch root junctions are shown to be structures of great significance. Besides connecting the branches to the axile systems, they serve also to connect the EMX and LMX vessels, and contain a tracheid barrier which prevents air embolisms entering the main vessels. These discoveries force us to revise the traditional view of water uptake by the root hair zone, and to suggest that much water must also enter bare roots, possibly via the laterals. There is some published evidence for this. The living LMX elements of the sheathed zone accumulate large concentrations of potassium which must join the transpiration water at the transition to the bare zone. Calculations suggest that this may be only a tenth of the requirement of a mature plant, and that the balance may enter the bare zones with the transpiration water.

Abbreviations. EMX, early metaxylem; LMX, late metaxylem

Pathways

The beautifully detailed drawing of two young wheat plants reproduced in Fig. 1 was published by Sachs in 1882. It shows a feature of field-grown grasses which has been neglected for 100 years. In the younger plant, all the roots are surrounded, except at their growing tips, by a thick, closely adhering sheath of soil. In the older plant, sheaths are absent from the proximal zones of the roots but present on the distal, younger zones. Clearly these sheaths are a developmental phenomenon; they form and are maintained in immature zones behind growing tips and are shed further back at a point where the underlying root has reached the requisite maturity. The soil sheaths are easily missed since they develop only to a limited extent in potted plants, and with field-grown plants careful excavation is required to find them at the tips of long roots. They are obscured by washing and by sampling from soil cores. Sand sheaths have occasionally been reported as peculiarities of the roots of some desert and dune grasses on which their development can be spectacular (*e.g.* see Wullstein and Pratt, 1981), but these sheaths were not recognized as developmental or as a feature in common with other grasses in other environments. Soil sheaths on cereals and other grasses have been rediscovered with their description in field-grown maize (references in McCully, 1987; Vermeer and McCully, 1982).

The axile root system of maize consists mainly of

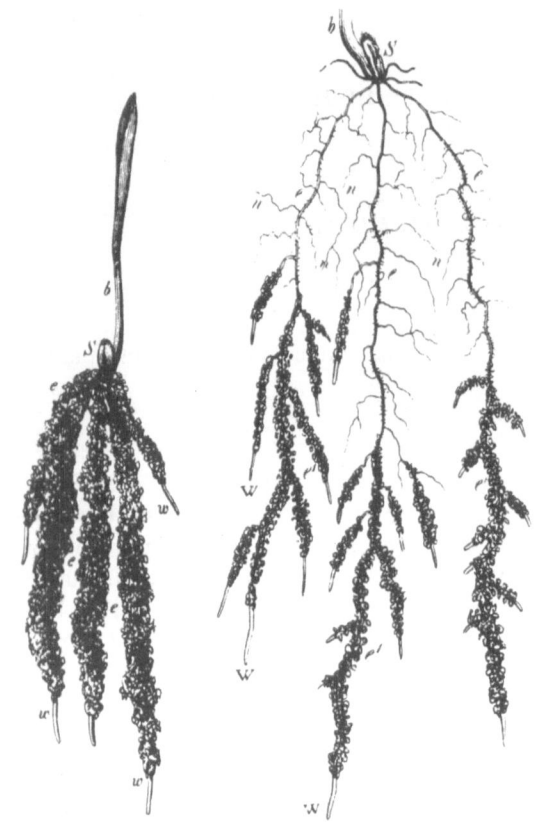

Fig. 1. Drawing of wheat plants showing soil sheaths. Reproduced from Sachs (1882).

tiers of roots developed throughout the season from increasingly younger stem nodes (Hoppe *et al.*, 1986). All these roots (and the primary root) develop soil sheaths. These are of variable length from about 5 cm on the very long oldest roots of mature plants to 30 + cm on young roots from the highest nodes (St. Aubin *et al.*, 1986). Sheathed

zones may be unbranched, or, particularly in the last-formed nodal roots, highly branched with sheath-bearing laterals. Bare zones, which can be two or more metres long in the oldest roots are often much branched (Fig. 2).

The sheath on maize roots is about 1 mm thick and in the youngest tiers approximately doubles root diameter. (Sheaths can be twice as thick relative to root diameter in grasses like wheat and oats.) The sheath is a complex of soil particles, microbes, mucilaginous substances and intertwined root hairs (Figs. 3 and 4). Under favourable moisture conditions living root-cap cells can persist in the sheath for a considerable distance behind the growing tip (Figs. 3 and 4). The mucilage, though probably of mixed origin coming from both the roots and the associated bacteria, shows similar staining properties to the mucilage produced by peripheral root-cap cells, and in part must be left behind along the root as the tip extends. The detached root-cap cells are always within regions of mucilage (Fig. 4). Many factors must maintain the sheath as a coherent entity (it, together with the underlying epidermis, forms a cylinder which can be easily slipped off corn roots — Fig. 1C, McCully and Canny, 1985). As suggested by Sachs (1882), the major anchorage seems to be the root hairs; specifically points where they are curled or distorted, for here vigorous washing and sonication does not remove adhering soil (Fig. 5). The stability and cohesion of the intervening soil aggregates must be effected by the mucilage and possibly by other root and bacterial exudates.

Sheaths are shed at their proximal ends during root development. Initially we thought that the basic cause of this loss must be senescence and/or

Fig. 2. Typical unwashed roots from field-grown maize plants at flowering stage. *Left*: older sheathed portion of root originating from stem node #6. *Centre*: tip of younger root also from a stem node #6. The relatively long, unsheathed tip indicates rapid elongation. *Right*: bare portion of long root which originated from stem node #3. × 1.0
Figs. 3–20 all show sections of field-grown maize roots.

Fig. 3. Transverse section showing root surface and adhering soil sheath from young nodal root. Root epidermis and underlying exodermis are clear. The section passes through parts of two root hairs and a detached cell (**R**) which has originated from the root cap. Plastic-embedded material, Nomarski optics. × 580.

Fig. 4. Electron micrograph of section similar to that in Fig. 3. Section passes through an epidermal (**E**) and an exodermal (**X**) cell and a detached root cap cell (**R**). Rhizosphere mucilage (**M**) and clay particles (**C**) are included. × 13 500.

Fig. 5. Surface of a sheathed portion of a root that was extensively washed and sonicated in water. Soil aggregates remain fastened to distorted regions of root hairs and in places to the root surface. Hand-cut section. Toluidine blue staining. × 90.

Fig. 6. Surface of bare region of root showing remnants of epidermal cells and a root hair. The exodermis (**X**) is now the outermost intact living tissue. Rhodamine B-induced fluorescence. × 170.

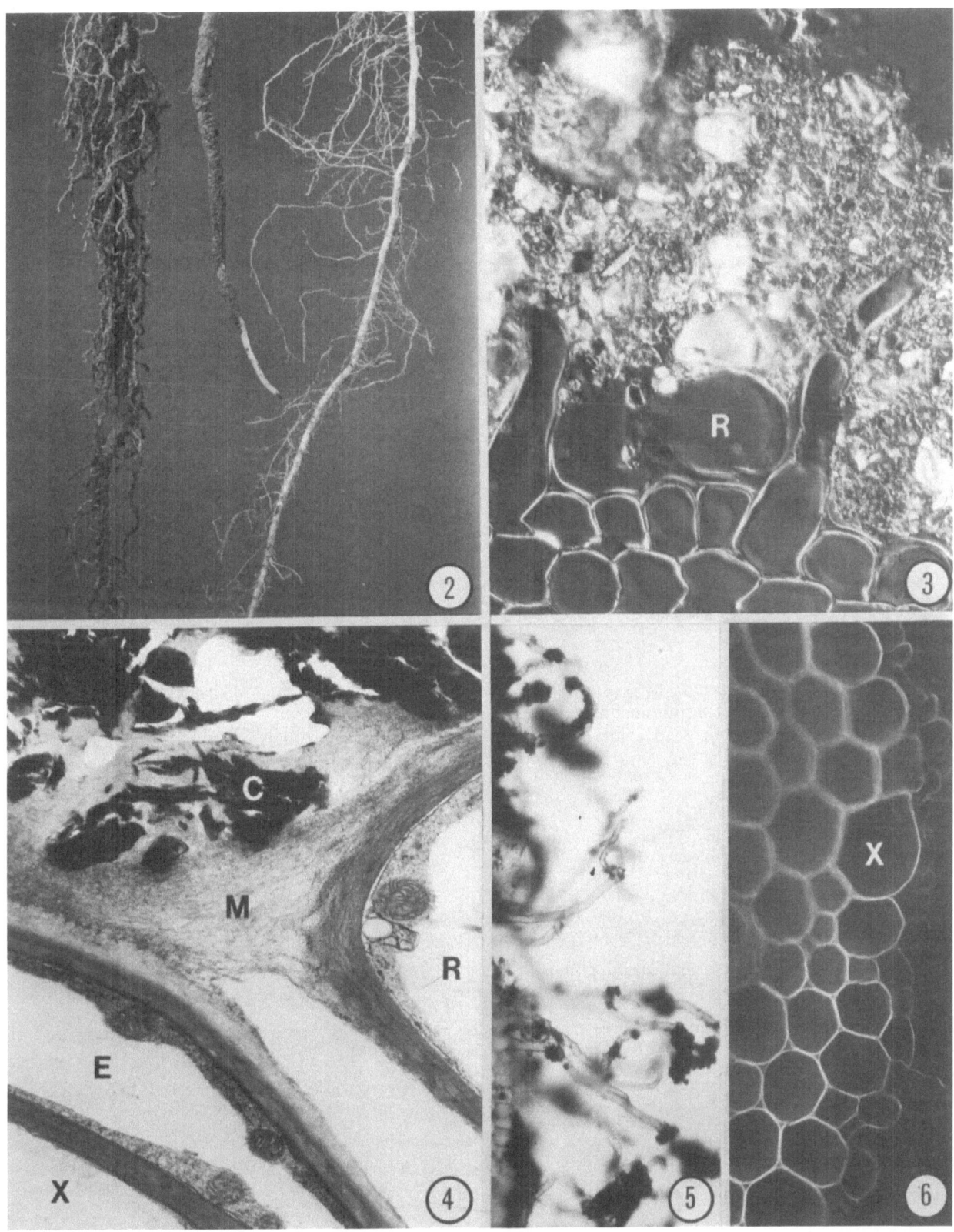

Table 1. Features of axile roots of maize

	Sheathed zone	Bare zone
Epidermis[1,2]	Alive. Root hairs curled and distorted, with clinging soil (Fig. 5)	Mostly dead, and root hairs shed (Fig. 6)
Exodermis[1,2]	Casparian strip, suberized lamellae in distal end of zone	Tertiary, lignified wall (Fig. 6)
Endodermis[1]	Similar to exodermis (Fig. 7)	Heavy, lignified tertiary walls (Fig. 11)
Remaining cortex[1,2]	Alive, thin, primary walls (Fig. 7)	Alive. Secondary walls, lignified and suberized (Fig. 11)
Stelar parenchyma[1,2]	Progressively developing secondary lignified walls (Fig. 7)	Fully lignified (Fig. 11)
Early metaxylem	Maturation in distal region of zone (Fig. 8)	Mature and open (Fig. 11)
Late metaxylem[3,4]	Immature, with living protoplasts and intact end walls. Lignified secondary lateral walls at distal end of zone (Figs. 9 and 10)	Mature and open (Fig. 11)
[14]C-labelled assimilate[1]		
a) in stele	Soluble and insoluble	Similar to sheathed zones
b) in cortex	Soluble and insoluble	Similar to sheathed zones
c) exuded to rhizosphere	About 7%	Similar to sheathed roots
Residual at root surface[5]		
a) root-cap mucilage	In sheaths (Fig. 4)	Absent
b) detached root cap cells	In sheaths (Figs. 3 and 4)	Absent
Water status	High	Low
Actinomycetes on surface[6]	Present	Absent

[1] McCully and Canny, 1985; [2] McCully, 1987; [3] St. Aubin *et al.*, 1986; [4] McCully *et al.*, 1987; [5] Vermeer and McCully, 1982; [6] Gochnauer, McCully and Labbé, submitted.

other changes in the immediately underlying tissues. And indeed, the sheath-anchoring root hairs are lost. Remnants of dead epidermal cells remain (Fig. 6), as well as a few living cells, some with straight root hairs which persist on even the oldest bare roots.

Results of our studies of structural and physiological characteristics of epidermal and cortical

Fig. 7. Transverse section of sheathed zone in which the developing LMX elements (**LX**) are beginning to lignify. No contents appear in these still-living cells because they were pulled out during the sectioning. EMX elements (∗) have thick, lignified walls. Hand-cut section of fresh tissue. Rhodamine B-induced fluorescence. × 180.

Fig. 8. Transverse section of a root similar to that shown in Fig. 7 but cut close to the distal end of the sheath. Protoxylem (∗) is probably mature, EMX has lignified secondary walls but the largest element, at least, still has contents. The LMX element (**LX**) has not begun to produce a secondary wall. Periodic acid-Schiff's reaction followed by aniline blue. Fluorescence optics. × 680.

Fig. 9. Transverse section of sheathed zone, 20 cm from the tip. All late metaxylem elements contain protoplasts with thin transvacuolar cytoplasmic strands. Pieces of root were fixed in formalin, then hand sectioned. Phase contrast optics. × 225.

Fig. 10. Longitudinal section of a part of a sheathed root similar to that in Fig. 8 showing intact end wall in a living late metaxylem element. Hand-cut section of fresh material. Rhodamine B-induced fluorescence. × 100.

Fig. 11. Transverse section of old bare region of a primary root. Stelar parenchyma is completely lignified. Cortical cells (except where aerenchyma is forming) have lignified, secondary walls. Hand-cut formalin-fixed tissue. Rhodamine B-induced fluorescence. × 120.

Fig. 12. Sheathed zone of root through which an aqueous solution of sulphorhodamine G was drawn with a hand vacuum pump. The tissue was then freeze-substituted and embedded in resin. Dye clearly moved up and out of early metaxylem elements, not the late metaxylem (∗) which is immature. Close to the base of a branch root, at the upper and lower right of the photomicrograph, the dye has moved through the connecting tracheary elements between the early and late metaxylem strands. Fluorescence optics. × 190.

Fig. 13. Bare zone of root through which an aqueous solution of basic fuchsin was drawn with a hand vacuum pump. The dye moved up and diffused laterally from the late metaxylem strands. Hand-cut section. × 120.

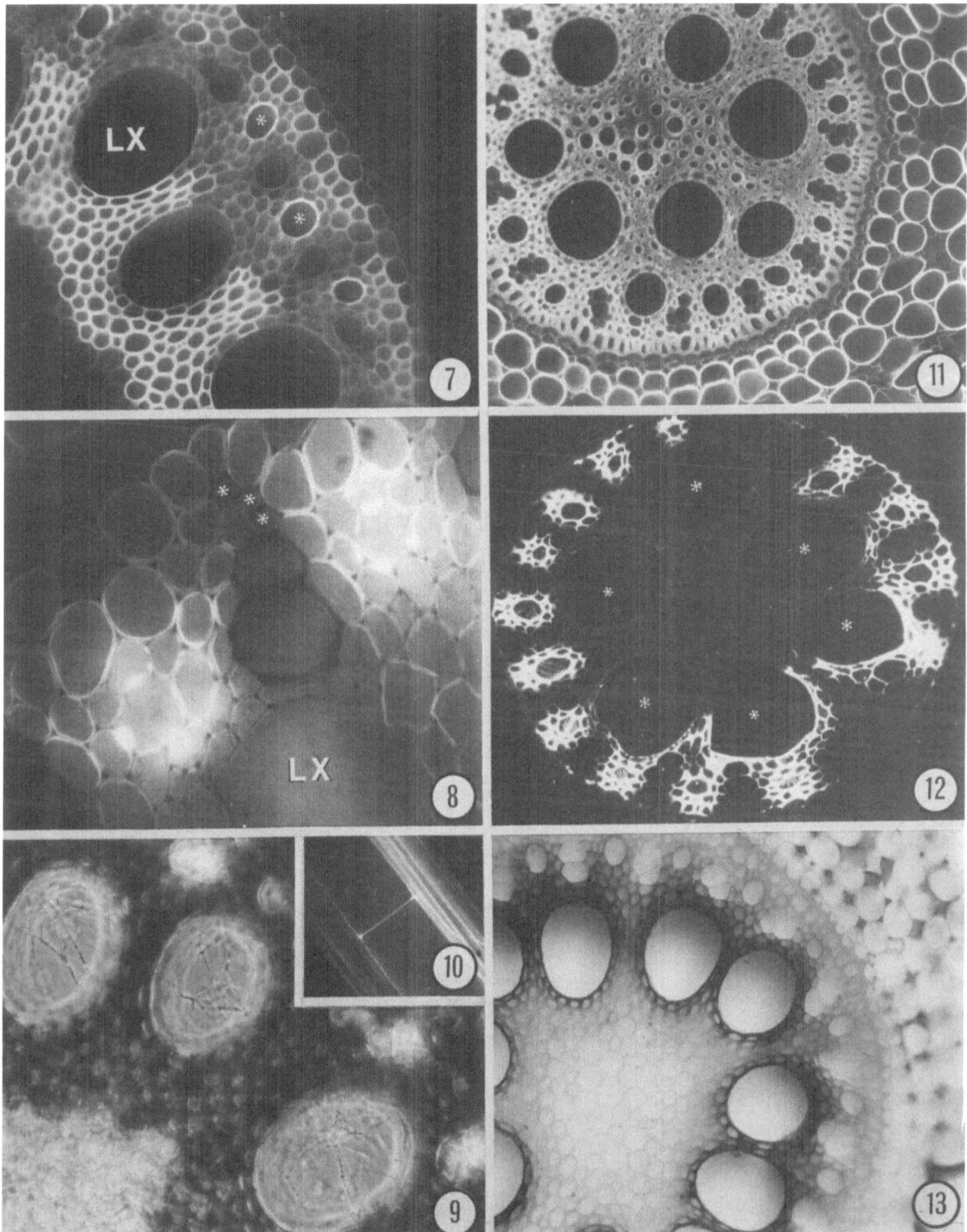

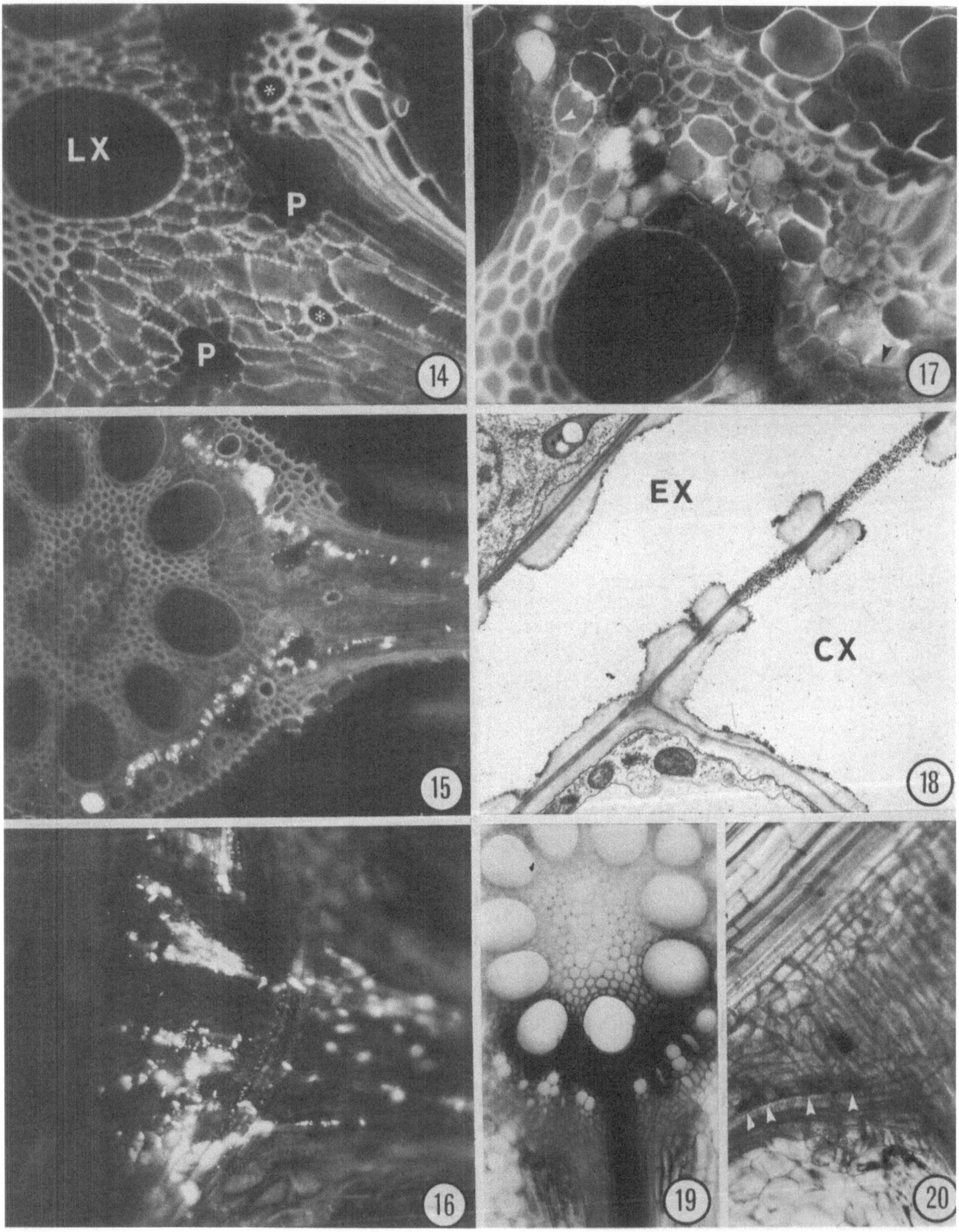
LX
P
P
14
17
15
EX
CX
18
16
19
20

tissues of sheathed and bare root zones in maize are included in Table 1. Two features in particular should be noted. Cells in all tissues outside the stele remain alive in bare zones, and about the same amount of organic carbon is exuded from bare zones as from sheathed zones. The sheathed zones of freshly excavated roots, like their white growing tips, are soft and juicy, in contrast to bare zones which are tougher and drier. Other observations (Table 1) are consistent with a marked difference in water status of the two zones. For example, expanded mucilage gel of root-cap origin and detached, living root-cap cells persist in soil sheaths (Fig. 4) though both have poor water-retaining properties (Guinel and McCully, 1986), while actinomycetes (which prefer dry environments) are commonly present on the surface of bare zones but absent from sheathed surfaces.

Some structural differences between stelar tissues of sheathed and bare zones are listed in Table 1. A most surprising discovery was that the late metaxylem (LMX) elements retain living protoplasts (Fig. 9) and intact end walls (Fig. 10) throughout the sheathed zone; their maturation coinciding with sheath loss (St. Aubin *et al.*, 1986). We have not determined the distances behind the root tips of field-grown plants at which the protoxylem and early metaxylem (EMX) vessels mature but figures for maize roots in liquid culture (Burley *et al.*, 1970; Higinbotham *et al.*, 1972) are approximately 5 and 10 cm respectively. This would place protoxylem maturation near the distal end of the sheath, with the EMX opening somewhere within the zone. The relative sizes of the three types of xylem close to the tip are shown in Fig. 8.

The difference in axial water pathways of roots having immature and mature LMX elements has been shown by the elegant dye transpiration work of Kozinka (this symposium). The same difference can be demonstrated by dye pulled by vacuum through sheathed and bare root zones (*i.e.* initial flow through EMX and LMX respectively—see Figs. 12 and 13).

The vascular connections between branch roots and the axile roots provide unexpectedly complex links not only between the fine root system and the main roots, but also between the vascular strands within individual axile roots. The extensive proliferation of stelar parenchyma which accompanies branch root initiation (*e.g.* Fig. 15, Bell and McCully, 1970) forms cells which differentiate to tracheary elements and sieve tubes. These connect the branch root xylem and phloem to 1/3 to 1/2 of the corresponding vascular poles of the main root (Figs. 12, 15, 19). Connecting elements extend acropetally and basipetally along the parent-root vascular strands beyond the insertion of the branch (Fig. 16). Connecting tracheary elements provide links not only between individual parent poles within the EMX and the LMX but they also form

Fig. 14. Transverse section of bare zone of a nodal root at the base of a branch root (*right of micrograph*). Phloem (**P**), and early (*) and late (**LX**) metaxylem of main root are joined by connecting tracheary elements. Hand-cut section of fresh tissue. Rhodamine B-induced fluorescence. × 180.

Fig. 15. A section similar to that of Fig. 14 but stained with aniline blue to show the position of connecting sieve tubes. Fluorescence optics. × 120.

Fig. 16. A longitudinal section through a region of a sheathed root passing through the base of a lateral root. Aniline-blue induced fluorescence shows the numerous strands of connecting phloem. Branch root is toward the right. Fluorescence optics. × 120.

Fig. 17. Transverse section of bare zone of a nodal root at base of branch (*toward lower right*). There are direct contacts between sieve tubes and xylem elements (*arrows*). Aniline blue-induced fluorescence. × 350.

Fig. 18. Electron micrograph of a longitudinal section of main root at the base of a branch showing junction of a connecting tracheid (**CX**) with early metaxylem (**EX**) of the main root. × 6500.

Fig. 19. Transverse section of bare portion of axile root from stem node 1. Aqueous basic fuchsin was drawn by vacuum through a branch into the main root. Hand-cut section of fresh tissue. × 100.

Fig. 20. Similar root and treatment to Fig. 19 except that an aqueous solution of a coloured latex paint was drawn through the branch into the main root. The water moved freely but the latex accumulated (*arrows*) at interfaces between connecting tracheids and main root vessels. This longitudinal section passes through the periphery of the base of the branch root (*right of micrograph*). Hand-cut, unstained section. × 120.

direct links between the two types of elements (Figs. 14 and 17). Elsewhere in the axile roots EMX and LMX strands are always separated by living parenchyma cells. Furthermore, within the bed of connecting xylem and phloem, tracheary elements and sieve tubes often adjoin (Figs. 14 and 17).

The important studies by Luxová (1986 and this proceedings) show that a zone of small diameter vessels lies at the base of barley roots. She has proposed that these are regions of high resistance, providing hydraulic protection to the shoot during drought. Such regions might function also by preventing air bubbles from moving out of the root system. The connecting elements which join main-root xylem vessels to the bed of tracheary elements at the base of branch roots in maize may provide similar protection. All of the tracheary elements in the bed have diameters considerably smaller than those of the main root LMX vessels. Also, the connecting elements immediately adjoining the main xylem strands are tracheids, with modified walls (Fig. 18) forming fine-porosity barriers between the branch and main root xylem systems. Water soluble dyes readily pass through these walls when sucked from a branch into the main root (Fig. 19). In contrast, the coloured latex particles accumulate in the connecting tracheids when an aqueous solution of latex paint (method of Zimmermann and Jeje, 1981) is pulled into the main root in the same manner (Fig. 20). Air bubbles would be similarly trapped. Thus hydraulic safety zones exist throughout the maize root system as shown also for barley (Luxová, this symposium).

Processes

This revised view of xylem anatomy of maize roots (which can probably be extended to include all the grasses) calls for a re-appraisal of how the roots are functioning, especially in their principal activities of absorbing water and nutrients. The traditional physiological view is that both water and ions enter roots close to the tip in the root hair zone and are conducted upwards in mature open vessels in the transpiration stream. Since only the protoxylem and EMX vessels are open for the first 25 cm or so of the root, only these few small tubes would be available for this traffic. The large LMX vessels are not available for carrying water, and, as

living cells, could be large reservoirs accumulating ions.

Let us look first at the capacity of the sheathed and bare zones to carry water. The diameter of the EMX elements is about $20\,\mu m$, while that of the LMX is about $100\,\mu m$ (*e.g.* Figs. 7, 9, 13 and 14). It is easy to forget how powerful a dependence of volume flow upon tube radius is expressed by the Hagen-Poisseuille relation. To emphasise this the relative flows are shown in Table 2 for tubes of different relative diameter. A 20% increase in diameter of a vessel may pass unnoticed, and yet allows twice the flow. Each LMX vessel, with five times the diameter of an EMX, would carry 625 times the volume with the same gradient of pressure. Allowing for their fewer number, there is still an enormous increase in longitudinal conductance when the root makes the transition from sheathed to bare. In a root with both systems open, virtually all the water must travel in the LMX. This is shown by dye movements in Dr Kozinka's paper in this volume and in our Figs. 12 and 13 and expressed in numbers in Fig. 21. The ratio of the two conductances in this root is 1800. Looking at it as a gradient of water potential: to draw the same volume of water through the sheathed zone as through the bare zone requires 1800 times the gradient. If a gradient of the usually accepted size of $0.2\,atm\,m^{-1}$ extends down to the end of the open LMX, and all the water is coming (as in the traditional view) from near the tip, the final 30 cm would need a gradient of $360\,atm\,m^{-1}$ or $3.6\,atm\,cm^{-1}$. The negative potential generated by the leaves would reach down little attenuated to the end of the bare zone. In the sheathed zone it would be sharply reduced (at the rate of something like $3.6\,atm\,cm^{-1}$) before the root hair zone was reached. It is worth asking whether the traditional view of water uptake near the tip may be wrong.

If the bare zone of the root contains the vessels which could conduct large volumes of water in a low resistance pathway, and if this is the region where the negative water potential of the shoot extends little diminished, it would be a more logical place for the root to take up water than would the

Table 2. Flow varies as (radius)⁴

Radius	1	1.2	1.5	2	5	10
Flow	1	2	5	16	625	10,000

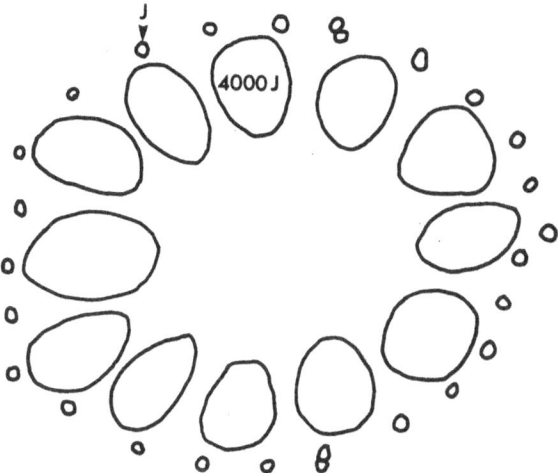

Fig. 21. Tracing of xylem vessels in transverse section of a nodal maize root from tier 2. The flux of water in an EMX vessel is denoted by J. Measurements of the mean diameters of individual vessels and the application of the (radius)[4] rule of Table 1 enable the calculation of fluxes in individual vessels relative to J. The labelled LMX vessel would carry 4000 J. The sum of the flows in the EMX is 27 J, and in LMX is 50 000 J. Total LMX/total EMX = 1800.

sheathed region. It has already been stressed that when a plant is dug up, bare roots are dry while sheathed roots are wet, as though the shoot were vigorously extracting water from the former, but not the latter. Perhaps the wiry, lignified tissues of Figs. 6 and 11 are not as impermeable as has been assumed. The Casparian bands at the exodermis and endodermis, the lignified cortex and stelar tissue, may have paths for water through them in spite of their thoroughly water-proofed appearance. Certainly these roots exude organic carbon to the rhizosphere (McCully and Canny, 1985). There have long been ambivalent views about the zone of water uptake, well exemplified by the contributions to this conference by Drs Steudle and Barber. The one, following the physiologists' tradition, details measurements of water uptake by the zone just behind the tip of barley roots; the other, following the agronomists' tradition, presents a model of plant growth which assumes uniform uptake of water and nutrients along the whole length of the roots.

If one asks, what do the published measurements show about the permeability of different zones of roots to water? The answer is rather one-sided and incomplete. For reasons of experimental convenience most measurements have been made on

quite small (mostly unbranched) root systems, and mostly on the rates of exudation by cut roots rather than the uptake drawn in by transpiration. The radial conductivity may be expressed either per unit length of root (P_L) or per unit surface (P_S). It has commonly been measured by the changes in rate of exudation when roots are transferred between solutions of different osmotic potential. It is not known whether such a system gives similar values to the intact plant drawing water from soil by the negative potential in the transpiration stream. The unit of P_L, $cm^3/cm_{root}/second/atm$, is often condensed to $\mu m^2 s^{-1} bar^{-1}$, which is 10^{-8} of the other. Newman (1973) has collected a number of published values for comparison (Table 3) and supplies several of his own measurements. All the values are less than $30 \mu m^2 s^{-1} bar^{-1}$, and many are under 10 of these units. This range may be taken as typical of values for seedling roots measured in this way, or for the root hair zone of roots in general.

We have found one set of published measurements of water permeability in older roots which might be comparable with the bare zone of Zea, those of Maertens (1971). Maertens grew maize plants in sandy soil to the age of $2\frac{1}{2}$ months, by which time the basal parts of their older roots would certainly have open LMX. The roots were washed free of sand and placed in culture solution. Different zones of the roots were enclosed in a tube between rubber stoppers 5 cm apart, and the uptake measured under the transpiration pull of the shoot. Maertens does not calculate a P_L as defined here, and it is necessary to assume a value for root mass per unit length to convert his figures to P_L. Our own measurements on bare roots of maize give 45 cm per gram, and, using this value, Maertens'

Table 3. Permeability of roots to water P_L ($\mu m^2 s^{-1} bar^{-1}$) (= $10^{-8} cm^3$/cm root/second/bar)

P_L	Species	Author (year)
5.7–21.4	*Zea mays*	Anderson and Collins (1969)
		House and Findlay (1966)
13.5–28.3	*Allium cepa*	Hay and Anderson (1972)
6.3–12.9	*Vicia faba*	Brouwer (1953)
0.5–4.3	*Phaseolus, Vicia, Helianthus, Zea, Lycopersicon*	Newman (1973)
	Zea mays	Maertens (1971)
46	apical	
290	basal	

figures convert to a P_L of 46 for the apical zone and 290 for the basal zone. Not only is the permeability of the old zone greater by a factor of 6 than that of the young zone, but its P_L value is 10 to 600 times the values quoted by Newman. Given the concurrence of the indications from the xylem anatomy and the scanty measurements, the time seems ripe for a careful study of the water permeability of older roots.

Such a study must include the part played in water uptake by the lateral roots. As shown above, these provide a path right through the lignified and suberized tissues of the cortex and endodermis, connecting to the LMX elements through a bed of tracheary elements. This, rather than the direct radial route which is the only one available near the tip, could provide easy entry. In bare zones it will be necessary to discover what proportion of the water enters by each of the two paths, as well as how the total compares with that in the sheathed zones. We are currently devising ways to make these comparisons.

It has been stressed that the transition of sheathed to bare zone is a quantum jump in longitudinal conductance. Distal to this, water moves in the EMX. But this is only the last of three such jumps in the maturation of the root (Higinbotham *et al.*, 1972). At less than 10 cm from the tip the EMX is also closed, and the only open vessels are the protoxylem, which have diameters 5-fold smaller again (Fig. 8). So at 10 cm there is another large change in longitudinal conductance of comparable magnitude. The protoxylem vessels themselves are open only after 5 cm, at which distance the first large change in conductance occurs. The connections between these three systems have been referred to above. Protoxylem and EMX are in direct contact (Fig. 8), and the connection at 10 cm is obvious. But the vessels of EMX and LMX are separated by parenchyma cells (Figs. 7, 8 and 13). The pathway by which water passes from EMX to newly-opened LMX vessels at the transition to the bare zone is found in the bed of tracheary elements that connects both sets of vessels to a lateral root (Fig. 14).

If the large LMX elements of a sheathed root are not conducting water, what are they doing? We hypothesised that they might be storing ions in their vacuoles for later release to the transpiration stream when their cross walls disappear. We chose

K^+ as a convenient ion to measure, and used the X-ray microprobe to measure $[K^+]$ in the vacuoles of LMX of sheathed zones (McCully, Canny and Van Steveninck, 1987). Field-grown roots were frozen in large pieces, snapped in the column of the scanning electron microscope, and their K^+ content estimated. While the open EMX vessels of sheathed roots and the open LMX of bare roots contained about 10 mM K^+, the vacuoles of the living LMX elements of sheathed roots contained 200 to 400 mM K^+. The balancing anion has not yet been identified. It is not chloride or phosphate, which were present in low concentrations. It may well be malate which is known to occur in maize root xylem sap (Butz and Long, 1979).

In the zone of transition from sheathed to bare root the cross walls of the LMX elements break down and the contents of their vacuoles are released to the transpiration stream in the vessels thus formed. This must supply part at least of the traffic of potassium to the shoot. Whether this mode of uptake and release is important for other inorganic nutrients remains to be determined. It would appear to provide about the observed amount of K^+ in exudates from severed roots (McCully *et al.*, 1987), forced out by root pressure.

These findings must return to prominence the mechanism proposed by Hylmö (1952) for the generation of root pressure. In a plant of high water status, at the top of the sheathed zone of a root, when the last cross wall in a file of LMX elements ruptures, it will release from constraint the protoplast below it. The protoplast contains, as we have seen, a high osmotic concentration. It will absorb water from the tissues around it, expand against the column of water above it in the vessel, and exert pressure on the column. Many experiments have been made on exudates coming from the cut end of grass roots, especially maize and barley, and sophisticated instrumentation is available to study their behaviour. It is time for an investigation carefully relating exudate physiology to root anatomy. Are the production of the exudate, its changes in concentration, pressure, *etc.*, consistent with Hylmö's hypothesis? Does the flow come from the EMX, which is alive back to 10 cm, or from the LMX which is alive back to the cut surface (if this is less than 25 cm from the tip)? It may even be possible to detect the rupturing of individual cross walls as small jumps in root pressure.

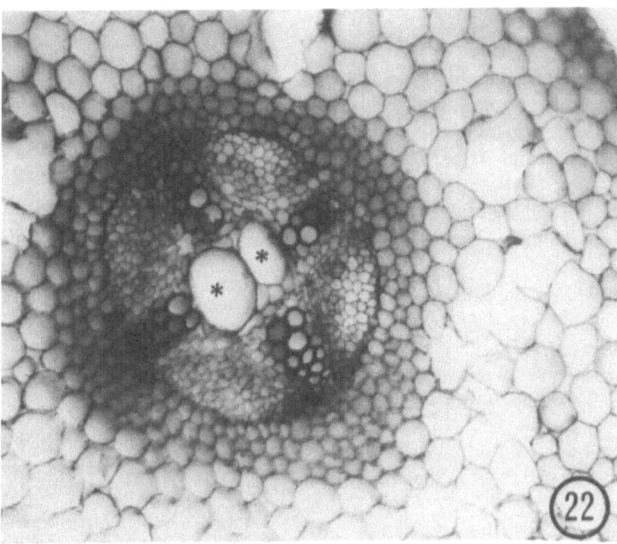

Fig. 22. Primary root of 10-day-old soybean plant, 18 cm from tip. The LMX elements (∗) have just matured, 10 cm proximal to the last living early metaxylem. Hand-cut section, toludine blue stain. × 180.

The question whether all the supply of nutrients for a mature maize plant could come in this way through the living LMX of the sheathed zones can be approached using the data we have collected about the architecture of the root system (Hoppe, McCully and Wenzel, 1986). At flowering our plants have about 70 axile roots, in seven tiers, whose numbers and complements of LMX vessels are given in that paper. The grand total of LMX vessels in these roots in one plant is about 2200; the radius of each is about 50 μm. Assuming a rate of extension of 3 cm per day, and so a breakdown of the cross walls in 3 cm of each of these vessels, the total volume of sap released would be $30 \times \pi$ $(0.05)^2 \times 2200 \, mm^3 = 500 \, \mu l \, day^{-1}$. With $250 \, mM$ K^+ in the sap, this mechanism would contribute $250 \times 500 \times 40 \, ng \, d^{-1} = 5 \, mg \, K^+ \, d^{-1}$ to the shoot. The needs of a mature crop for K^+ are quoted as $2 \, Kg \, ha^{-1} \, d^{-1}$ (Kuijper, personal communication). Assuming 4 plants to a square metre, this amounts to $50 \, mg \, d^{-1}$ per plant, or 10 times that calculated as coming from the sheathed zones. Though the factors assumed in this calculation are fairly rough, it does not seem likely that any of the assumptions are wrong enough, singly or collectively, to account for a ten-fold difference. Thus it seems likely that, along with a large proportion of the water needed by the shoot, 90% of the K^+ might enter the bare zones through laterals or the main axes. This would accord again with Maertens'

measurements, where in one of his experiments 15 times as much K^+ entered the basal zone as the apical zone (Maertens, 1971, Table 1). Studies of the possibility of such a traffic are planned along with those of water uptake.

Impelled by these discoveries and new ways of looking at corn roots, we have begun to explore dicotyledon roots to see if they hold similar surprises. They do (Kevekordes *et al.*, 1988). In Fig. 22 a section is shown of a soybean root 18 cm behind the tip. The four EMX poles surround a central pith in which have developed a pair of large LMX elements. At this distance they have just matured to become vessels. From here to 10 cm behind the tip they are alive and closed, just like the LMX in sheathed zones of corn roots. They first appear in the central pith at about 10 cm. Again it will be noticed, the ratios of diameters LMX/EMX is about 5, and a quantum jump in hydraulic conductance will result from LMX maturation. This late development of large LMX vessels in the centre of dicotyledon roots appears to be widespread, and must have profound consequences for the study of water and nutrient uptake.

Acknowledgements

We thank Janet Vermeer for the preparation used in Fig. 3 and for Fig. 4 and Joan Mallett for Fig. 18 and for making the plates.

References

Anderson W P and Collins J C 1969 The exudation from maize roots bathed in sulphate solution. J. Exp. Bot. 20, 72–80.

Bell J K and McCully M E 1970 A histological study of lateral root initiation in *Zea mays*. Protoplasma 70, 179–205.

Brouwer R 1953 Water absorption by the roots of *Vicia faba* at various transpiration strengths. I. Analysis of the uptake and the factors determining it. Proc. Koninkl. Nederl. Akad. v. Wetensch. C56, 106–115.

Burley J W A, Nwoke F I O, Leister G L and Popham R A 1970 The relationship of xylem maturation to the absorption and translocation of P^{32}. Am. J. Bot. 57, 504–511.

Butz R G and Long R C 1979 L-malate as an essential component of the xylem fluid of corn seedling roots. Plant Physiol. 64, 684–689.

Guinel F C and McCully M E 1986 Some water-related physical properties of maize root cap mucilage. Plant Cell and Envir. 9, 657–666.

Hay R K M and Anderson W P 1972 Characterization of exudation from excised roots of onion. J. Exp. Bot. 23, 577–584.

Higinbotham N, Davis R F, Mertz S M and Shumway L K 1972 Some evidence that radial transport in maize roots is into living cells. *In* Ion Transport in Plants. Ed. W P Anderson, pp 493–506. Academic Press, London.

Hoppe D, McCully M E and Wenzel C L 1986 The nodal roots of Zea: Their development in relation to structural features of the stem. Can. J. Bot. 64, 2524–2537.

House C R and Findlay N 1966 Analysis of transient changes in fluid exudation from isolated maize roots. J. Exp. Bot. 17, 627–640.

Hylmö B 1953 Transpiration and ion absorption. Physiol. Plant. 6, 333–405.

Kevekordes K G, McCully M E and Canny M J 1988 Late maturation of large metaxylem vessels in soybean roots: Significance for water and nutrient supply to the shoot. Ann. Bot. *In press.*

Luxová M 1986 The hydraulic safety zone at the base of barley roots. Planta 169, 465–470.

Maertens C 1971 Etude expérimentale de l'alimentation minérale et hydrique du Maïs. Capacité d'absorption des partie basales et apicales de racines de *Zea Maïs*. C.R. Acad. Sci. (Paris) 273, 730–732.

McCully M E 1987 Selected aspects of the structure and development of field-grown roots with special reference to maize. *In* Root Development and Function. Eds. Gregory P J, Lake J V and Rose D A, pp 53–70. Cambridge University Press, Cambridge.

McCully M E and Canny M J 1985 Localization of translocated ^{14}C in root exudates of field grown maize. Physiol. Plant. 65, 380–392.

McCully M E, Canny M J and Van Steveninck R F M 1987 Accumulation of potassium by differentiating metaxylem elements of maize roots. Physiol. Plant. 69, 73–80.

Newman E I 1973 Permeability to water of the roots of five herbaceous species. New Phytol. 72, 547–555.

Sachs J 1882 Vorlesungen über Pflanzenphysiologie. Englemann, Leipzig.

St. Aubin G, Canny M J and McCully M E 1986 Living vessel elements in the late metaxylem of sheathed maize roots. Ann. Bot. 58, 577–588.

Vermeer J and McCully M E 1982 The rhizosphere in Zea: New insight into its structure and development. Planta 156, 45–61.

Wullstein L H and Pratt S A 1981 Scanning electron microscopy of rhizosheaths of *Oryzopsis hymenoides*. Am. J. Bot. 68, 408–419.

Zimmermann M H and Jeje A 1981 Vessel length distribution in stems of some American woody plants. Can. J. Bot. 59, 1882–1892.

Note added in proof

Huang and Van Steveninck (1988) show that the maturation of the single, large LMX vessel in barley seminal roots is delayed until 10 to 15 cm proximal to the apex. As with maize, the living LMX elements are reservoirs of accumulated ions. These are released to the transpiration stream when the elements mature. In salt-stressed plants, the immature LMX elements may effect salt tolerance by their initial accumulation of a high concentration of salt which is subsequently reabsorbed by surrounding cells before the contents of the dying LMX cells are released into the open xylem.

Huang C X and Van Steveninck R F M 1988 Effect of moderate salinity on patterns of potassium, sodium and chloride accumulation in cells near the root tip of barley: role of differentiating metaxylem vessels. Physiol. Plant. *In press.*

B. C. Loughman et al. (Eds.), Structural and functional aspects of transport in roots, 15–20.
© 1989 by Kluwer Academic Publishers.

The vascular system in the roots of barley and its hydraulic aspects

MÁRIA LUXOVÁ

Institute of Experimental Biology and Ecology, CBES, Slovak Academy of Sciences, CS-814 34 Bratislava, Dúbravská 14, Czechoslovakia

Key words: barley, *Hordeum vulgare*, roots, stele, structural gradients, vascular system

Abbreviations: nodal root, NR; seminal root, SR.

Introduction

Comparatively few data are available about the hydraulic architecture of roots compared with the above-ground organs. The results presented in the literature were mostly obtained by an analysis of root secondary xylem. The roots possess the widest vessels; as distinct from stems, the diameter of vessels in the secondary root xylem usually increases in an acropetal direction (Fayle, 1968; Lebedenko, 1962; Riedl, 1937). The present paper deals with proximal-distal changes of the vascular pattern in barley roots and their hydraulic consequences.

The most basal region of barley roots, being *ca* 0.5 mm long, is formed by the hydraulic safety zone (Luxová, 1986): this is an anatomical adaptation which ensures hydraulic protection and at the same time makes possible an efficient axial transport of solutions from the roots to shoots and vice versa. The acropetal continuation of the safety zone is the basal root region which in barley has been the subject of several reports (Hagemann, 1957, Heimsch, 1951; Jackson, 1922). The former author summarized the available data and characterized the individual barley root types according to structural properties of their basal parts as follows: a) roots formed in the embryo, most frequently designated as seminal roots (Merry, 1941), are comparatively thin with a comparatively wide stele.

The vascular system consists of 6–9 alternating xylem and phloem strands; one narrow peripheral metaxylem vessel is adjacent centrally to each individual pole of protoxylem. A single wide metaxylem vessel is situated in the centre of the stele. b) Crown roots (nodal roots) are thicker and their branching starts later than in seminal roots. On the periphery of their stele there are 12–17 alternating xylem and phloem strands, and 3–6 vessels of the internal wide metaxylem are located around the centre. c) Lateral roots are thin and their vascular system can be reduced to four xylem and phloem strands and one central metaxylem vessel.

From the point of view of axial water flow in barley roots, the most efficient vessels are those of the inner wide metaxylem. Based on the Hagen-Poiseuille equation for paraboloid flow through capillaries, *e.g.* in the primary seminal barley root, the relative conductivity of a central vessel with a $30\,\mu m$ diameter amounts to 93%, while that of the 7 narrow peripheral vessels with a diameter of 9–$10.5\,\mu m$, amounts to a total of 7% (Luxová, 1986). These conductivity values were calculated for ideal capillaries. Although the properties of vessels are not those of ideal capillaries (Taylor and Klepper 1978) the number and diameter of vessels is of decisive importance in establishing the value of conductivity in plants which may largely control water flow in the xylem (Gibson *et al.*, 1984). This aspect served as a basis in evaluating the longitudinal histological gradient and consequently the hydraulic properties of roots.

Materials and methods

Quantitative histological analyses were carried out on roots of spring barley (*Hordeum vulgare* L. cv. Karát). The plants were grown under field conditions in deep vessels filled with soil. Seminal roots (SRs) were obtained from 2 month old plants and nodal roots (NRs) from plants whose grains were at the dough stage, when the SRs were partially damaged or dead. The number of vessels in the inner wide metaxylem in the base of NRs had a broader range (from 4 to 9) than that reported by Hagemann (1957). Paraffin cross-sections of the roots were prepared at a distance of 1, 15 and 30 cm, and occasionally 45 cm from the base and treated with tannic acid and ferric chloride. At these levels the stelar tissues were mature and the cortical tissues were partially shrivelled. In this way, 7 SRs and NRs each were examined; samples from individual roots being treated separately.

The examination of the cross sections included the determination of : a) the stele diameter; b) the number of peripheral xylem and phloem strands; c) the number of vessels of the wide metaxylem and the mean value of their radial-tangential internal diameter; d) the proportion of the wide metaxylem (in %) of the stelar cross-sectional area; e) the proportion (in %) of the relative conductivity of the wide metaxylem; the relative conductivity of wide metaxylem at a distance of 1 cm from the base was taken as 100%. In SRs, the number of cell layers of the parenchyma sheath surrounding the central metaxylem was also determined.

Results

Changes in the histological pattern along the roots are the result of a complex interaction of internal and external factors affecting the activity of root meristems during ontogenic development of the plant. It is not surprising, therefore, that the structural manifestations along individual roots may be locally and quantitatively specific. To give a clear picture of this variability (observed in plant roots also under constant conditions; unpublished results), values for some selected roots will be presented.

The results of a quantitative analysis of three SRs are summarized in Table 1 and Figs. 1A, B, C illustrate cross sections of the stele of the SR No. 1.

The structure of SRs at a distance of 1 cm from the base corresponded to the known properties mentioned above (Hagemann, 1957). However, in the barley cultivar under investigation some SRs possess two adjacent vessels located in the centre of

Table 1. Seminal root

	No.	Distance from the base of the root			
		1 cm	15 cm	30 cm	45 cm
Stele diameter (μm)	1	108.0	81.0	84.0	—
	2	112.5	90.0	82.5	—
	3	105.0	112.5	75.0	78.0
Number of xylem strands	1	7	7	7	—
	2	7	7	7	—
	3	7	7	6	6
Number of parenchyma layers around the central metaxylem vessel	1	3–2	2–1	1	—
	2	4–2	2–1	2–1	—
	3	4–2	3–2	2–1	1
Central metaxylem diameter (μm)	1	33.0	36.0	37.5	—
	2	30.7	39.0	34.5	—
	3	34.5	31.5 + 10.5	31.5	31.5
Proportion of the central metaxylem (in %) of the stelar cross-section area	1	9.3	19.7	19.9	—
	2	7.5	18.8	17.5	—
	3	10.8	8.7	17.6	16.3
Relative conductivity of the central metaxylem (in %)	1	100	142	167	—
	2	100	258	158	—
	3	100	70	69	69

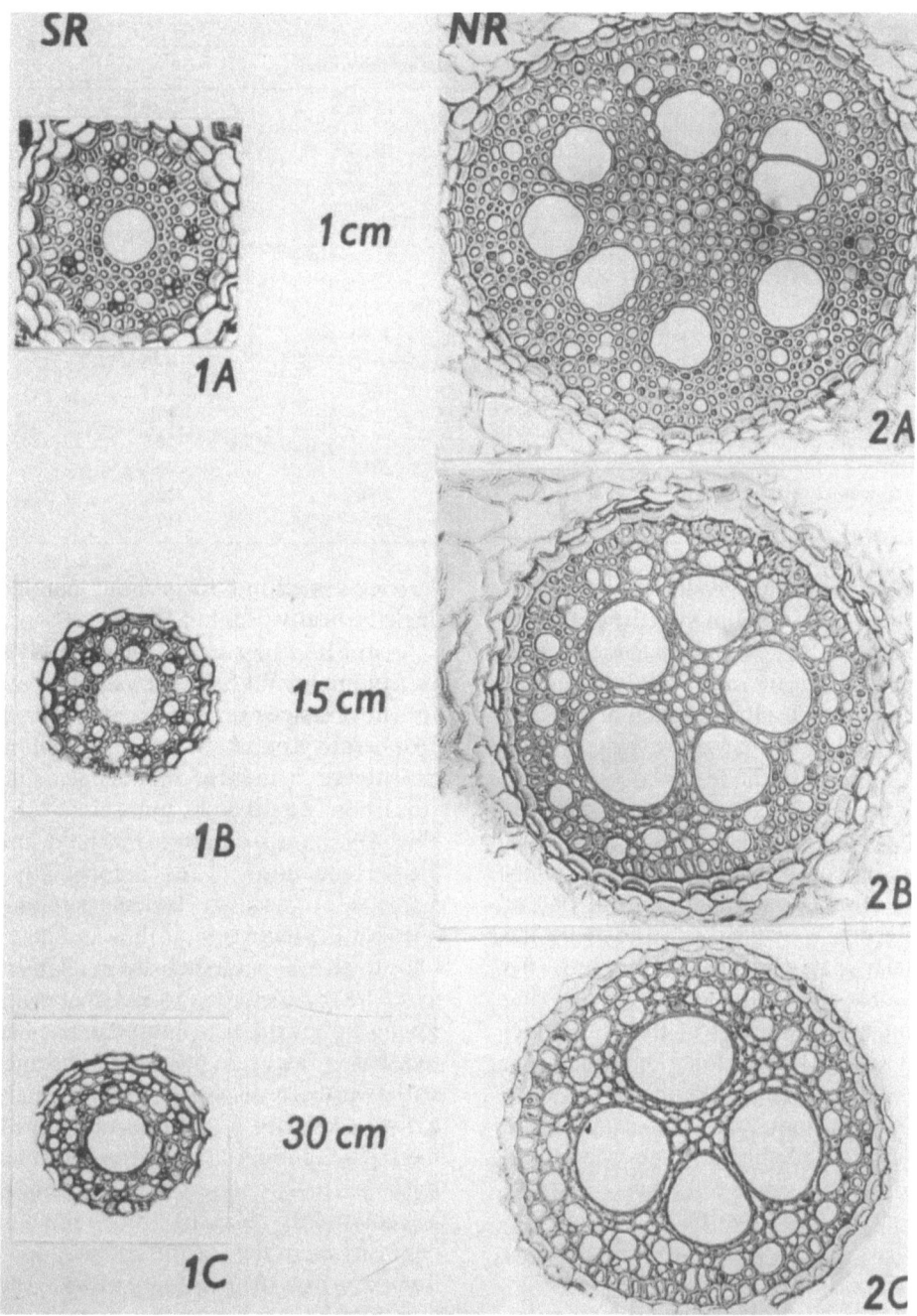

Figs. 1 and 2. Cross sections of stele of seminal (Fig. 1) and nodal (Fig. 2) barley root at 1 cm (**A**), 15 cm (**B**) and 30 cm (**C**) from the base. (× 170).

the stele instead of one vessel. Stele diameter, after an initial moderate increase, decreased in the acropetal direction. Nevertheless the number of peripheral vascular strands remained unchanged or, in some roots, it decreased at most by 1–2

strands. The diameter of the central vessel, equal to 30–35 μm at a distance of 1 cm from the base, increased in the acropetal direction, sometimes fluctuating, but in distally more remote parts of the root it was regularly greater than at the base. The

Table 2. Nodal root

	No.	Distance from the base of the root			
		1 cm	15 cm	30 cm	45 cm
Stele diameter (μm)	1	216.0	204.0	210.0	—
	2	255.0	225.0	213.0	—
	3	235.5	240.0	180.0	150.0
Number of xylem and phloem strands	1	15 + 15	13 + 13	13 + 13	—
	2	16 + 18	14 + 14	14 + 14	—
	3	15 + 15	15 + 15	13 + 13	11 + 11
Wide metaxylem vessels— number and max.–min. diameter (μm)	1	8 40.5–30.0	5 55.5–42.0	4 57.0–51.0	—
	2	7 36.0–27.0	5 58.5–44.0	4 66.0–48.0	—
	3	6 42.0–30.0	6 45.0–37.5	3 49.5–45.0	2 45.0
Proportion of the wide metaxylem (in %) of the stelar cross-section area	1	22	27	27	—
	2	13	26	29	—
	3	15	16	20	18
Relative conductivity of the wide metaxylem (in %)	1	100	201	266	—
	2	100	360	452	—
	3	100	115	117	65

relative conductivity of the central vessel changed slightly along the root, depending on the changing diameter of the vessel. Profound changes in relative conductivity result from fusion of the two central vessels as well as from local longitudinal division of a wide central vessel into two narrow vessels (root No. 3). The proportion of the central vessel on the cross-sectional area of the stele increased acropetally. Its isolated position might appear dangerous because in case of its disfunction there is no alternative pathway of water movement in the SRs. But hydraulic protection is evidently provided by the sheath of stelar parenchyma surrounding the vessels in the form of a continuous layer of living cells. The number of cell layers of this sheath decreased in the acropetal direction. The decreasing diameter of the stele is in accordance with it. In distal parts of the root, the central vessel is separated from the peripheral metaxylem and metaphloem only by a single cell layer. This very regular gradient is suggestive of a gradual longitudinal decrease of resistance to radial water flow.

Table 2 presents the results obtained by a quantitative analysis of NRs, Figs. 2A, B, C illustrate cross-sections of the stele of NR No. 1.

The basal diameter of the stele of NRs is approximately twice that of SRs. The vascular pattern of NRs is richer than that of SRs. Vessels of the inner metaxylem are wider than the solitary central vessel at the base of the SRs. In relation to their number and diameter, the relative conductivity of the wide vessels in NRs is about tenfold that of the single central vessel in SRs.

As in SRs, the stele diameter in NRs decreases longitudinally. Whereas the vascular pattern in SRs does not change or changes only little, in NRs both peripheral vascular strands and wide inner vessels are subject to marked reduction in the acropetal direction. The different number of the xylem and phloem strands is connected with the mechanism of their reduction. The acropetally increasing diameter of vascular elements not only compensates for the reduction of the vascular pattern, but leads to an acropetal increase in relative conductivity. The diameter of wide metaxylem vessels, reaching a maximum of 40–50 μm at 1 cm from the root base, was up to 60 μm at a distance of 30 cm, so that at this level the relative conductivity was up to fivefold higher. The proportion of wide vessels in the cross-sectional area of the stele increases, and this is partially connected with a changed proportion of tissues resulting from reduction of stelar parenchyma in the acropetal direction. This reduction occurs on the periphery of wide vessels which, as in SRs, suggests a decreasing resistance to radial water flow in more distal parts of the stele. The number of parenchyma cells in the root centre and between wide vessels, which thus approach each other, also decreases. The more distal regions of both types of barley roots are thus hydraulically favoured.

The formation of structural gradients is due to different intensities of growth processes, suggesting their relative independence and different requirements. Cell divisions responsible for pattern formation in barley root apices are inhibited with progressing ontogeny, and the pattern of the stele becomes simpler. In contrast, cell enlargement is stimulated, the diameters of vascular elements increase so much that in NRs the relative conductivity of wide metaxylem vessels increases acropetally in spite of a decreasing number of vessels. In terminal parts of the roots cell enlargement is finally inhibited. This trend is more or less affected by external conditions which may lead to local quantitative changes. The character of structural gradients and the values of relative conductivity of the hydraulically most efficient wide metaxylem cells indicate an acropetal decrease of resistance to radial and axial water flow and thus an acropetally increasing ability of the roots to extract the available water contained in the soil.

Discussion

A quantitative analysis elucidated the regularities of the incidence of structural changes appearing along the SRs and NRs of barley. In SRs, possessing a simple vascular pattern, there occur acropetally decreasing gradients of stele diameter and of the relative proportion of stelar parenchyma; in some SRs there is also an acropetally decreasing gradient of the number of vascular strands. In NRs with a more complicated vascular pattern there occurs regularly an acropetal decrease in the number of vascular strands, accompanied by an acropetally decreasing number of wide metaxylem cells. In addition, both root types are characterized by an acropetally increasing gradient of the diameter of vascular elements. Due to the latter, the axial relative conductivity increases longitudinally in spite of the simplification of the vascular pattern of NRs. In the most distal parts of the roots the vessel diameter decreases. The basic trends established in the development of structures along barley roots evidently are not limited to the species under investigation but appear to be of a rather general importance. This confirmed also our unpublished results obtained by quantitative analysis of maize roots. Ponsana (1975)(c.f. Greacen *et al.*, 1976), observed a similar trend in seminal roots of wheat.

A hydraulic protective role of stelar parenchyma surrounding metaxylem vessels in the roots can be anticipated based on the known function of paratracheal contact parenchyma occuring in the secondary xylem of numerous woody plants (Braun, 1970). Sheaths formed by this paratracheal parenchyma have long been known to protect vessels from air embolism. Even if embolisms were to occur, the sheaths can substantially contribute to their removal.

Earlier authors already noticed changes in the vascular pattern along roots of barley (Jackson 1922) and other Monocotyledons (Jost, 1932). Fahn (1967) described an acropetally increasing gradient of the width and length of tracheal elements of horizontal roots of *Retama raetam* and assumed that the greater width and length of tracheal elements in distal parts of the roots secures a more profuse water flow. These relations were experimentally confirmed by Altus *et al.* (1985) in wheat leaves.

The present study was limited to a quantification of stelar root structures and conclusions about their hydraulic properties were inferred from calculated conductivity of wide metaxylem vessels. Along SRs with simple root geometry, characterized by a single wide central vessel, changes in conductivity following changes in vessel diameter were less pronounced. For NRs with several fold longitudinal fusions of wide metaxylem vessels it was important from the point of view of hydraulics that the relative conductivity values did not decrease longitudinally but, in contrast, that they increased with increasing diameter of the fewer vessels.

References

Altus D P, Canny M J and Blackman D R 1985 Water pathways in wheat leaves. II. Water-conducting capacities and vessel diameters of different vein types, and the behaviour of the integrated vein network. Aust. J. Plant Physiol. 12, 183–199.
Braun H J 1970 Funktionelle Histologie der sekundären Sprossachse. I. Das Holz. Handbuch der Pflanzenanatomie Bd. IX Teil 1. Gebr. Borntraeger, Berlin.
Fahn A 1967 Plant Anatomy. Pergamon Press Oxford.
Fayle D C F 1968 Radial growth in tree roots. Techn. Rep. No. 9. Faculty of Forestry Univ. Toronto.
Gibson A C, Calkin H W and Nobel P D 1984 Xylem anatomy, water flow, and hydraulic conductance in the fern *Cyrtomium falcatum*. Am. J. Bot. 71, 564–574.

Greacen E L, Ponsana P and Barley K P 1976 Resistance to water flow in the roots of cereals. *In* Water and Plant Life, Eds. O L Lange *et al.* pp 86–100. Springer Verlag Berlin.

Hagemann R 1957 Anatomische Untersuchungen an Gerstenwurzeln. Kulturpflanze 5, 75–107.

Heimsch Ch 1951 Development of vascular tissues in barley roots. Am. J. Bot. 38, 523–537.

Jackson V 1922 Anatomical structure of the roots of barley. Ann. Bot. 37, 21–39.

Jost L 1932 Die Determinierung der Wurzelstruktur. Z. Bot. 25, 481–522.

Lebedenko L A 1962 Comparative anatomical analysis of the mature wood of roots and stems of some woody plants. (In Russ.) Trudy Inst. lesa i drev. AN SSSR, Sib. Otd. 51, 124–134.

Luxová M 1986 The hydraulic safety zone at the base of barley roots. Planta 169, 465–470.

Merry J 1941 Studies on the embryo of *Hordeum sativum*. I. The development of the embryo. Bull. Torr. Bot. Club 68, 585–598.

Riedl H 1937 Bau und Leistungen des Wurzelholzes. Jb. Wiss. Bot. 85, 1–75.

Taylor H M and Klepper B 1978 The role of rooting characteristics in the supply of water plants. *In* Advances in Agronomy Vol. 30. Ed. N C Brady, pp 99–128, Academic Press, New York.

B. C. Loughman et al. (Eds.), Structural and functional aspects of transport in roots, 21–24.
© 1989 by Kluwer Academic Publishers.

Experimental control of cellular patterns in the cortex of tomato roots

P.W. BARLOW and J.S. ADAM
Department of Agricultural Sciences, University of Bristol, AFRC Institute of Arable Crops Research, Long Ashton Research Station, Bristol BS18 9AF, UK

Key words: cell division, computer simulation, roots, tomato

Each species of plant has a characteristic pattern of cells underlying the structure of its roots. This pattern has two aspects: the first relates to the anatomy of the root and results from differential gene activity within the community of cells generated by the meristem (Barlow, 1982; Goldberg, 1986); the second aspect of cellular patterning is evident in the arrangement of its cell-files and is a consequence of cellular ancestry (Barlow, 1987). These two views of cellular patterns — based on histology and cell lineages — are complementary (though not necessarily mutually exclusive) features of root morphology.

The number of cell-files that comprise a root not only determines its diameter but also helps maintain a particular relationship between the volumes of cortex and stele (Barlow and Rathfelder, 1984). Presumably, this last-mentioned relationship has significance in optimizing the efficiency of root function within the limits of the resources available for root growth. Within an established root system, roots of different diameter occupy specific locations and comprise the axes of different order. Fine roots with small cortices and large surface-to-volume ratios probably contribute more to the efficiency of the root system in capturing water and mineral nutrients than thicker roots with larger cortices (*cf.* Hackett, 1969).

We shall describe experimentally induced variations of the number of cell-rows across the cortex of tomato roots cultured *in vitro* (*cf.* Street and McGregor, 1952) and suggest ways in which this may be regulated in the intact plant. In connection with this, we have explored, with the aid of computer simulation, a hypothetical 'rule' which we propose dividing cells obey in order to specify the orientation of their partition walls and which could, therefore, also regulate the number of cell-rows. In a wider context, the discovery of such developmental rules is one of the aims of students of plant morphogenesis (*cf.* Lindenmayer, 1978), and the testing of such hypothetical rules against the realities of development, is an integral way of establishing a rational basis for morphogenesis.

Our material was a clone of tomato roots (*Lycopersicon esculentum* cv. Ailsa Craig) which had been maintained for two years *in vitro* in White's medium containing 1.5% sucrose. In order to alter the pattern of cellular activity in the meristem, the sucrose in the medium at the start of one of the weekly subculture periods was adjusted to either 0.5%, 1.0% or 2.0%. At intervals following such a routine transfer of 1 cm root tips, samples of apices were fixed, embedded in paraffin wax, and sectioned longitudinally at 6 μm. Sections were then dewaxed, stained by the PAS reaction (Jensen, 1962), and mounted under a coverglass. Observations and drawings were made of median and near-median sections.

Following their transfer to fresh medium, the root tips show a triphasic pattern of growth: an initial period of slow growth is followed by a phase of rapid elongation of the root axis which then gives way to a decline in elongation rate; subculturing the tips reinitiates. Most observations were made on roots 7 days after the initial subculture into the new sucrose concentration. At this time they were either just entering the final phase (those roots in 1.0% and 2.0% sucrose) or still rapidly elongating (those in 0.5% sucrose). In either case, the cortex had achieved a stable pattern of cellular organization. The number of cell-rows constituting

Table 1. The mean ($\pm$ SE) number of cell rows across the cortex, and width of the root and its cortex, in excised tomato roots grown for 7 days in three different concentrations of sucrose

Sucrose (%)	Number of cell rows	Width (μm)[a]	
		Cortex	Root
0.5	4.1 $\pm$ 0.1	46 $\pm$ 2	170 $\pm$ 10
1.0	5.4 $\pm$ 0.3	71 $\pm$ 3	236 $\pm$ 4
2.0	7.8 $\pm$ 0.2	83 $\pm$ 3	257 $\pm$ 11

[a] Measured 300 μm behind the root/cap junction.

Table 2. The frequency of periclinal divisions in different cell-rows of the cortex in roots grown for 7 days in three different concentrations of sucrose

Sucrose (%)	Frequency (%)			
	I[a]	I-1	I-2	I-3
0.5	100.0	0	0	0
1.0	94.0	6.0	0	0
2.0	92.9	4.6	1.5	1.0

[a] I—Innermost row of the cortex; I-1—row next to I; I-2—row next to I-1, *etc.*

the width of the cortex was related to the sucrose concentration (Table 1). Thus, a cortex four rows wide was characteristic of roots grown in 0.5% sucrose, while 2.0% sucrose supported a cortex with double this number of cell-rows (Fig. 1).

The branching of the files of cells and hence the number of cell-rows seen across the cortex, is related to the number of periclinal divisions. Such divisions are located close to the apex and are most frequently found in the innermost row of cortical cells (Fig. 1; Table 2). Where there are four cortical rows (*e.g.* roots grown in 0.5% sucrose), these arise from three periclinal divisions invariably located in the innermost row. Where there are eight rows (*e.g.* roots grown in 2.0% sucrose) there are seven periclinal divisions; usually, these all occur in the

innermost row, but occasionally some are found in the adjacent one or two rows.

Numbering the cells in the inner cortical row basipetally, starting from the most distal cell (cell number 1) (Fig. 1), and considering the cell just distal to each of the branch positions as the site of a periclinal division, shows that the majority of such divisions occur in regular locations along the cell-row (Table 3). They are particularly frequent close to the apex of the cortical complex. However, the final periclinal division, which may be the third (occasionally the fourth) such division in 0.5% sucrose-grown roots, or the seventh in 2.0% sucrose-grown roots, is always separated from the site of the preceding periclinal division by a few cells in which divisions occur transversely. Thus, if one considers the probability of encountering periclinal divisions in the innermost cortical row of, say the 2.0% sucrose-grown roots, it is highest close to cell number 1 and declines to zero by about cell number 15. The probability of encountering transverse divisions varies inversely.

Occasionally, additional periclinal divisions occur in the innermost cortical row beyond the above-mentioned proximal limit. But they are sporadic since the additional cell-rows which they generate extend for only relatively short distances.

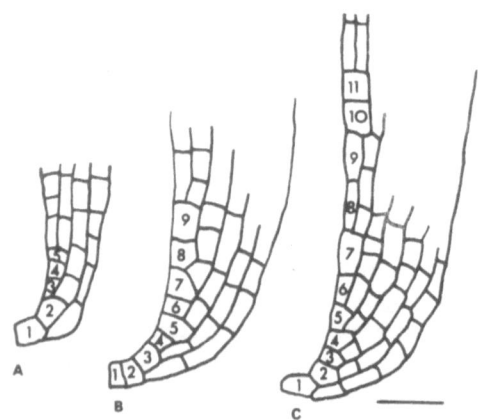

Fig. 1. Typical cell patterns in the cortex of tomato roots grown for seven days in (**A**) 0.5% sucrose, (**B**) 1.0% sucrose, and (**C**) 2.0% sucrose. Cells of the innermost cortical row are numbered sequentially starting from the most apical cell of the cortical cell-complex (cell number 1). Cells considered to be the next to divide longitudinally in (**A**) are: 1, 2 and 5; in (**B**) they are 2, 3, 5, 7 and 9; and in (**C**) they are 1, 2, 3, 4, 5, 7 and 11. *Scale bar* represents 25 μm.

Table 3. Mean position of successive periclinal divisions in the innermost cortical row recorded as the number of the cell proximal to the origin (designated as cell number 1) of the set of cortical cell-rows (see Fig. 1)

Sucrose (%)	Cell position for periclinal division number:						
	1	2	3	4	5	6	7
0.5	1.2	2.6	5.6	8.0	—	—	—
1.0	1.1	3.0	4.6	6.0	8.5	—	—
2.0	1.4	2.7	4.1	5.7	7.5	9.1	12.3

However, an additional periclinal division in the cortex regularly occurs opposite each of the two protophloem poles (also described by Heimsch, 1960).

Besides periclinal divisions, the cortex supports frequent radial longitudinal divisions. These increase the number of cell-rows in the circumferential plane. Although we have not investigated these divisions, we assume that they coexist with the periclinal divisions. In fact, it is likely that at certain locations in the distal portion of the innermost cortical row, periclinal and radial divisions may alternate in successive cell generations. However, the occasional transverse division would also need to be interspersed with these two types of longitudinal division in order to accommodate the gradual increase in the length of the file. Clearly, there is much to understand concerning the simple 3-dimensional process of generating files of cells such as exist in the cortex.

Morphogenesis depends upon the co-ordination of the rates and planes of cell growth and division (Lloyd and Barlow, 1982). One outcome of the present work was the possibility of utilizing the variation of division patterns in the cortex induced by the cultural conditions as a means of inferring how cells are determined to divide either longitudinally or transversely. In many systems, the plane of cell division seems to be oriented perpendicular to the principal direction of cell growth. Thus, in more proximal locations along the inner cortex the principal direction of cell growth is parallel to the axis of the cell-rows and here the cells divide transversely. However, in more distal locations, radial and circumferential growth are more in evidence, and in consequence longitudinal divisions are more frequent. We explored, therefore, the possibility that should the relative amounts of longitudinal and radial growth determine the plane of division, then these might find expression in the ratio of the length and breadth of a dividing cell. We term this ratio the 'cellular aspect ratio'.

Aspect ratio reflects the past pattern of cell growth. But whether the plane of division is determined prior to division by the integration of the amount of growth in each of its three principal directions (longitudinal, radial and circumferential), or whether it is determined simply by the value of the aspect ratio at the time of commitment to

division, is not known. For the sake of simplicity (and because we lack further information), we assume the latter. Also, for the same reasons, we apply the aspect ratio to only two of the three possible orientations of division, namely, periclinal and transverse.

Given that cell shape at the time of division determines the orientation of the new cell wall, it is necessary to define numerical values of the aspect ratio that permit one or other of the two types of divisions. A 'critical aspect ratio' is taken as the value which, if exceeded, would cause a cell to divide transversely, but if not exceeded would result in its longitudinal division. In practice, cells which divide transversely tend to be longer than they are wide and hence have a relatively high value for their aspect ratio, whereas cells that divide longitudinally tend to be wider and hence have a lower aspect ratio. The critical aspect ratio must therefore lie between these two limits. When estimates are made of the width and length of cells expected to divide in either of the two planes, there is quite a sharp difference between their ratios: they lie in the range 0.9–1.9 for longitudinal divisions, and 2.3–3.0 for transverse divisions. The critical aspect ratio must therefore lie between 1.9 and 2.3. This applies irrespective of the sucrose concentration in which the roots are grown.

The concept of the aspect ratio makes possible a computer simulation of cell division patterns. In one such simulation model, realized with the aid of a program originally devised by Miss Alison Thurlbeck, the rates of cell elongation and cell division can be varied independently along a row of cells. As a consequence, the aspect ratio of a cell at the time of its division varies along the length of the row. By setting a suitable value for the critical aspect ratio, the position beyond which a file of

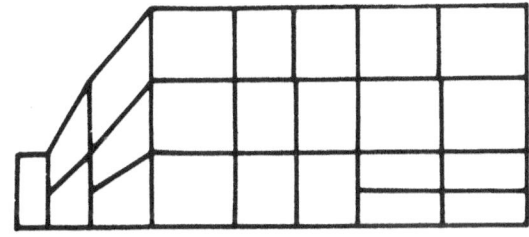

Fig. 2. Computer simulation of a simple cellular branching pattern such as found in the cortex of tomato roots grown in 0.5% sucrose.

cells will no longer branch by longitudinal division can be determined. The result of one such simulation that mimics quite closely the cellular pattern seen in the cortex of roots grown in 0.5% sucrose is shown in Fig. 2.

Further work is obviously necessary to test the validity of cell shape as a determinant of the plane of cell division. An alternative hypothesis is that the division plane is specified by the principal direction of cell growth, in which case aspect ratio is only a derivative of this and therefore has no intrinsic significance as a regulator of division.

Whatever the true explanation, the orientation of cell growth is clearly of importance for morphogenesis. Observations of sections of tomato roots show that the innermost row of cortical cells is virtually the only one in which radial growth occurs. The immediate question must be why this is so. We speculate that the innermost row grows radially rather than longitudinally as a response to sucrose (or some other substance whose distribution is regulated by sucrose) leaking from the stele. The additional periclinal divisions that consistently occur opposite the phloem of roots grown both *in vivo* and *in vitro* support the view that the stele could be the source of sucrose, even though in the *in vitro* situation sucrose presumably permeates the whole of the cortex. Accepting this possibility, we further suggest that the degree of development of the endodermis and/or pericycle might regulate the passage of sucrose to the cortex and hence confine its effects to cortical cells located close to the apex. Thus, it would be of interest to chart the development of endodermis, as well as the phloem, in relation to the longitudinal distribution of periclinal divisions, and also to observe whether any differences exist in the development of these cells in roots grown in different sucrose concentrations.

Another important problem is the means by which growth rate in each of the three planes is regulated. Is this a feature of cytoskeletal structures (microtubules), or is it due to some subtle polarization of membranes and cytoplasm (including plasmodesmata)? And what then is the link between these subcellular structures and the growth response engendered by available sucrose? Only when answers to these questions are forthcoming will anything be understood about the fundamental processes of morphogenesis, not only of roots but also of other plant organs.

References

Barlow P W 1982 Root development. *In* The Molecular Biology of Plant Development. Eds. H Smith and D Grierson, pp 185–222. Blackwells Scientific Publications, Oxford.

Barlow P W 1987 Cellular packets, cell division and morphogenesis in the primary root meristem of *Zea mays* L. New Phytol. 105, 27–56.

Barlow P W and Rathfelder E L 1984 Correlations between the dimensions of different zones of grass root apices, and their implications for morphogenesis and differentiation in roots. Ann. Bot. 52, 249–260.

Goldberg R 1986 Regulation of plant gene expression. Phil. Trans. Roy. Soc. Lond. B 314, 343–353.

Hackett C 1969 A study of the root system of barley. II. Relationships between root dimensions and nutrient uptake. New Phytol. 68, 1023–1030.

Heimsch C 1960 A new aspect of cortical development in roots. Am. J. Bot. 47, 195–201.

Jensen W A 1962 Botanical Histochemistry: Principles and Practice. W H Freeman and Co., San Francisco.

Lindenmayer A 1978 Algorithms for plant morphogenesis. *In* Theoretical Plant Morphology. Ed. R. Sattler, pp 37–81. Leiden University Press, Leiden.

Lloyd C W and Barlow P W 1982 The co-ordination of cell division and elongation: The role of the cytoskeleton. *In* The Cytoskeleton in Plant Growth and Development. Ed. C W Lloyd, pp 203–228.

Street H E and McGregor S M 1952 The carbohydrate nutrition of tomato roots. III. The effects of external sucrose concentration on the growth and anatomy of excised roots. Ann. Bot. 16, 185–205.

B. C. Loughman et al. (Eds.), Structural and functional aspects of transport in roots, 25–28.
© 1989 by Kluwer Academic Publishers.

Determination of vascular tissue in roots of dicotyledonous plants

P. B. GAHAN and D. F. CARMIGNAC

Biology Department, King's College London, Campden Hill Road, London W8 7AH, UK

Key words: cytochemistry, determination, dicotyledonous plants, roots, tissue culture, vascular tissue

Introduction

Quantitative enzyme cytochemistry together with *in vivo* and *in vitro* studies of primary and secondary meristems have permitted the identification of a precocious marker enzyme for the determination of meristematic cells to form vascular tissue, namely, carboxylesterase (Gahan, 1981). The quantitative cytochemical test (Gahan, 1984; Rana and Gahan, 1983) involves the use of the following reaction:

Naphthol AS-D Acetate +   Fast Blue BB
   (substrate)   (diazonium salt)
   (colourless)   (yellow)
   (water-soluble)   (water-soluble)

ESTERASE
(pH 6.5)

NAPHTHOL AS-D - FAST BLUE BB
COMPLEX
(water-insoluble ppt; blue)

The carboxylesterase activity can be distinguished from the other esterases that can act on this broad-spectrum substrate, being inhibited by either $10^{-4} M$ diisofluoropropylphosphate (DFP) or $10^{-4} M$ diethylparanitrophenylphosphate (E600), but resistant to either $10^{-4} M$ eserine or $10^{-4} M$ parachloromercuriphenylsulphate (PCMPS) can break between phenyl and sulphate if necessary (Gahan, 1981, 1984; Holmes and Masters, 1967; Veerhabhandrappa and Montgomery, 1971). This reaction permits the identification of a central column of cells in the meristem regions of roots of dicotyledonous species, which are determined to

form vascular tissues. This column of cells is contiguous with the vascular tissue of the root and terminates at the edge of the quiescent centre (Gahan, 1981; Rana and Gahan, 1982).

Evidence for the determination of cells to form vascular tissue elements in root apices

If the central column of meristematic cells are truly determined, then removal of the cells from the environment containing the determination initiator(s) should leave the cells still in the determined state, *e.g.* they should retain the carboxylesterase activity. In order to test this possibility, the three terminal 0.5-mm segments of the root tip of *Pisum sativum*, after removal of the root cap, were explanted into basal culture medium (Murashige and Skoog, 1962) containing two per cent sucrose as a carbohydrate source, but no added growth regulators. Three to six days later, the explants were frozen and serially sectioned transversally (Gahan, 1984) and reacted cytochemically for carboxylesterase activity. In all of the segments examined the carboxylesterase activity was not only retained, but had increased, in the absence of the initiating substance(s), so confirming the true determination of these cells. In addition, all of the cells in the zone of the procambium which will form the xylem in segments two and three, had developed secondary cell walls (birefringent on viewing with polarized light). In the first segment, the tissue nearest the root tip had only carboxylesterase activity, but at the end farthest from the root tip, protoxylem cells with secondarily thickened walls could be seen. So far, it has not been possible to induce this cell-wall change down to the edge of the

Table 1. Some features of the carboxylesterase identified as a marker of vascular tissue differentiation

1. Sensitivity to DFP and E600 (both at $10^{-4}M$), but resistance to eserine ($10^{-4}M$) and PCMPS ($10^{-4}M$).
2. A substantial proportion is linked within the cell wall (Gahan and McLean, 1969; Carmignac and Gahan, 1988).
3. Electron microscopy indicates that it is formed in the RER and is transferred to the cell wall via vesicles formed from the SER (Gahan and McLean, 1969).
4. It is rapidly exported to and through the plasmalemmae of protoplasts prepared from suspension culture cells of *Daucus carota*.
5. Inhibition *in vivo* of the esterase in roots of *Pisum sativum* by E600 results in a shortening of the longitudinal cell wall (Gahan, Amakiri and Maple, in preparation).

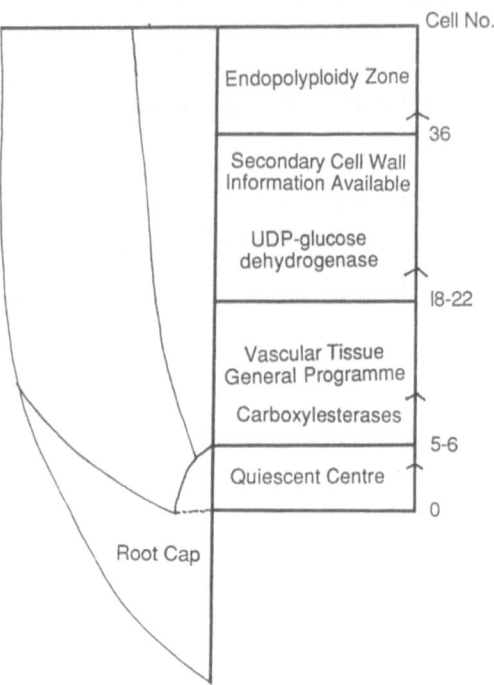

Fig. 1. Diagram of a median longitudinal section of a half root of *Pisum sativum*, indicating the cell number back from the root cap/root tip junction at which enzyme activities are either initiated or sharply increased. The availability of parts of the genetic programme for the formation of vascular tissue elements is also indicated.

quiescent centre. This implies the presence of either a barrier to complete differentiation into the meristem or lack of part of the programme for such a development. Either way, this is of importance since there appears to be a mechanism preventing the loss of the meristematic activity through differentiation, and hence from loss of growth by increased cell mass.

Some features of the carboxylesterase

Although it has not yet been possible to biochemically characterize this enzyme, preliminary studies would indicate that it is involved with cell wall changes (Table 1).

Is the programming in roots multi-step?

It does not seem possible to induce cell differentiation right down to the quiescent centre in roots of either *Pisum sativum* or *Vicia faba*. The barrier occurs at 18–22 cells back from the root tip/root cap junction. A general programming of the central procambial cells occurs from cells 5–6 back to cells 18–22 when the secondary cell wall information becomes available and UDP-D-glucose dehydrogenase shows increased rates of activity (McGarry and Gahan, 1986) as does glucose-6-phosphate dehydrogenase (G6PD) activity (Fig. 1). This implies a multi-step programming, an idea which was further tested by studying the de-

velopment of a vascular bridge after severing the central vascular bundle in roots of *Pisum sativum* (Robbertse and McCully, 1979).

A quantitative cytochemical study was made to determine the earliest moment that the carboxylesterase activity was present in the cortical parenchymal cells that will form the vascular bridge, *i.e.* how early can the general programming be observed? Unfixed, frozen sections of the wound region were reacted for carboxylesterase activity, and by 18–20 hours after wounding, the first esterase activity was observed in the cortical cells which will form the vascular elements (Fig. 2). At 24 hours, mitosis was observed adjacent to the severed ends of the vascular bundle, *i.e.* at the top and bottom ends of the bridge. The first sign of xylem elements was at about 72 hours after wounding (Fig. 2).

A further series of tests were made to check (a) if the enzyme activity was evidence of determination, and (b) when the first-determined cells were programmed. This involved wounding the roots, leaving them for a period of time (x hours) isolating the whole of the wounded region and leaving them

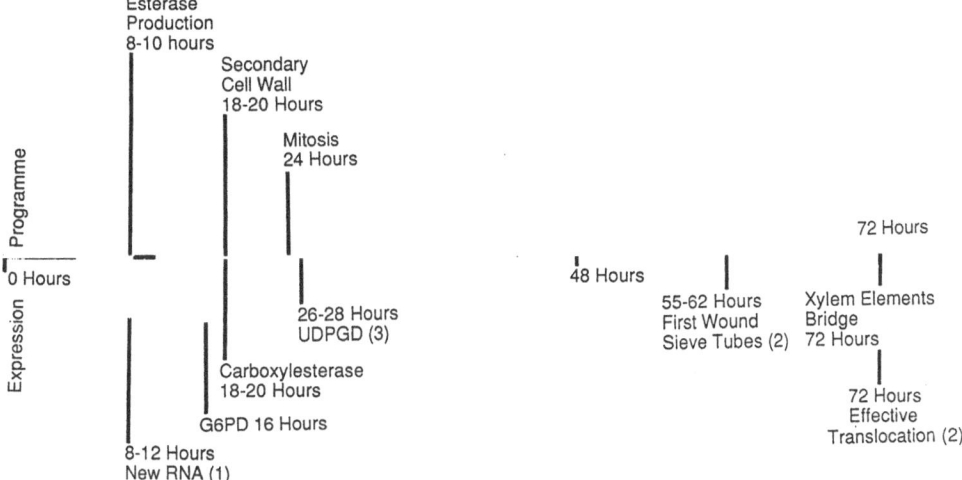

Fig. 2. Timings, in hours, of events observed to occur in the region of the vascular bridge-forming, cortical parenchyma on severing the vascular bundle of the main root of *Pisum sativum*. At 72 hours after wounding, a viable vascular bridge has been formed. (1) = Shininger, 1979; (2) = Schulz, 1987; (3) = McGarry and Gahan, 1986.

in basal culture conditions for a further period (y hours) where x + y = 96 hours. Cytochemical examination of the bridge regions indicated that the first carboxylesterase activity appeared in the probridge segments isolated 8–10 hours after wounding and left for a further 86–88 hours, but no elements with secondarily thickened cell walls were observed. The latter event occurred only if the plants were left for 18–20 hours after wounding and before removal of the bridge zone. These results indicate that (a) on removing the stimulus, the carboxylesterase activity remains and hence that the cells are determined; (b) although carboxylesterase programming is present 8–10 hours after wounding, the information to make secondary cell walls is not available until about 10 hours later; and (c) although the esterase programme is available at 8–10 hours after wounding, it is not expressed until 8–10 hours later. This confirms the multi-step nature of the programming.

Further confirmation of this state has been gained from analysis of the metabolic pathways as assessed by quantitative cytochemical analysis of G6PD activity as a flux-generating step in the pentose phosphate pathway (Turner and Turner, 1980; Gahan *et al.*, 1979) and of UDP-D-glucose dehydrogenase (UDPGD) activity as a measure of flux through the pathway generating hemicellulose precursors (McGarry and Gahan, 1986). G6PD activity was markedly increased by 16–18 hours after wounding, but UDPGD activity did not increase

substantially until about 30 hours after wounding (Fig. 2).

Initiation of the programme

Clearly, the initiation of the programme must occur normally during embryogenesis, this basic plan being maintained and extended as the embryo, and subsequently the plant, develops. The cytochemical analysis for the identification of the first procambial cells of the embryo to be determined (and so show carboxylesterase activity) can be made either *in vitro* or *in vivo*. In the first case, esterase-positive cells indicating the procambium were first observed after six days of culture in somatic embryos of *Daucus carota* at the polarized globular stage of development (Caligo *et al.*, 1986). Esterase-positive procambial cells of embryos of *Pisum sativum*, *in vivo*, can be observed by the third day after the sepals are fully-reflexed (Gahan and Rana, in preparation) whilst King and Heyes (1986) reported cytological identification of such cells only five days after the sepals are fully-reflexed. Further, if embryos of *Pisum sativum* from the smallest to the largest in a pod are explanted into Murashige and Skoog (1962) basal medium containing two per cent sucrose, and in the case of the larger embryos, sliced transversally, the procambial cells not only maintain their carboxylesterase activity, but produce secondary cell walls as well. Thus, all or a major part of the programme for vascular tissue production is likely to be present already in the smallest embryo in the pod (Gahan and Rana, in preparation).

Possible mechanisms of initiation

When segments of roots of *Pisum sativum* are explanted into Murashige and Skoog (1962) medium containing $1 \, \text{mg} \, l^{-1}$ 2,4-dichlorophenoxy-acetic acid (2,4-D), $5 \, \text{mg} \, l^{-1}$ benzylaminopurine and 2% sucrose at 30°C in the dark, the cortical parenchyma cells convert to tracheids (Phillips and Torrey, 1973). Normally, no carboxylesterase is detectable in the cortical parenchyma cells, yet under the above mentioned conditions of culture, its activity is initiated prior to the cell wall changes. 2,4-D alone gives a higher activity than that observed with either benzylaminopurine alone or in combination with it (Gahan *et al.*, 1983). However, although both cytokinin and auxin are involved in such tracheid production and at a stage when the early marker enzyme is initiated, unpublished data (Gahan, Roberts and Welbourne) show the carboxylesterase being switched-on in lettuce pith cells by both ethylene and GA. Thus, more data are needed to decide on the basic mechanism of initiation of vascular tissue differentiation, although a model based upon simple diffusion patterns can be indicated from the root wounding experiments (Rana and Gahan, 1983).

Nematode resistance

Preliminary studies indicate that the resistance factors to nematode infection in tomato roots have an early involvement of carboxylesterase activity in the cortical parenchymal cells surrounding the nematode after its entry into the root (Melillo *et al.*, 1988).

Acknowledgements

P B G wishes to thank the Science Research Council (U.K.), the University of Geneva (Plant Physiology) and the Central Research Fund (University of London) for partial financial assistance.

References

Caligo M A, Nuti Ronchi V and Nozzolini M 1986 Proline and serine affect polarity and development of carrot somatic embryos. Cell Diff. 17, 193–198.

Fahn A 1974 Plant Anatomy, 2nd Edition. Pergamon Press, Oxford.

Gahan P B 1981 An early cytochemical marker of commitment to stelar differentiation in meristems of dicotyledonous plants. Ann. Bot. 48, 769–775.

Gahan P B 1984 Plant Histochemistry and Cytochemistry: An Introduction. Academic Press, London.

Gahan P B and McLean J 1969 Subcellular localization and possible function of acid β-glycerophosphatases and naphthol esterases in plant cells. Planta (Berl.) 89, 126–135.

Gahan P B, Auderset G and Greppin H 1979 Pentose phosphate pathway activity during floral induction in spinach. Ann. Bot. 44, 121–124

Gahan P B, Rana M A and Phillips R 1983 Activation of carboxylesterase in root cortical parenchyma cells of *Pisum sativum* during xylem induction, *in vitro*. Cell Biochem. Function 1, 109–111.

Holmes R S and Masters C J 1967 The developmental multiplicity and isozyme status of Cavian esterases. Biochem. Biophys. Res. Comm. 132, 379–399.

King G A and Heyes J K 1986 Morphology and cytology of pea embryo during histogenesis. Ann. Bot. 58, 633–640.

McGarry A and Gahan P B 1986 A quantitative cytochemical study of UDP-D-glucose dehydrogenase:NAD oxidoreductase (EC.1.1.1.22) activity during stelar differentiation in *Pisum sativum* L. cv. Meteor. Histochemistry 83, 551–554.

Melillo M T *et al.* 1988 Histochemical localization of carboxylesterases in roots of *Lycopersicon esculentum* in response to *Meloidogyne incognita* infection. Ann. Appl. Biol. *In press*.

Murashige T and Skoog F 1962 A revised medium for rapid growth and bioassays with tobacco tissue cultures. Physiol. Plant. 15, 473–479.

Phillips R and Torrey J G 1973 DNA synthesis, cell division and specific cytodifferentiation in cultured pea root cortical explants. Develop. Biol. 31, 336–347.

Rana M A and Gahan P B 1982 Determination of stelar elements in roots of *Pisum sativum*. Ann. Bot. 50, 757–762.

Rana M A and Gahan P B 1983 A quantitative cytochemical study of determination for xylem element formation in response to wounding in roots of *Pisum sativum*. Planta (Berl.) 157, 307–316.

Robbertse P J and McCully M E 1979 Regeneration of vascular tissue in wounded pea roots. Planta (Berl.) 145, 167–173.

Schulze A 1987 Sieve-element differentiation and fluoresceine translocation in wound phloem of pea roots after complete severance of the stele. Planta (Berl.) 170, 289–299.

Shininger T L 1979 The control of vascular development. Plant Physiol. 30, 313–337.

Turner J F and Turner D H 1980 The regulation of glycolysis and pentose phosphate pathway. *In* The Biochemistry of Plants 2, Ed. D D Davies, pp 279–316. Academic Press, London.

B. C. Loughman et al. (Eds.), Structural and functional aspects of transport in roots, 29–33.
© 1989 by Kluwer Academic Publishers.

The heterogeneity of root tip cells and tissues as revealed by *in situ* enzyme localization and isoenzyme patterns: A comparison of proteo- and esterolytic activities

KAREL BENEŠ[1], VĚRA HADAČOVÁ[1] and JAN HLAVÁČEK[2]
[1]Institute of Experimental Botany, Czechoslovak Academy of Science, Ke dvoru 15, CS-166 30 Praha 6, Czechoslovakia and [2]Institute of Organic chemistry and Biochemistry, Czechoslovak Academy of Science, Flemingovo nám. 2., CS-166 10 Praha 6, Czechoslovakia

Key words: esterase, isoenzymes, localization, protease, root tip, *Vicia faba, Zea mays*

Introduction

The root tip is a relatively simple and well defined developmental system. Moreover, germination is one of the best-provided biological processes. Taking genetically homogeneous material like seeds of a certified cultivar of crop plants, we obtain fairly standard objects. This led us to the decision to use seedling root tips in our research on enzymes as markers of cell and tissue differentiation, which is the main aspect of our studies on enzyme heterogeneity within the root tip.

Proceeding from the works of Brown and his pupils (see Brown, 1963), two approaches are used in our laboratory: *in situ* histochemistry which is, with a few exceptions, the only technique that makes possible studies on differential enzyme activity at tissue level and the comparison of isoenzyme patterns in particular regions of the root tip. We limited ourselves to the study of hydrolases.

A series of papers has been published from our laboratory on the localization of carboxyl esterases (Beneš, 1971; 1977), phosphatases (Beneš and Opatrná, 1964) and glycosidases (Beneš and Hadačová, 1980, Beneš, Ivanov and Hadačová, 1981). The results obtained hitherto concerning 1) the presence of hydrolases in root tips, 2) their localization, and 3) some aspects of the regulatory mechanisms involved may be summarized as follows. 1. Many hydrolases are present in the root tip and the activity of some of them is fairly high. The knowledge of the role of hydrolases in cell metabolism is still so fragmentary that it is not possible to adequately explain their presence in root tips. 2. Whereas many efforts have been devoted to studies of the subcellular distribution of enzymes, we have looked for regularity in enzyme localization at tissue level. It appeared that the localization of particular hydrolases is usually not equal both concerning the same enzyme in root tips of different species and different enzymes in the same root tip. In some roots there are quantitative gradients in enzyme activity, in others, rarely, the differences in activity are regarded as qualitative. The localization patterns correspond to the common structure of the root and may be described in terms of plant anatomy. However, the particular tissue complex (*e.g.* pericycle) need not reveal the same degree of activity: it depends on the level of differentiation or on the functional state of the tissue. 3. In differentiating tissue complexes the differences in enzyme activity appear in most cases only shortly before or simultaneously with the appearance of the relevant structural features. If we distinguish 3 phases, or levels, of plant differentiation (a, organogenesis; b, histogenesis; and c, the differentiation of the histogen) then, leaving aside organogenesis the enzyme localization patterns concern the differentiation of the histogen, if the functional state of the given cell complex is not concerned.

Using polyacrylamide gel electrophoresis (PAGE) we tried to compare isoenzyme patterns in extracts of dissected parts of the root tip corre-

30   *Beneš* et al.

sponding to particular root growth zones. We obtained not only differences in staining intensity and width of particular bands, which are regarded as quantitative, but also qualitative differences *i.e.* unequal number and disparate position of the bands. However, we were not able to discern any general conclusion concerning the relation between isoenzynme patterns and the progress of cell growth and maturation (Hadačová and Beneš, 1977).

The present paper is a continuation of our above mentioned studies and is devoted to the comparison of the localization and of the isoenzyme patterns of some peptidases and esterases.

In situ studies

The objects of our studies are seedling root tips of *Vicia faba* L. cv Chlumecký and *Zea mays* L. cv. Český bílý koňský zub obtained as described elsewhere (Beneš and Kutík, 1978). With respect to our preliminary results, in the present paper the *in situ* histochemistry concerns maize root tips only whereas the isoenzyme studies were performed with root tips of *Vicia faba*.

Since the histochemistry of proteases has received little attention in plants, it was necessary to standardize the procedure (Beneš, in press). We use free floating sections of cold Ca formol fixed root tips and a simultaneous azo-coupling procedure (2.5 mg substrate in 0.5 ml dimethylformamide + 9.5 ml phosphate buffer pH 6.55 containing 10 mg Fast blue B salt or Fast Garnet GBC salt). As substrates the naphthyl amides (NA) of the following amino acids were used: alanine, leucine, histidine, arginine, tryptophane, tyrosine, glycine, proline, hydroxyproline, phenylalanine, α-and β-

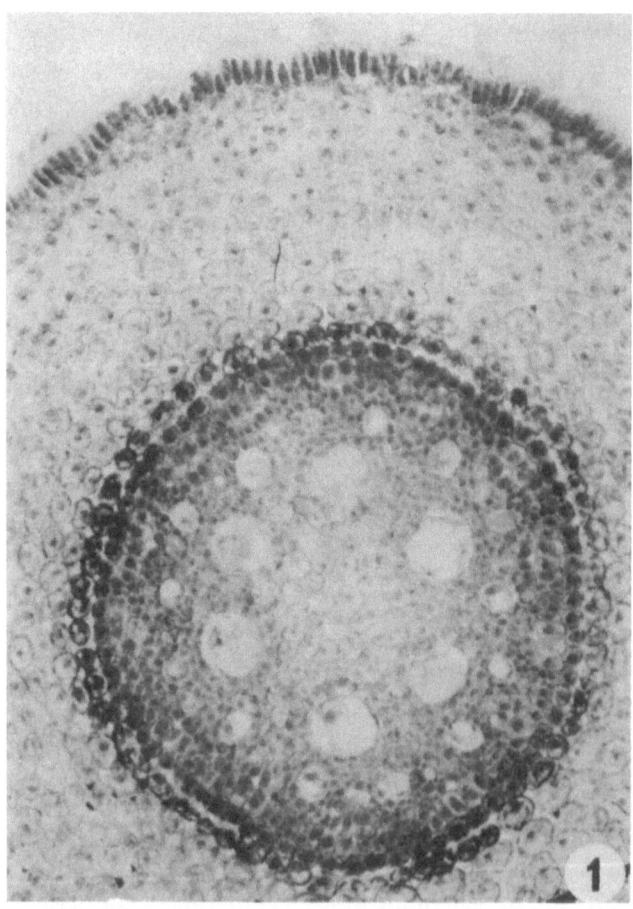

Fig. 1. The localization of hydrolytic activities in transverse sections of maize root tips. Substrate: 1-naphthyl butyrate. Mag. 40 ×.

glutamic acid. Several hours incubation is necessary. The resulting azo-dye must be carefully distinguished from the background staining, which is of similar shade and of high intensity under these conditions. Since the incubation medium becomes coloured, there is the danger of false positive results.

Clearly positive results were achieved with NAs of leucine, alanine and phenyalanine. Rhizodermis is the most positive part of the section, cortex is weak or negative and the central cylinder positive. There are no substantial differences in the localization where the three mentioned substrates are compared. In the case of proline and hydroxyproline NAs it was not possible to exclude the false positive reaction. Intense staining with histidine NA is definitely a false positive. With other NAs the results are very weak or negative.

If we consider the results of esterolytic activity, evaluating other sections from the same roots incubated in corresponding media, then, with 1- and 2-naphthyl acetate and 1-naphthyl propionate the whole section is positive within 15 to 30 min. The incubation in 1-naphthyl butyrate must be prolonged to 1 to 2 h. Here, rhizodermis becomes positive, cortex very weak or negative, but endodermis and adjacent cell layer are positive (Fig. 1). In the central cylinder, the periphery is more coloured than the central part. With naphthol AS acetate the results are similar to those obtained with 1-naphthyl butyrate, but the two layers are not conspicuous.

In further experiments we tried some other amide and ester substrates. The material was handled as mentioned before and the media — with the exception of substrates — were the same. Plain N-acyl NAs (see Table 1) were not hydrolysed at all. From several N-acetylated aminoacid esters applied (see Table 1), a positive reaction appeared only with N-acetyl-L-glycine-2-naphthyl ester. Again, the rhizodermis and the periphery of the central cylinder were positive (Fig. 2).

To evaluate the results achieved in the present work it is necessary to mention that the substrate specificity of some proteases is broad because they hydrolyse not only different peptides but also some esters; on the other hand, some enzymes denoted as esterases also hydrolyze peptidic substrates. In some of these the term esteroprotease has been used and usually concerns the serine proteases. Another

Table 1. The comparison of hydrolytic activity of particular PAGE protein bands. Note their position (Rm), number, width and intensity (black — high, double hatched — medium, hatched — low; the dotted lines represent very weak bands not always detected). Four groups of substrates were used

meaning of this concept, but not contradictory, is based on the ability of esteroproteases to slit esters of amino acids. Thus, the term esteroprotease is not quite unambiguous and is not always used (von Deimling and Bocking, 1976; McDonald, 1985 in animal material; in plants Mikola and Mikola, 1986; Storey and Wagner, 1986). However, some relevant data from plant material are available, see *e.g.* Aducci, Ascenzie and Ballio, 1986; Burger, Prentice, Moeller and Kasten-Schmidt 1970; Mikola and Pietilä, 1972).

If we now compare the results of proteo- and esterolytic activity, we may conclude that all sites in the sections, positive with animoacyl NAs are also positive with ester substrates. Nevertheless, the coincidence of the localization of proteolysis and esterolysis is only a prerequisite for the ambiguous

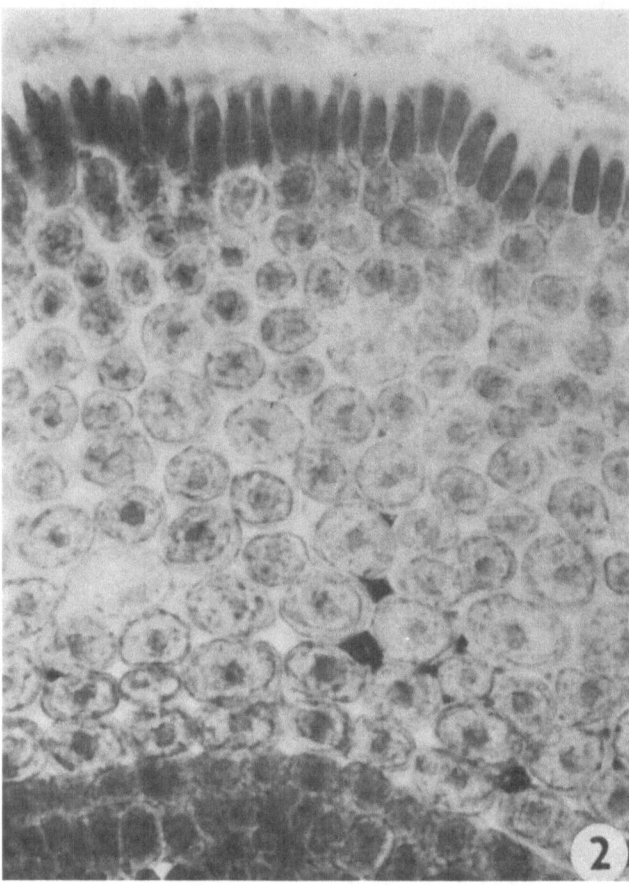

Fig. 2. The localization of hydrolytic activities in transverse section of maize root tips. Substrate: N-acetyl-L-glycine 2-naphthyl ester. Mag. 100 ×.

esteroprotease activity, not proof of its presence. Positive reaction was achieved with one of the N protected aminoacid esters suggesting that its localization was also similar to that of the peptidase.

PAGE patterns

A step further in our research is the comparison of the activity of particular bands separated by PAGE using different amide and ester substrates. Again, four groups of substrates (Table 1) were tested: 1) esters of lower fatty acids of different chain length with various naphthols, 2) naphthylamides of amino acids of various chain length, 3) naphthylamides of plain 2C and 4C fatty acids, 4) naphthyl esters of N-acetylated amino acids. The question is whether the naphthyl esters and naphthylamides applied are hydrolysed by PAGE identical or different enzyme proteins.

The material used in this part of our work were 12 mm tips of seedling roots of *V. faba*. An homogenate (1:1) was made in phosphate buffer containing 0.9% NaCl and dithiothreitol (1 mg ml^{-1}). The supernatant (22 000 g, 20 min, +4°C) was about four times concentrated by the addition of Sephadex G-25. After electrophoresis (Davis, 1964) the gels were washed and processed to reveal the isoenzyme bands. The same media were used as in our histochemical studies. The number, position, width and colour intensity of bands were evaluated both visually and by means of a densitometer.

The substrates used and the results obtained are summarized in Table 1. Concerning esters of lower fatty acids, both the total activity and the number and position of bands depend on the chain length (compare 1-naphthyl acetate and butyrate) and on the naphtholic moiety (see acetates of 1- and AS naphthols).

Several bands, localized in the region of the

faster moving esterases, were positive with amino-acyl NAs. A lesser number of bands was detected with glycine NA than with alanine and leucine NAs. Negative results were obtained with plain acyl NAs.

Similarly, as in the case of aminoacyl NAs, only some of the faster moving bands were positive using esters of N-acetylated amino acids. The total activity was weak. The differences revealed by different substrates of this type are regarded as insubstantial.

It follows that some protein fractions of higher Rm display activity towards both esters of aliphatic acids and aminoacyl NAs and esters of N-protected amino acids. This fact, and the coincidence in the *in situ* localization using ester and amide substrates, support the possibility of the existence of estero-proteases in our material. However, the purpose of the present study was to localize the enzyme activity and to visualize the isoenzyme patterns using the simple aminoacyl NAs and naphthyl esters of lower fatty acids and of aminoacids as substrates with the final aim of evaluating the results from the standpoint of cell and tissue heterogeneity within the root tip. Work is in progress to perform the inhibition tests, to apply other types of visualizing reactions, and to use some more complex substrates as has been done in animal material (Gossrau, 1983; Lojda, 1984).

References

Aducci, Ascenzi P and Ballio A 1986 Esterolytic properties of leucine proteinase, the leucine specific serine proteinase from spinach (*Spinacia oleracea* L) leaves: A steady state and pre-steady state study. Plant Physiol. 82, 591–593.

Beneš K 1971 Histogenesis and localization of non-specific esterase in root tip. Biol. Plant. 13, 110–121.

Beneš K 1977 Histochemistry of carboxyl esterases in the broad bean root tip with indoxyl substrates. Histochemistry 63, 79–87.

Beneš K, Ivanov V B and Hadačová V 1981 Glycosidases in the root tip. *In* Structure and Function of Plant Roots. Eds. R Brouwer *et al.* pp 137–139. Nijhoff and Junk, The Hague.

Beneš K and Kutík J 1978 The localization of starch in root tips. Biol. Plant. 20, 458–463.

Beneš K and Opatrná J 1964 Localization of acid phosphatase in the differentiating root meristem. Biol. Plant. 6, 8–16.

Brown R 1963 Cellular differentiation in the root. *In* Cell differentiation (17th Symp. Soc. Exp. Biol.). Ed. G E Fugg, pp 1–17. Cambridge Univ. Press, London.

Burger W C, Prentice N, Moeller M and Kasten-Schmidt J 1970 Hydrolysis of α-napthylacetate and L-leucyl naphthylamide by barley enzymes. Phytochemistry 9, 33–40.

Davis B J 1964 Disc electrophoresis. II. Method and application to human serum proteins. Ann. N. Y. Acad. Sci. 121, 407–427.

Gossrau R 1983 Fluorescence histochemical detection of hydrolases in tissue sections and culture cells. Histochemistry 79, 87–94.

Hadačová V and Beneš K 1977 Investigation of isoenzyme patterns of some oxidoreductases, hydrolases and transferases in different growth zones of broad bean (*Vicia faba* L.) roots. Physiol. Vég. 15, 735–745.

Lojda Z 1984 Die Histochemie der Proteasen. Acta Histochemica Suppl. 30, 9–29.

McDonald J K 1985 An overview of protease specificity and catalytic mechanisms: aspects related to nomenclature and classification. Histochem. J. 17, 773–785.

Mikola J and Pietilä K 1972 Hydrolysis of ester substrates of trypsin and chymotrypsin by barley carboxypeptidase. Phytochemistry 11, 2977–2980.

Mikola L and Mikola J 1986 Occurrence and properties of different types of peptidases in higher plants. *In* Plant proteolytic enzymes. Ed. M J Daling pp 97–117. CRC Press, Boca Raton.

Storey R D and Wagner F W 1986 Plant proteases: A need for uniformity. Phytochemistry 25, 2701–2709.

von Deimling O and Böcking A 1976 Esterases in histochemistry and ultrahistochemistry. Histochem. J. 8, 215–252.

B. C. Loughman et al. (Eds.), Structural and functional aspects of transport in roots, 35–40.
© 1989 by Kluwer Academic Publishers.

Significance of the exodermis in root function

CAROL A. PETERSON

Department of Biology, University of Waterloo, Waterloo, Ontario, Canada N2L 3G1

Key words: apoplast, Casparian band, exodermis, hypodermis, roots

Introduction

An exodermis is a specialized type of hypodermis which often occurs in roots. The hypodermis, like the endodermis, is part of the cortex since it is derived from the ground meristem. The hypodermis and endodermis form the outer and inner boundaries of the cortex, respectively. The hypo-dermis is a uni- or multiseriate layer of cells which are morphologically distinct from those of the neighbouring cortex. External to the hypodermis lies the epidermis derived from the protoderm. In roots, cells of the hypodermis are frequently modified by the presence of suberized walls; in this case the layer is termed an exodermis according to von Guttenberg (1968). A recent survey of 202 species

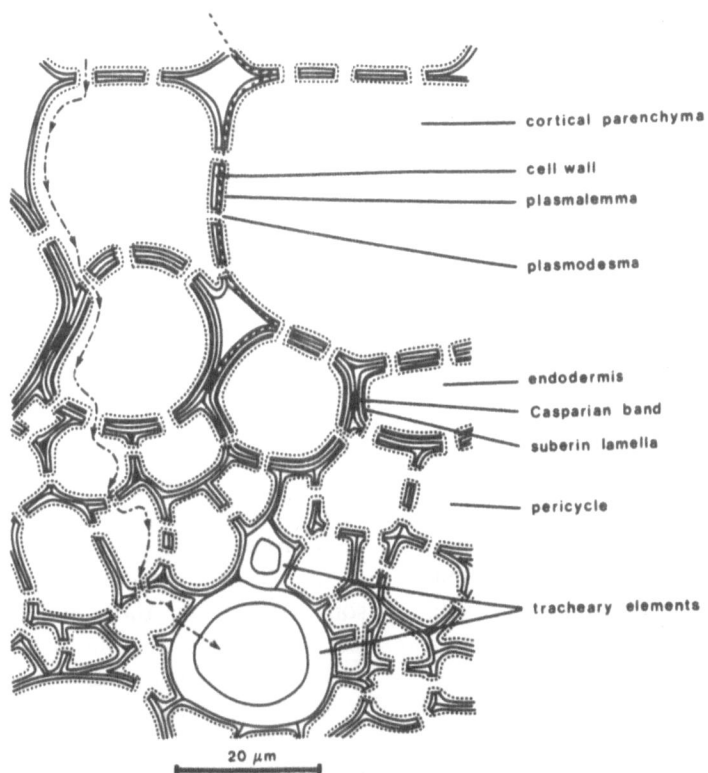

Fig. 1. Diagram of portion of a nonexodermal root in cross section including (*top to bottom*) cells of the inner cortical parenchyma, endodermis, pericycle, xylem parenchyma and tracheary elements. Symplasmic transport (–·–·–) of ions occurs from cell to cell through plasmodesmata. Apoplastic transport (--------) from the cortical parenchyma to the pericycle is prevented by the endodermal Casparian band.

35

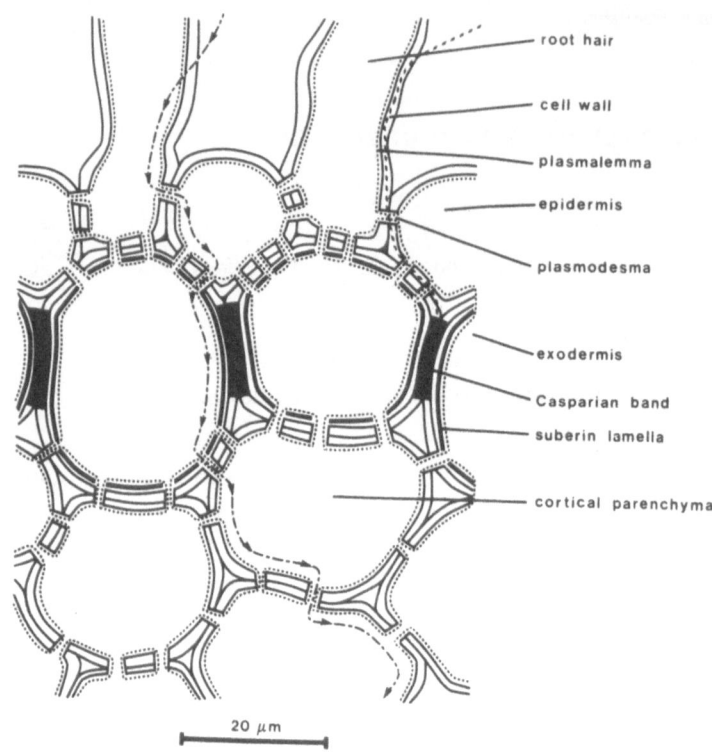

Fig. 2. Diagram of a portion of an exodermal root in cross section including (*top to bottom*) cells of the epidermis, exodermis and outer cortical parenchyma. Symplasmic transport (–·–·–) of ions occurs from the epidermis to the cortical parenchyma through plasmodesmata. Apoplastic transport (--------) is prevented by the exodermal Casparian band. The suberin lamella between the exodermal wall and plasmalemma prevents ion uptake by the exodermis and contributes to the formation of an impermeable layer after death of the epidermis.

from 54 families of angiosperms has shown that 91% of the species have a suberized hypodermis and that they invariably also have Casparian bands. Conversely, 9% of the species either have no hypodermis or have a hypodermis with un-suberized walls. None of these have Casparian bands. Therefore, it seems logical to extend von Guttenberg's definition of an exodermis to include the presence of a Casparian band in the cells comprising the layer.

The existence of two sets of Casparian bands, the outermost adjacent to the epidermis, necessitates a reexamination of root structure and function. These topics are treated in turn in the following sections.

Comparison of the exodermis with the endodermis

Casparian bands

The structure (and function) of the endodermal Casparian band is well known and has been the subject of several reviews (Clarkson and Robards, 1975; Van Fleet, 1961). Briefly, the band consists of a deposit of suberin, lignin, or a mixture of both which fills the intermicrofibrillar spaces (normally filled with water) in a discrete zone in the anticlinal walls of the cells (Fig. 1). The band encircles each endodermal cell and is continuous through its middle lamella with the bands of adjacent endodermal cells. Virtually every plant root has Casparian bands in its endodermis. Some reports of the absence of Casparian bands have proved incorrect upon reexamination (see Jorns, 1987).

The Casparian bands of the exodermis are in some respects like those of the endodermis. They are composed of suberin or a mixture of suberin and lignin and also infiltrate the intermicrofibrillar spaces so that the modification of the wall does not increase its width. The deposit encircles the cell and is continuous with deposits in adjacent cells forming a cylinder of modified wall in the root. Usually the Casparian band of the exodermis occupies the entire width of the radial wall (Fig. 2) and is thus more massive than its counterpart in the endodermis. With an improved staining technique, we see that in some exodermal cells of some species, there are two bands in each cell, one at the outer edge of the anticlinal wall and the other at the inner edge (Brundrett, Enstone and Peterson, 1987). Whereas the endodermis is virtually always uniseriate, the exodermis is sometimes multiseriate (*e.g.* in 21 of the 202 species examined by Perumalla, 1986). In the 3 species so far investigated in detail, Casparian bands occur in all suberized layers of the exodermis (Perumalla, 1986).

Ontogeny

A complete description of the development of the endodermis can be found in the review by Clarkson and Robards (1975). Only the salient features for comparison with the exodermis will be included here. Endodermal cells can pass through 3 stages of development commonly called State I, II and III. In State I, Casparian bands have developed. This occurs within a few millimeters of the root tip, the actual distance from the tip being related to the species and the growth rate of the root (Wilcox, 1962). For some species, differentiation of the endodermis is complete with the formation of the Casparian band. In others, the cells may progress to State II by forming a suberin lamella on all faces of the wall. This lamella does not sever the plasmodesmata connecting the protoplasts of the cells (Clarkson *et al.*, 1971). Still later, the cells may develop to State III by laying down an additional cellulosic wall which may itself become suberized and/or lignified. The position in the root at which States II and III development occur are highly variable; there is evidence that State II development occurs closer to the root tip in slowly growing roots than in rapidly growing roots (Wilcox, 1962). States II and III development often occur asynchronously in the root, being delayed in endodermal cells opposite the xylem poles. Such endodermal cells are called passage cells despite the fact that they have Casparian bands. The mature endodermal cells are uniformly elongate.

The development of the exodermis has not been extensively studied so that the following summary has of necessity been compiled from descriptions of only a few species. Further investigation is required to determine whether the picture presented here is a general one. The cells of the exodermis can also

be described in terms of State I, II and III development. In rapidly growing roots, Casparian bands mature a considerable distance from the apex, *i.e.* 30–40 mm in onion and 40–50 mm in corn (Peterson and Perumalla, 1984). However, in slowly growing roots, the band can mature very close to the apex (within 5 mm in corn according to Perumalla and Peterson, 1986). In the exodermis of corn and onion, Casparian band (State I) development is quickly followed by deposition of a suberin lamella (State II). In his survey, Perumalla (1986) found that all species which have an exodermal Casparian band also have suberin in their tangential walls. This indicates that the cells of the exodermis invariably progress at least to State II. In corn and onion, State III development is also achieved and it remains to be seen whether or not development to this state is a general feature of the exodermis. Some species, like corn, have a uniform exodermis in which the cells are uniformly elongate (Fig. 3a). Others, like onion, have a dimorphic exodermis in which some cells are conspicuously shorter than others (Fig. 3b). The arrangement of long and short cells can be rather irregular (as illustrated for onion) or regular so that long and short cells alternate in longitudinal files. Shishkoff (1987) has identified 98 species which have a dimorphic exodermis and has listed an additional 97 on the basis of the previous literature. The short cells are called passage cells and in fact resemble the passage cells of the endodermis in that the development to State II and III is delayed (von Guttenberg, 1968). With this anatomical background, we can now consider the physiological significance of the exodermis.

Function in absorption of water and ions

Nonstressed roots. The current explanation of water and ion uptake by plant roots (which can be found in many sources, *e.g.* Marschner, 1986), portrays this process in roots lacking an exodermis *i.e.* in roots having one Casparian band which is located in the endodermis. In such roots, water and ions can potentially move apoplastically through the walls of the epidermis and cortex as far as the endodermal Casparian band (Fig. 1). This hydrophobic structure prevents further apoplastic movement of the water and ions. To the extent that these substances can permeate membranes, either passively or by active carriers, symplasmic movement will also occur (Fig. 1). Symplasmic movement inward occurs via plasmodesmata which connect the cells of the epidermis, cortex (including the endodermis), pericycle and xylem parenchyma (Walter *et al.*, 1984; Warmbrodt, 1985). Water and ions could be taken up by the symplast at any point external to the endodermal Casparian band *i.e.* by the cells of the epidermis and cortex. Thus, a large membrane surface can be used for absorption.

In roots with an exodermis, apoplastic movement of the soil solution occurs as far as the outer boundary of the exodermal Casparian band (Fig. 2). Evidence that the Casparian band of the exodermis (and endodermis) limits apoplastic movement has been obtained by observing the movement of apoplastic fluorescent dyes (Peterson *et al.*, 1978; Peterson *et al.*, 1982), iron salts (de Rufz de Lavison, 1910) and by measuring the free space of roots to sulfate ions (Peterson, 1987). The dyes and ions penetrate the walls of the epidermis and the outer tangential walls of the exodermis. The permeability of the exodermal Casparian band to water has not been tested. However, based on the hydrophobic nature of the barrier, it is likely that the apoplastic movement of water is also impeded by this structure. Ions and water are thus absorbed from the soil solution by the epidermal cells. Recall that in the exodermis, suberin lamellae are also present (Fig. 2). This hydrophobic layer presumably prevents contact between the soil solution and the plasmalemmae of the exodermal cells. An exception to this general rule occurs in the case of the dimorphic exodermis in which suberin lamella formation in the short cells is delayed. Until these lamellae form, the short cells of the exodermis, in addition to the cells of the epidermis, are potentially absorptive. Water and ions in the cytoplasm of the epidermis would be transported through the exodermis symplasmically; plasmodesmata link the cells of each layer (Warmbrodt, 1985). The pathway of transport of water and ions from the exodermis to the stele is unknown. Symplasmic connections, which occur between all the intervening cell types, provide one potential route.

Stressed roots. Little is known regarding the effect(s) of stress on root structure. Stresses such as heat, drought and freezing are experienced by plant root systems in various climates. Of course, extreme stress will lead to death of the entire plant

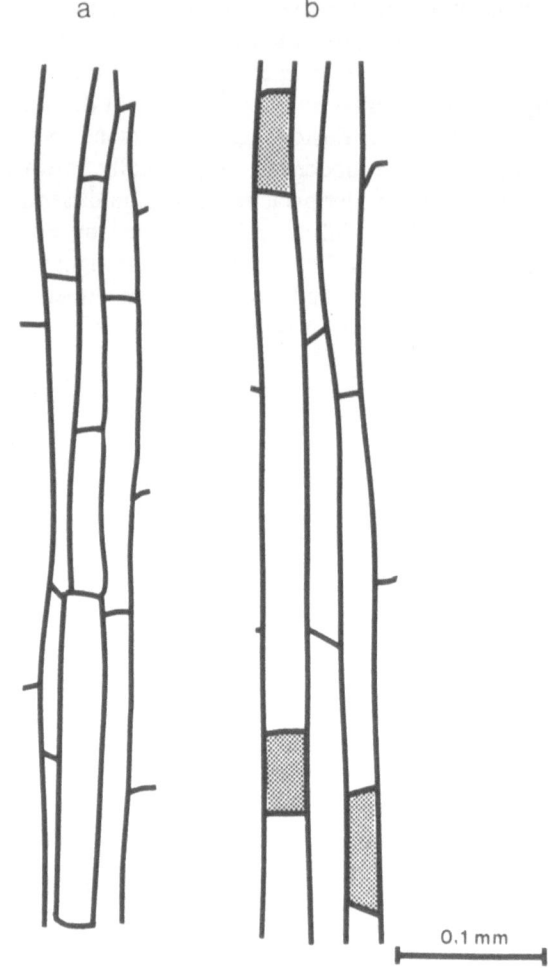

Fig. 3. Exodermal cells in face view (tangential longitudinal section). (a) uniform exodermis of corn, (b) short-celled exodermis of onion (short cells stippled).

including its root system, but less extreme or less prolonged stresses lead to changes which are presumably adaptive. In droughted roots without an exodermis (*e.g.* barley and perennial ryegrass) the cells of the epidermis and cortex die (Clarkson *et al.*, 1968; Jupp and Newman, 1987). The suberized endodermis is the outermost intact layer and the stele, with its vascular tissues and potentially meristematic pericycle, is preserved. In stressed roots with an exodermis, usually the epidermis dies and the exodermis is the outermost intact layer. Thus, the cortex as well as the stele is preserved in these roots. Onion roots react to stress in this fashion but even in severely droughted roots the short cells of the exodermis remain absorptive

(Stasovski and Peterson, 1987). Despite the presence of an exodermis, seedling roots of corn die back to the endodermis when desiccated (Stasovski and Peterson, 1987). However, older field-grown corn roots which have developed sclerenchymatous layers in the cortex adjacent to the exodermis retain their cortices (Hoppe *et al.*, 1986).

It is possible that in some species death of certain cells is a normal consequence of development. For example, Walker *et al.* (1984) reort that in Citrus (which has a dimorphic exodermis) the long cells die shortly after they are suberized. This does not occur in onion (Stasovski, unpublished results) and further work is needed to clarify these observations.

Adaptive significance of the exodermis

At first glance, the presence of an exodermal Casparian band so near the edge of the root appears to be disadvantageous because the root system has a much-reduced plasmalemma surface area with which to absorb ions compared to root systems of nonexodermal species. However, in view of the fact that the large majority of flowering plants so far examined do have an exodermis, it is of interest to speculate on the possible advantages such a layer may confer.

One consequence of Casparian band function is that molecules which do not cross the plasmalemma, or do so at a slower rate than water, tend to accumulate in the walls next to the band. If the first Casparian band encountered by these molecules were in the exodermis, they would have a shorter diffusion path to traverse in returning to the soil than if the first Casparian band encountered were deep within the root in the endodermis. The presence of an exodermis could thus mediate against the build-up of some ions within the apoplast of the cortex, an important consideration in saline soils.

Exodermal and nonexodermal roots also differ in their reaction to drought. In general, dieback is arrested at cells with suberized walls, which apparently reduce water loss from the root to the soil, so that the cells internal to the layer survive. In nonexodermal roots, the cortex is lost but in exodermal roots this layer, with its food reserves, is retained. Survival of this layer is especially impor-

tant for perennial monocotyledons which die back to the root system each winter since it constitutes the site of food storage for the plant.

Suberized cell walls in general represent a common defense for plants against attack by microorganisms (Kolattukudy, 1984). In roots with an exodermis, such a layer is located very close to the root surface.

Finally, the cortex of most field-grown plants is populated with a mycorrhizal fungus which assumes many nutrient absorption functions of the root (Harley and Smith, 1983). Preservation of the cortex provides a haven for these fungi when conditions are unfavourable for the growth of hyphae outside the roots (Brundrett, personal communication). The widespread occurrence of the exodermis in angiosperm roots suggests that it was a feature of the ancestral taxon for this group, and that because the advantages of possessing this layer outweigh the disadvantages, it has been retained by most members.

References

Brundrett M C, Enstone D E and Peterson C A 1988 A berberine-aniline blue fluorescent staining procedure for suberin, lignen and callose in plant tissue. Protoplasma. *In press*

Clarkson D T and Robards A W 1975 The endodermis, its structural development and physiological role. *In* The Development and Function of Roots. Eds. J G Torrey and D T Clarkson. pp 415–436. Academic Press, New York.

Clarkson D T, Robards A W and Sanderson J 1971 The tertiary endodermis in barley roots: Fine structure in relation to radial transport of ions and water. Planta (Berl.) 96, 292–305.

Clarkson D T, Sanderson J and Russell R S 1968 Ion uptake and root age. Nature 220, 805–806.

Guttenberg H von 1968 Der primäre Bau der Angiospermenwurzel. *In* Handbuch der Pflanzenanatomie Vol. 8 part 5. Ed. K Linsbauer Gebrüder Borntraeger, Berlin.

Harley J L and Smith S E 1983 Mycorrhizal Symbiosis. Academic Press, London.

Hoppe D C, McCully M E and Wenzel C L 1986 The nodal roots of Zea: their development in relation to structural features of the stem. Can. J. Bot. 64, 2524–2537.

Jorns A C 1987 Presence and function of the Casparian band in roots of Norway spruce [*Picea abies* (L.) Karst.]. J. Plant Physiol. 129, 493–496.

Jupp A P and Newman E I 1987 Morphological and anatomical effects of severe drought on the roots of *Lolium perenne* L. New Phytol. 105, 393–402.

Kolattukudy P E 1984 Biochemistry and function of cutin and suberin. Can. J. Bot. 62, 2918–2933.

Marschner H 1986 Mineral Nutrition in Higher Plants. Academic Press, London.

Perumalla C J 1986 Studies on the Hypodermis of Roots and Rhizomes of Various Angiosperm Species. PhD. Thesis, University of Waterloo, Canada.

Perumalla C J and Peterson C A 1986 Deposition of Casparian bands and suberin lamellae in the exodermis and endodermis of young corn and onion roots. Can. J. Bot. 64, 1873–1878.

Peterson C A 1987 The exodermal Casparian band of onion roots blocks the apoplastic movement of sulphate ions. J. Exp. Bot. 38, 2068–2081.

Peterson C A, Peterson R L and Robards A W 1978 A correlated histochemical and ultrastructural study of the epidermis and hypodermis of onion roots. Protoplasma 96, 1–21.

Peterson C A, Emanuel M E and Wilson C 1982 Identification of a Casparian band in the hypodermis of onion and corn roots. Can. J. Bot. 60, 1529–1535.

Peterson C A and Perumalla C J 1984 Development of the hypodermal Casparian band in corn and onion roots. J. Exp. Bot. 35, 51–57.

Rufz de Lavison J de 1910 Du mode de pénétration de quelques sels dans la plante vivante. Role de l'endoderme. Revue Génerale de Botanique 22, 225–241.

Shishkoff N 1987 Distribution of the dimorphic hypodermis of roots in angiosperm families. Ann. Bot. 60, 1–15.

Stasovski E and Peterson C A 1987 The vitality of root tissues experiencing drought. Proceedings of The Canadian Society of Plant Physiologists, Eastern Regional Meeting. University of Guelph, Canada.

Van Fleet D S 1961 Histochemistry and function of the endodermis. Bot. Rev. 27, 165–220.

Walker R R, Sedgley M, Blesing M A and Douglas T J 1984 Anatomy, ultrastructure and assimilate concentrations of roots of Citrus genotypes differing in ability for salt exclusion. J. Exp. Bot. 35, 1481–1494.

Warmbrodt R D 1985 Studies on the root of *Hordeum vulgare* L. — ultrastructure of the seminal root with special reference to the phloem. Am. J. Bot. 72, 414–432.

Wilcox H 1962 Growth studies of the root of incense cedar. *Libocedrus decurrens.* I. The origin and development of primary tissues. Am. J. Bot. 49, 221–236.

B. C. Loughman et al. (Eds.), Structural and functional aspects of transport in roots, 41–44.
© 1989 by Kluwer Academic Publishers.

Seed hydration as a trigger of cell elongation in bean hypocotyl and radicle

N. V. OBROUCHEVA and O. V. ANTIPOVA
Institute of Plant Physiology, Academy of Sciences of USSR, Botanical Street 35, SU-127 276, Moscow, USSR

Key words: acid growth, cell elongation, hydration, osmotic potential, radical protrusion, seed germination

Abbreviations: PEG — polyethylene glycol, m.w. 6000

Introduction

When studying root growth one usually considers roots rapidly growing by active cell division and elongation. But another situation is very attractive too — an activation of growth occurring, for example, in imbibing seeds where the cells start their growth during a rather short period measured in hours. We have shown earlier (Obroucheva, 1981) that radicle emergence, the first visible growth phenomenon in germinating seeds, is a result of cell elongation. Cell division in many seed species begins later, after radicle protrusion. It is elongation that provides rapid root protrusion and its contact with soil water. In the seeds of *Vicia faba minor* radical emergence and further growth of hypocotyl and root occur only by cell elongation until the axis is 1 cm in length. Thus it is a good model to look for some preliminary processes preparing the axial organs for elongation and developing along with their hydration.

Results and discussion

The general view of the hydration curve is shown in Fig. 1. The axial organs of the killed seeds imbibe water up to 60% moisture. This first rapid stage of hydration occurs mainly by matric forces and partially by osmotic forces, as at 60% moisture osmotic potential ($-2.5\,\text{MPa}$) represents only a small part of water potential ($-11.4\,\text{MPa}$).

The processes preparing cell elongation appear to develop in a moisture range from 60%, at which the physical water absorption is complete, to 72–73%, at which the bean radicles protrude.

The first such physiological process can be a rapid additional accumulation of osmotically active substances observed at 65–68% moisture and increased as further hydration proceeded. Three lines of evidence were obtained.

1. The direct measurement of osmotic potential in cell sap has shown that it contributed 36% of water potential ($-4.7\,\text{MPa}$) at 65% moisture and 62% of water potential ($-2.4\,\text{MPa}$) at 68% moisture. Osmotic forces became dominant in water uptake by axial tissues.

2. The estimation of the main components of osmotic potential in the cell sap of axes, hypocotyl and radicle is shown in Table 1. Up to 65% moisture, the main osmotic components were K^+ and Cl^- ions, but at higher moisture levels sugars (particularly glucose) and K^+ dominated. Osmotic potential values result from two processes, namely water uptake and changes in content of the substances. If we compare their absolute amounts in axes or radicles at various moisture levels (Table 1, below), we can observe at 68% and higher moisture the increasing amounts of sugars and K^+ appearing, perhaps after the commencement of starch and phytin

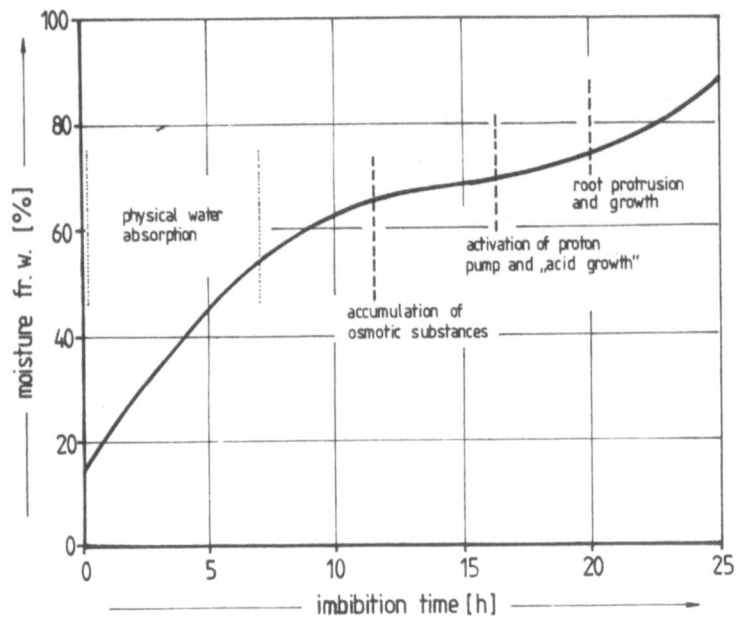

Fig. 1. Hydration curve of seeds of *Vicia faba minor*.

Table 1. The components of osmotic potential in imbibing bean axes and radicles

Moisture of organ, % fresh weight	Imbibition time, h	Osmotic potential developed by each component, -MPa			
		Sugars	K$^+$ + coanions	Cl$^-$	NO$_3^-$
Axial organs					
60	9	0.32	0.60	0.61	0.29
65	12	0.23	0.46	0.73	0.11
68	16	0.30	0.45	0.13	0.05
72	18	0.29	0.44	0.05	0.04
80	24	0.29	0.55	0.09	0.03
Radicles					
63	12	0.24	0.44	0.84	0.14
66	16	0.28	0.44	0.13	0.05
70	18	0.30	0.49	0.02	0.04
74	24	0.45	0.54	0.12	0.03
Content of osmotic substances in cell sap, μg per axis or per radicle					
Axial organs					
60	9	249	39	72	60
65	12	239	38	107	29
68	16	300	43	22	14
72	18	408	59	11	16
80	24	563	99	29	15
Radicles					
63	12	141	22	77	22
66	16	168	23	12	8
70	18	281	39	3	11
74	24	334	53	21	8

Table 2. The effect of dehydration-rehydration on the rate of rehydration of bean axes in imbibing and germinating seeds

Hydration		Rehydration		Hydration time of control seeds within the same range of axis moisture increase, h
Time, h	Moisture of axes, %	Range of increase of axis moisture (%) during rehydration	Time, h	
4	40	20–40	3	2.5
6	50	20–50	5	4
8	55	20–55	5	5
9	60	20–60	8	7
12	65	20–65	10	10
16	68	50–68	3	10
16	68	40–68	4	12
16	68	30–68	5	13
16	68	20–68	7	14
Radicle protrusion				
18	72	60–72	1	9
18	72	40–72	2	12
18	72	20–72	4	14
Germinating seeds				
29	84	70–84	2	13
29	84	30–84	7	27

mobilization. Their accumulation provides the additional hydration of axial organs.

3. We have performed experiments with seed hydration-dehydration-rehydration. The seeds were hydrated up to fixed levels of moisture, then dessicated to various levels of moisture and rehydrated back to the same fixed levels (Obroucheva and Antipova, 1985). Then seeds

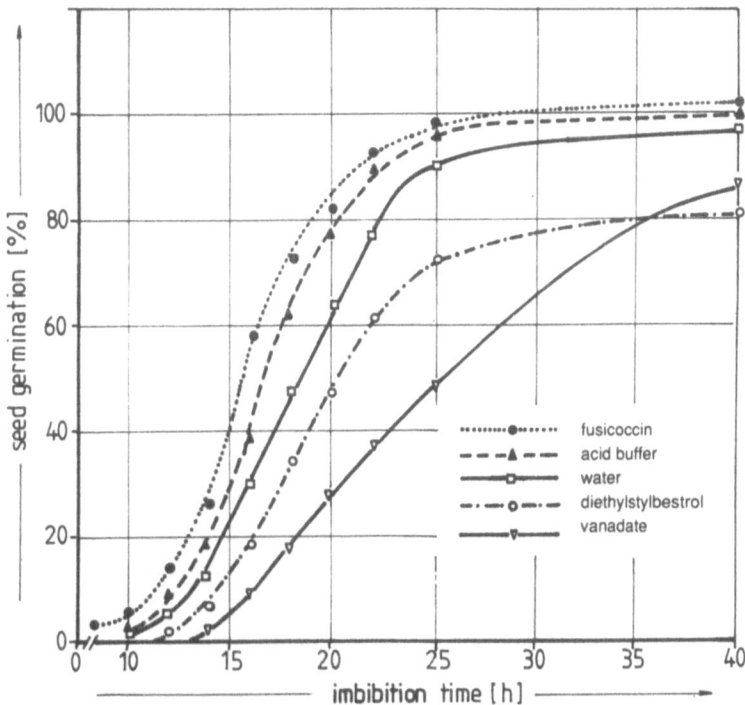

Fig. 2. The effects of fusicoccin (20 μM); phosphate-citrate buffer pH 3.9 (1 mM), o-vanadate (5 mM) and diethyl stilbestrol (0.5 M) on radicle emergence of *Vicia faba minor*.

Table 3. Acidification of ambient solution by bean axes

Imbibition time, h	Moisture % fr.wt	Δ pH (100 axes in 10 ml $10^{-3} M$ KCl 30 min)		
		no treatment	+ diethyl-stylbestrol $5.10^{-4} M$	+ fusicoccin $2.10^{-5} M$
9	60	0.01	0	0.01
12	65	0.09	0	0.11
16	68	0.34	0.05	0.49
20	75	0.65	0.31	0.86

proceed to absorb water, germinate, and grow on. The same scheme was applied to the seeds with protruding radicles. Control seeds were neither dessicated nor rehydrated. The only observed effect (Table 2) was the acceleration of the rehydration rate as compared to control seeds, but it occurred only in seeds whose hydration level was 68% and higher. We explain these results by an additional accumulation of osmotic substances within 65–68% moisture range providing the enhanced water uptake during rehydration.

The increased content of osmotic substances at 65–68% and higher moisture levels is not enough to radicles to protrude. When we exposed such seeds to PEG solutions at concentrations which do not affect their moisture levels, the seeds did not take up more water and no root protrusion occurred. It indicates the operation of another physiological mechanism necessary for cells to initiate their elongation.

We suppose this mechanism to be an acidification of cell walls as a result of proton pump activation at 68–70% moisture. This suggestion follows from the following observations:

1) fusicoccin ($2.10^{-5} M$) stimulated radicle emergence, water uptake and cell elongation as did 1 mMM phosphate-citrate buffer pH 3.9 while o-vanadate ($5.10^{-3} M$) and diethylstylbestrol ($5.10^{-4} m$), the inhibitors of plasmalemma H^+-ATPase, inhibited the radicle emergence (Fig.2);
2) axial organs of 68% and higher moisture became able to acidify the surrounding solution (Table 3); this acidification was inhibited by diethylstylbestrol and stimulated by fusicoccin.

So we assume that activation of plasmalemma H^+-ATPase results in an acidification of cellwalls, their loosening, additional water uptake and initiation of cell elongation.

Figure 1 shows the general scheme of processes preparing the cell for elongation. The first is the accumulation of osmotic substances commencing at 65–68% moisture; the second one is an activation of plasmalemma H^+-ATPase beginning at 68–70% moisture. They result in uptake of additional water and initiation of cell elongation at first in the hypocotyl, then in the radicle. The radicle lags behind the hypocotyl in its rate of hydration and reaching the moisture levels necessary to trigger both processes required for elongation. Seeds in PEG solutions were unable to take up more water and germinate so we can consider hydration levels to be the triggers of physiological processes preparing elongation. At critical hydration levels the necessary conformations of enzymes permit them to start functioning, *i.e.* hydrolases to provide the accumulation of osmotica, and H^+-ATPase to operate the proton pump.

References

Obroucheva N V 1981 Development of mature root growth pattern in germinating seeds. *In* Structure and Function of Plant Roots. Eds. R. Brouwer *et al.* pp 23–27. Nijhoff, Dordrecht, The Netherlands.

Obroucheva N V and Antipova O V 1985 The level of seed hydration that controls the events preceding cell elongation in germinating beans. Soviet Plant Physiol. 32, 932–941.

B. C. Loughman et al. (Eds.), Structural and functional aspects of transport in roots, 45–48.
© 1989 by Kluwer Academic Publishers.

Ultrastructure of cortical cells in the primary root of barley during germination

MILOŠ MIKUŠ
Department of Plant Physiology and Biotechnology, Comenius University, Bratislava, Czechoslovakia

Key words: barley, germination, lipid bodies, protein bodies, root, ultrastructure

Abbreviations. LB, lipid body; M, mitochondria; P, plastid; PB, protein body; V, vacuole

Introduction

Most of the previous ultrastructural studies of hydration during seed germination have primarily involved changes in specialized storage tissues. During germination the hydrolytic processes in these tissues lead to their degeneration and senescence. In contrast, cells in the embryo proper also contain reserve material, but hydration during germination results in reactivation of their growth and differentiation.

We examined the primary seminal root of barley, which comprises root meristem covered with the root cap and a histologically determined part that has completed cell division. During germination the restoration of the growth processes begins in this non-meristematic region by cell elongation with the exception of the zone adjacent to scutellar node where growth is absent (Luxová, 1986). This zone was studied here. Observations were made with respect to the storage structures and to the organelles participating in degradation and utilization of storage materials.

Material and methods

Seeds of *Hordeum distichum* L. cv. Slovenský dunajský trh were germinated in rolls of wet filter paper and the samples were taken after 0, 12, 24 and 48 hours of cultivation at 23°C in dark. Thin cross-sections of the embryos containing the basal part of the primary root were excised with a razor blade, fixed with 5% glutaraldehyde in 0.1 M phosphate buffer and post-fixed with 2% osmium tetroxide in the same buffer. After dehydration in acetone, the samples were embedded in Durcupan ACM (Fluka)

For correct orientation in the blocks, semi-thin sections, stained with toluidine blue and basic fuchsin, were used (Lux, 1981). Thin sections were stained with uranyl acetate and lead citrate. Cytochemical staining for catalase activity was performed with diamino benzidine as described by Fahimi (1969).

Quantification of changes in the cell ultrastructure involved stereology. Weibel grids were used for derivation of volume density, surface density and numerical density (Toth, 1982) in two levels: level 1 (micrograph magnification × 7500) 'cytoplasmic' volume fractions were determined by point counting and level 2 (micrograph magnification × 26 000) point and intersection counts were made on protein bodies, lipid bodies, plastids and mitochondria.

Volume densities of the cell structures were expressed in two ways: 1. relative volume per cell (cell volume excluding nucleus) and 2. relative volume per cytoplasm (cell volume excluding nucleus and vacuoles).

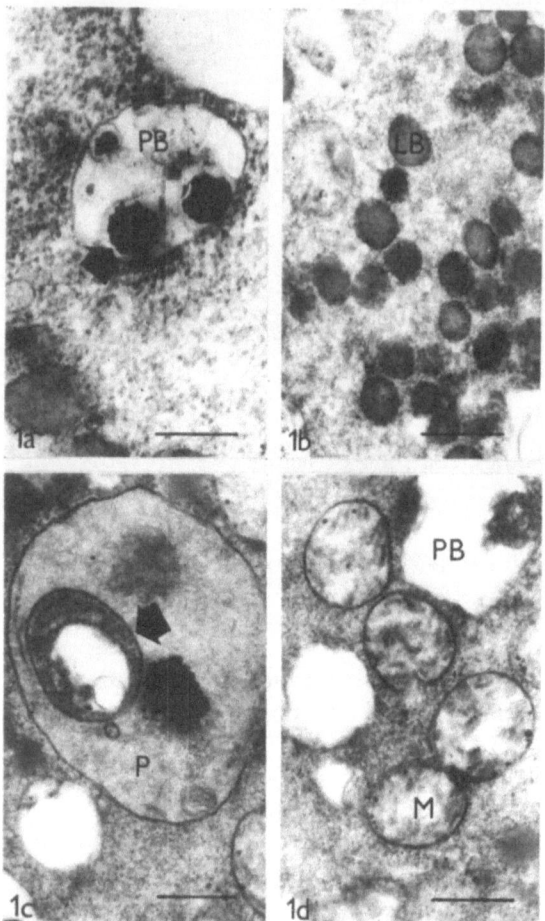

Fig. 1. Cortical cells of barley root of dry seeds. *Bars* = 0.5 μm.
a) PBs with dense globoids (*arrow*). b) Numerous LBs are
present in cytoplasm. c) A typical plastid of a dry seed which has
an invagination of its envelope (*arrow*) and group of plastoglobuli. d) Mitochondria with distinct envelopes and dense
granules.

Results and discussion

Dry seeds

Electron microscopic observations of cortical
cells in the basal part of primary roots showed
mainly two types of storage structures: protein
bodies (Fig. 1a) and lipid bodies (Fig. 1b). Starch
was found only rarely in plastids. PBs had narrow
peripheral zones of proteinaceous material and
most of them contained electron dense structures
corresponding to globoids (Fig. 1a). Globoids are
often present in PBs of various species (Pernollet,
1978) and consist of phytate. LBs were present in
cytoplasm mostly forming layers along cell walls
and nuclei. Some of them showed a thin electron-

dense boundary (Fig. 1b) which might correspond
to the half unit membrane described by Yatsu and
Jacks (1972) and others.

Plastids contained plastoglobuli, crystalline-like
material (probably phytoferritin), short thylakoid
lamellae, rarely small starch grains and invaginations of the plastid envelope (Fig. 1c). Similar invaginations of the plastid envelope were found in
cells of dry seeds of rice by Bechtel and Pomeranz
(1978) and it is possible that these invaginations
represent a response to dessication. Mitochondria
show distinct cristae, translucent areas and several
electron dense granules in the matrix (Fig. 1d).

Changes during germination

Changes in fine structure during germination are
more rapid in root cells than in all other parts of the
embryo. Proteinaceous material of PBs was already
digested after 12 hours, but phytate (globoids) was
degraded less intensively. PBs fused gradually and
they formed a central vacuole after 24 hours of
germination. Generally, proteolysis is more rapid
than lipolysis during germination. We also observed that the relative volume of LBs in the cell
significantly decreased only after 24 hours and the
relative volume of LBs in the cytoplasm decreased
only after 48 hours of germination (Figs. 3a, 3b).

The first significant decrease of relative volume
of LBs was concomitant with the onset of digestion
of LBs in vacuoles (formerly PBs) shown in Fig. 2a.
A similar mode of degradation of LBs in vacuoles
was described by Buvat and Robert (1983),
Čiamporová (1983) and others. This provides
further evidence that PBs are not only compartments for the storage and hydrolysis of protein
reserves, but PBs and vacuoles play an important
role as a part of the lytic compartment of the cell.
This role of plant vacuoles was described in different objects (Bobák *et al.*, 1983).

A significant decrease in the relative volume of
LBs in cytoplasm (Fig. 3b) was the result of their *in
situ* degradation (Fig. 2b). The first 48 hours of
germination, without *in situ* degradation of LBs in
cytoplasm, could be necessary for transfer of
lipases from PBs to LBs as described by Fernandez
and Staehelin (1987). Although they proposed the
transfer of lipases *via* membrane fusion, we
observed contact of PBs and LBs only rarely.

Protoplastids were transformed to the amylo-

involved in the degradation of storage lipids in cortical cells of the root during the early stages of germination in barley.

Storage lipids are apparently metabolized in a different way in the embryo, where they are re-utilized within a cell, as compared to the storage tissues, where glyoxysomes may play an important role in lipid metabolism (Huang, 1975) and metabolites are transported to the embryo.

Conclusion

Ultrastructural changes in cortical cells of the basal part of primary root during germination have

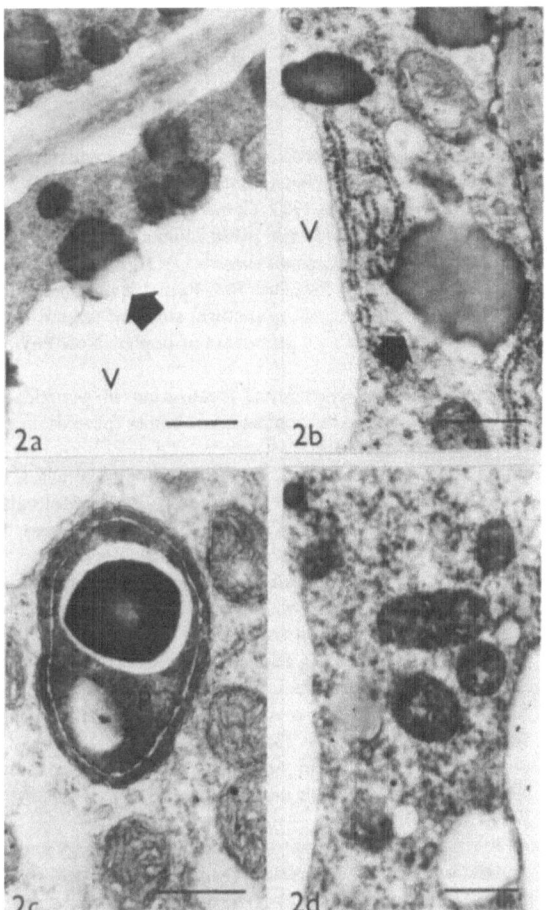

Fig. 2. Cortical cells of barley root after 48 hours of germination. *Bars* = 0.5 μm. a) Digestion of LB inside vacuole (*arrow*). b) *In situ* degradation of LB in cytoplasm (*arrow*). c) Plastoglobuli disappeared from plastids and starch was accumulated. d) Mitochondria have well defined internal structure.

plasts after the onset of germination. After 48 hours of germination they contained one or several starch grains, no plastoglobuli and no phytoferritin in their stroma (Fig. 2c). At the same time, mitochondria had a better defined internal structure and only a few electron dense granules were present in the matrix (Fig. 2d). Figures 3a and 3b show changes of the relative volume of mitochondria per cell and per cytoplasm, respectively, during germination.

As no glyoxysomes were observed, and an almost linear relationship was calculated between relative volume changes of mitochondria and LBs in the cell (Fig. 3a), it appears that mitochondrial β-oxidation rather than glyoxisomal β-oxidation is

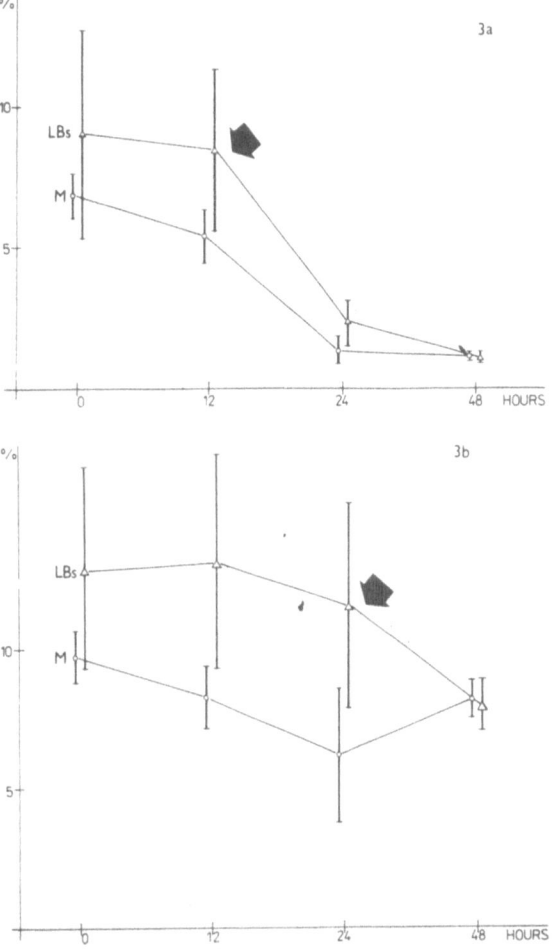

Fig. 3. Changes of relative volume of LBs (△—△) and mitochondria (○—○) in cell (3a) and in cytoplasm (3b) during germination. *Arrows* indicate the onset of significant decrease of LBs.

been investigated. These cells neither divide nor elongate, but continue to differentiate after the onset of germination.

Numerous PBs fuse and form a central vacuole after 24 hours of germination. Other storage structures are LBs (9.11% relative volume per cell). During germination the relative volume of LBs per cell decreased after 24 hours while the relative volume of these particles per cytoplasm significantly declined only after 48 hours. This ultrastructural study reveals two types of degradation of LBs: 1. Digestion inside the vacuoles which are formed from PBs and 2. *In situ* degradation of LBs in the cytoplasm.

During germination, ultrastructure of mitochondria became more complicated and plastids accumulated starch.

We suppose that the linear relationship between volume changes of mitochondria and lipid bodies in the cell, observed during germination, results from their biochemical linking. Since glyoxysomes were not observed, mitochondrial rather than glyoxysomal β-oxidation is involved in the degradation of reserve lipids in cortical cells during early germination of barley.

Acknowledgements

The TEM observations have been made with the kind assistance of V Polák. I express special acknowledgements to Dr M Čiamporová for valuable discussions during the work and helpful comments on the manuscript.

References

Bechtel D B and Pomeranz Y 1978 Ultrastructure of the mature ungerminated rice (*Oryza sativa*) caryopsis. The Germ. Am. J. Bot. 65, 75–85.

Bobák M, Herich R and Hudák J 1983 Lysosomal apparatus of a plant cell (in Slovak) Veda, Bratislava.

Buvat R and Robert G 1983 Cytologie ultrastructurale de l'hydration germinative des primordiums de racines des semences de l'Orge (*Hordeum vulgare* L.). II. La réhydration proprement dite. Ann. Sci. Nat. Bot. Paris 13e série 5, 49–70.

Čiamporová M 1983 An ultrastructural study of reserve lipid mobilization in stem root primordia of poplar. New Phytol. 95, 19–27.

Fahimi H D 1969 Cytochemical localization of peroxidatic activity of catalase in rat hepatic microbodies (peroxisomes). J. Cell Biol. 43, 275–288.

Fernandez D E and Staehelin L A 1987 Effect of gibberellic acid on lipid degradation in barley aleurone layers. *In* Molecular Biology of Plant Growth Control. Eds. J E Fox and M Jacobs. pp.323–334. Alan R. Liss Inc., New York.

Huang A H C 1975 Comparative studies of glyoxisomes from various fatty seedlings. Plant Physiol. 55, 870–874.

Lux A 1981 A rapid method for staining semi-thin sections of plant material (in Slovak), Biológia (Bratislava) 36, 753–756.

Luxová M 1986 The seminal root primordia in barley and the participation of their non-meristematic cells in root constructions. Biol. Plant. (Praha) 28, 161–167.

Pernollet J C 1987 Protein bodies of seeds: Ultrastructure, biochemistry, biosynthesis and degradation. Phytochemistry 17, 1473–1480.

Toth R 1982 An introduction to morphometric cytology and its application to botanical research. Am. J. Bot. 69, 1694–1706.

Yatsu L Y and Jacks T J 1972 Spherosome membranes. Plant Physiol. 49, 937–947.

B. C. Loughman et al. (Eds.), Structural and functional aspects of transport in roots, 49–52.
© 1989 by Kluwer Academic Publishers.

Chromatin structure in various tissues of the primary root of *Zea mays* L.

FRANTIŠEK BALUŠKA

Institute of Experimental Biology and Ecology, CBES, Slovak Academy of Sciences, CS-814 34 Bratislava, Czechoslovakia

Key words: chromatin structure, DNA content, maize root, radial ion transport

Introduction

The main function of the plant root, besides its rôle in fixing the plant in the soil, is to take up inorganic compounds from the soil and to transport them to the shoots. In comparison with the longitudinal transport, the routes and mechanisms of which are clear, the radial transport of ions in roots has for a long time been a matter of discussion. A widespread idea is that the transport of ions to the endodermis is realized not only *via* the symplasmic, but also through the apoplasmic pathway. The large surface area of cortical cells is considered to be the main site of the passage of ions from the apoplasm into the symplasm (Anderson, 1972; Clarkson, 1974). On the contrary, other authors (Bange, 1973; Moon *et al.*, 1986; Peterson and Perumalla, 1984; Vakhimstrov, 1967; Van Iren and Boer-Van der Sluijs, 1980) support a view according to which the main sites of ion entry into the symplasm are plasmalemmae of the root periphery (root hairs, epidermis, hypodermis) and subsequent transport continues mainly *via* the symplasm. Similarly the transfer of ions from symplasm into the xylem apoplasm has been discussed for a long time. But in recent years it has been proved that the parenchyma cells of the stele adjoining the vessels are active in releasing ions from the symplasm into xylem apoplasm (Clarkson and Hanson, 1986; De Boer and Prins, 1985).

In earlier papers (Baluška and Kubica, 1987a, b) we followed changes in the DNA content and degree of chromatin condensation during growth and differentiation of maize root cells. We used the method of squash preparations which enabled us to measure a great number of nuclei in a particular developmental stage. As a great diminution of the DNA content was found to be connected with chromatin decondensation in the basal region of the root, we decided to characterize the individual tissues of the maize root separately. It can be safely presumed that tissues capable of the active transfer of ions across the plasmalemma should also preserve their nuclear apparatus intact.

Results and discussion

The experimental material was grown and handled as in the preceding paper (Baluška and Kubica, 1987a). 10 μm thick paraffin sections of the 5 cm long primary root of *Zea mays* L. cv. CE-380 were examined in the Quantanal AMP 20 cytophotometer.

It was found that all tissues of the maize primary root examined were characterized by the enlargement of nuclei after completion of the mitotic cycles; this accompanied cell maturation at the base of the root. An interesting observation was that the size of nuclei did not correlate with the DNA content of the nuclei. In this respect it is remarkable that the nuclei of pericycle cells reached an average volume of $421,15 \pm 172,58 \mu m^3$ (Fig. 1A) at the root base where the majority of nuclei had 2C and 4C DNA content (Fig. 1B). On the other hand, for example, the nuclei of epidermis or root hairs, with DNA content of 8C and 16C, achieved an average volume of only $295,94 \pm 206,01 \mu m^3$ and $261,60 \pm 126,60 \mu m^3$ respectively, in this part of the maize root (Fig. 1A). Also, in a great number of stele parenchyma, endodermis and above all of cortex nuclei, a diminution of the DNA content

50   *Baluška*

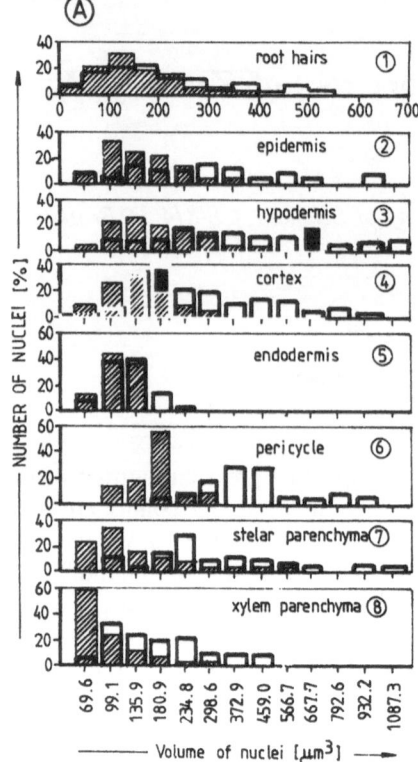

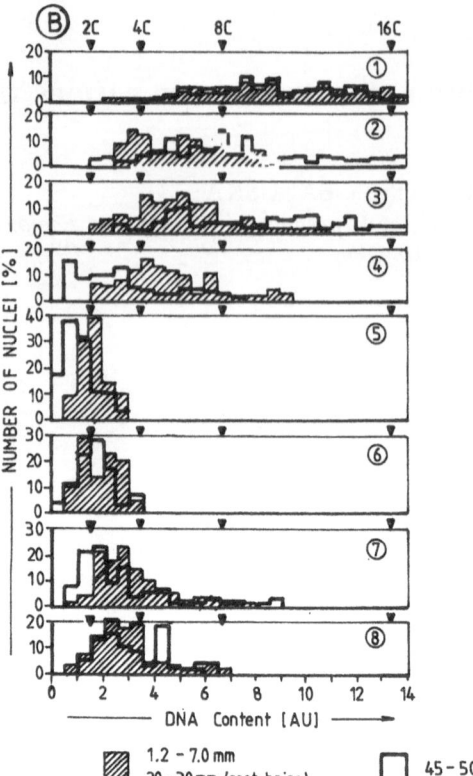

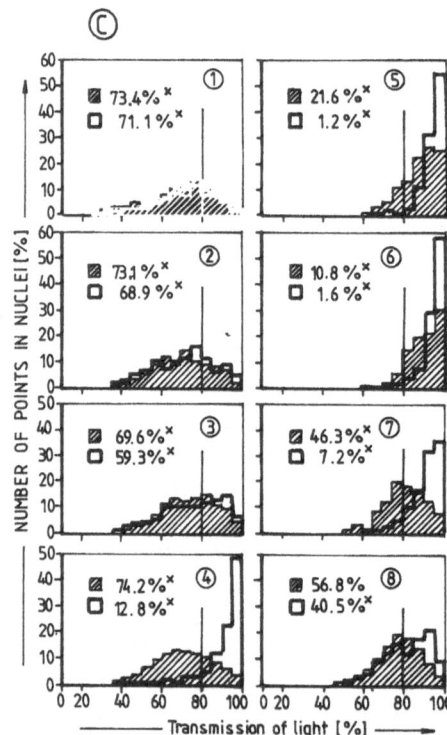

× of condensed chromatin

Fig. 1A. Volume of nuclei in cell elongation region (■) and basal region (□) of 5 cm long *Zea mays* primary root.

Fig. 1B. DNA content in cell elongation region (■) and basal region (□) of 5 cm long *Zea mays* primary root.

Fig. 1C. Chromatin condensation in cell elongation region (■) and basal region (□) of 5 cm long *Zea mays* primary root.

occurred without decline in nuclear volume after completion of cell growth (Fig. 1A, 1B). The factors determining the size of nuclei are not known, but an important rôle is probably performed by those factors which influence the concentration of individual proteins in the nucleus, or the strength of their binding to the DNA. Phytohormones are important candidates, particularly IAA, because of its ability to loosen the binding between histones and DNA (Grieshaber-Scheubel and Fellenberg, 1972). It is also known that IAA stimulates the loosening of the DNA double-helix from crown-gall tissues (Le Goff *et al.*, 1985). The tissues of the stele, with the exception of xylem parenchyma, exhibited very large nuclear volumes, in spite of the lower DNA content in comparison with cortical tissues (Fig. 1A, 1B). In connection with this observation, it is of interest that the IAA content is five times higher in the stele than in the cortex of maize root (Saugy and Pilet, 1984). The most prominent disproportion between nuclear size and DNA content is observed in the pericycle cells; it is probably connected with a high IAA content in these cells. It is known that high IAA concentration induces formation of root apical meristems not only in tissue cultures but also in intact plants in which the first mitotic divisions begin just in the pericycle cells (Wareing, 1982).

The cells of xylem parenchyma have an exceptional position among tissues of the stele. Apart from the xylem elements which were not included in our study, they have the greatest amount of condensed chromatin of all stelar tissues in the basal region of the maize primary root because of small changes in the DNA content and nuclear size during their growth and differentiation (Fig. 1A, 1B). With their stability of chromatin structure, the cells of xylem parenchyma are similar to the cells of root hairs, epidermis and hypodermis (Fig. 1C). As an active rôle of the xylem parenchyma cells in radial transfer of ions across roots has been proved in recent years (Clarkson and Hanson, 1986; De Boer and Prins, 1985), we assume that a stable chromatin structure is important for preservation of some active functions, especially those involved in active passage of ions across plasmalemmae. Much recent data has indicated that the entry of ions into the symplasm is realized mainly at the root periphery (Bange, 1973; Moon *et al.*, 1986; Peterson and Perumalla, 1984; Vakhamistrov, 1967). Van Iren and Boer-Van der Sluijs (1980)

found that, whereas during the first developmental phases of the cortical cells of barley root these were capable of ion absorption (as was also true of the epidermis and hypodermis cells) during further maturation the cortical cells lose this ability. Our observation that in some cells of the endodermis, stelar parenchyma, and especially of the cortex, a diminution of the DNA content occurs during the maturation (Fig. 1B) is quite interesting in this connection. Probably cells of the cortex, and also of some other tissues, lose their ability of active absorption or secretion of ions across the plasmalemma during the process of cell maturation due to the disintegration of their nuclear apparatus. It is known that the mitochondrial proteins are largely coded in the nucleus. We can then easily imagine that it is impossible to maintain for a long time a large number of functional mitochondria, which are needed as a source of energy for the process of active ion transport, without an intact nucleus. On the contrary, the cells of those tissues which are supposed to be active in transporting ions across the plasmalemma (root hairs, epidermis, hypodermis and xylem parenchyma) are characterized by a large number of mitochondria (Läuchli *et al.*, 1974), by high ATP-ase activity (Malone *et al.*, 197; Winter-Sluiter *et al.*, 1977) and, as shown in this study, also by the stable structure of their chromatin complex during cell growth and cell diffentiation. The functional similarity between the cells of the epidermis and the xylem parenchyma cells is based on the supposition that, in the conditions of higher demands for active ion transport, they both can be transformed into transfer cells (Letvenuk and Peterson, 1976; Kramer *et al.*, 1977, 1978).

In conclusion, our analysis of the DNA content and chromatin structure in the individual tissues of 5 cm long maize primary roots supports the view that in the proximal parts of higher plant roots the main sites of ion entry into the symplasm are plasmalemmae at the root periphery. Further radial transfer of ions across the root tissues continues *via* the symplasm to the xylem parenchyma cells which are responsible for active secretion of ions from the cytoplasm of the xylem parenchyma cells into the vessels.

References

Anderson W P 1972 Ion transport in the cells of higher plant tissues. Annu. Rev. Pl. Physiol. 23, 51–72.

Baluška F and Kubica Š 1987a Changes in chromatin condensation during growth and differentiation of maize primary root cells. Biologia (Bratislava) 42, 9–16.

Baluška F and Kubica Š 1987b DNA content, nucleus size and chromatin structure in cells of the maize primary root. Biológia (Bratislava) 42, 409–417.

Bange G G J 1973 Diffusion and absorption of ions in plant tissue. III. The role of root cortex in ion absorption. Acta Bot. Neerl. 22, 529–542.

Clarkson D T 1974 Ion transport and cell structure in plants. London, McGraw-Hill.

Clarkson D T and Hanson J B 1986 Proton fluxes and the activity of a stelar proton pump in onion roots. J. Exp. Bot. 37, 1136–1150.

De Boer A H and Prins H B A 1985 Xylem perfusion of tap root segments of *Plantago maritima*: The physiological significance of electrogenic xylem pumps. Plant Cell Environm. 8, 587–594.

Grieshaber-Scheubel D and Fellenberg G 1972 Beeinflussung der Bindung von lysinreichen und argininreichen Histon an DNS durch Wuchsstoffe. Z. Pflanzenphysiol. 66, 106–112.

Kramer D, Läuchli A, Yeo A R and Gullasch J 1977 Transfer cells in roots of *Phaseolus coccineus*: Ultrastructure and possible function in exclusion of sodium from the shoot. Ann. Bot. 41, 1031–1040.

Kramer D, Römheld V, Landsberg E and Marschner H 1980 Induction of transfer cell formation by iron deficiency in root epidermis of *Helianthus annuus* L. Planta 147, 335–339.

Le Goff L, Roussaux J, Aaron-da Cunha M I and Beljanski M 1985 Growth inhibition of crown-gall tissues in relation to the structure and activity of DNA. Physiol. Plant. 64, 177–184.

Letvenuk L J and Peterson R L 1976 Occurrence of transfer cells in vascular parenchyma of *Hieracium florentinum* roots. Can. J. Bot. 54, 1458–147-.

Läuchli A, Kramer D, Pitman M G and Lüttge U 1974 Ultrastructure of xylem parenchyma cells of barley roots in relation to ion transport to the xylem. Planta 119, 85–99.

Malone C P, Burke J J and Hanson J B 1977 Histochemical evidence for the occurrence of oligomycin-sensitive plasmalemma ATPase in corn roots. Pl. Physiol. 60, 916–922.

Moon G J, Clough B F, Peterson C A and Allaway W G 1986 Apoplastic and symplastic pathways in *Avicennia marina* Forsk. Vierh. roots revealed by fluorescence tracer dyes. Aust. J. Pl. Physiol. 13, 637–648.

Peterson C A and Perumalla C J 1984 Development of the hypodermal Casparian band in corn and onion roots. J. Exp. Bot. 35, 51–57.

Saugy M and Pilet P E 1984 Endogenous indol-3yl-acetic acid in the stele and cortex of gravistimulated maize roots. Plant Sci. Lett. 37, 93–99.

Vakhimstrov D B 1967 On the function of the apparent free space in plant roots: A study of the absorbing power of epidermal and cortical cells in barley roots. Sov. Pl. Physiol. 14, 103–107.

Van Iren F and Boer-Van der Sluijs P 1980 Symplastic and apoplastic radial ion transport in plant roots: Cortical plasmalemmas lose absorption capacity during differentiation. Planta 148, 130–137.

Wareing P F 1982 Determination and related aspects of plant development. *In* The Molecular Biology of Plant Development. Eds. H Smith and G Grierson pp 517–541, Blackwell Scientific Publications.

Winter-Sluiter E, Läuchli A and Kramer D 1977 Cytochemical localization of K^+-stimulated adenosine triphosphatase activity in xylem parenchyma cells of barley roots. Pl. Physiol. 60, 923–927.

B. C. Loughman et al. (Eds.), Structural and functional aspects of transport in roots, 53–56.
© 1989 by Kluwer Academic Publishers.

Activity and polymorphism of enzymes in different root tissues

OTILIA GAŠPARÍKOVÁ

Institute of Experimental Biology and Ecology, CBES, Slovak Academy of Sciences, Obrancov mieru 3, CS-814 34 Bratislava, Czechoslovakia

Key words: cytodifferentiation, enzyme activities, isoenzymes, root, *Zea mays*

Introduction

The manner in which cells of higher organisms acquire structural and functional specificity during differentiation and development is of general biological interest. Most authors assume differential gene expression and the level of regulation would be either at the gene, or in the pathway between the gene and its final product, the functional enzyme. One approach to the problem is to study the ontogeny of enzymes characteristic of a particular system since this provides a sensitive index of the basic changes occuring during differentiation. Therefore, enzyme activities, protein and isoenzyme patterns in different root regions have been compared many times (Beneš and Hadačová, 1988; Beneš et al., 1981; Hadačová and Beneš, 1977; Khavkin, 1977; Obrucheva, 1975; Polter and Müller-Stoll, 1969; Steward et al., 1965; Sutcliffe and Sexton, 1974). The available quantitative evidence in some cases relates to metabolic gradients from the apex to the mature root cells. Although even qualitative differences in the isoenzyme patterns of different root regions were observed (Beneš et al., 1981; Gašparíková, 1985; Hadačová and Beneš, 1977), we are still not able to discern any general conclusion concerning the relation between enzyme and isoenzyme patterns and the progress of cell growth and maturation.

The present paper is a continuation of our research on enzymology of different regions of the maize root. It is devoted to the comparison of the activity and isoenzyme patterns of lipoxygenase (LOX), superoxide dismutase (SOD), catalase and peroxidase in different root tip regions as well as in peripheral tissues (epidermis plus cortex) and stele of maize root.

Material and methods

Primary root of 3-day-old seedlings of *Zea mays* L., cv. CE 330 were used as an experimental material. The 70–80 mm long roots were selected and the following segments were dissected: 0–1 mm, 1–3 mm, 5–7 mm from the root cap boundary which represent the region of cell division, elongation and maturation respectively (Luxová, 1981). Peripheral tissues and stele were separated mechanically over the 10–70 mm region. Root segments were homogenized in $2 . 10^{-1}$ mol $. 1^{-1}$ phosphate buffer, pH 7.4, containing $5 . 10^{-3}$ mol $. 1^{-1}$ cysteine hydrochloride. Extracts were centrifuged (20,000 g; 20 min., 4°C) and enzyme activities in the supernatant were tested.

LOX activity was assayed according to the method of Surrey (1964) using a Clark oxygen electrode unit. The reaction was initiated by the addition of linoleic acid. SOD was determined spectrophotometrically by its ability to inhibit reduction of ferricytochrome c, using alkaline

Table 1. Activity of oxidative enzymes in different root segments

Enzyme activity	M	E	MC	C	CC[a]
	units . mg protein^{-1}[b]				
Lipoxygenase	53.5	87.7	148.2	1944.5	130.4
Superoxide dismutase	59.4	90.7	182.0	170.5	70.3
Catalase	6.6	5.0	4.6	3.4	13.3
Peroxidase	147.9	259.6	1000.0	1162.7	871.0

[a] M – meristem, E – region of elongation, MC – maturing cells, C – cortex plus epidermis, CC – stele.

[b] A unit of lipoxygenase is defined as the amount of enzyme consuming 1 μmol O_2 min^{-1} at 25°C. A unit of dismutase activity is defined as that amount which inhibits the DMSO-mediated cytochrome c reduction by 50%. Activities of catalase and peroxidase are expressed as in Frič and Fuchs (1970).

dimethylsulphoxide as the superoxide anion-generating system (Hyland *et al.*, 1983).

Catalase and peroxidase assays were performed according to Frič and Fuchs (1970). Proteins were determined by the method of Bradford (1976).

Isoenzyme patterns of SOD, catalase and peroxidase were performed as detailed previously (Gašparíková, 1985).

Results and discussion

The results of the quantitative enzyme analyses performed on the various root segments are summarized in Table 1. They indicate that root differentiation and maturation are associated with a gradual increase in specific activity of all enzymes observed apart from catalase. The peripheral

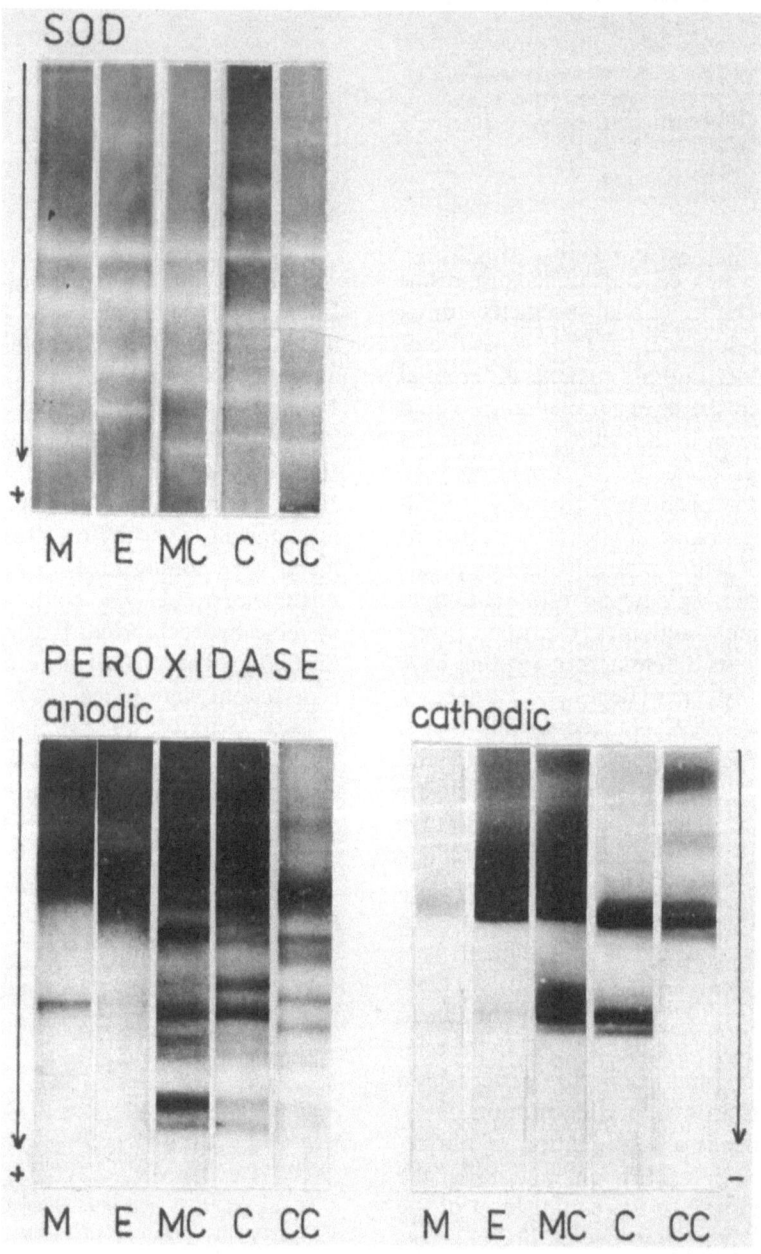

Fig. 1. Polyacrylamide gel electrophoresis patterns for superoxide dismutase and peroxidases in different region of maize root (*Symbols* as in the Table 1).

tissues (epidermis and cortex) present the final step of activity gradient of LOX, SOD and peroxidase; catalase is mostly localized in the stele.

Enormously high *in vitro* activity of LOX in the complex of peripheral root tissues could indicate the possibility of a high rate of phospholipid turnover and catabolism in these tissues. If we consider that LOX specifically catalyses the oxidation of polyunsaturated fatty acids including linoleic and linolenic acid, which usually constitute the sn-2 acyl chains in membrane phospholipid (Leshem, 1987) then the high LOX activity could be connected with the rebuilding of membranes in the mature cortex.

SOD, which catalyzes the disproportionation of superoxide and hydroxyl free radicals, products of LOX and other oxidases, seems to present almost the same pattern of activity distribution along the root tip. However, a small decrease was observed in the mature tissues. On the other hand the activity of catalase is rather low in the root tip, a two peaks of activity were observed in the meristem and in the stele.

As in the case of other plants, peroxidases exhibit a conspicuous age related activity gradient. The specific activity is highest in the differentiated root cells, whereby the activity in cortex exceeds that in the stele. Recently work suggests that peroxidases are organ and tissue specific (Shannon, 1968) and we have tried to find out if the rise in enzyme activity associated with root differentation is due to quantitative or qualitative changes in isoenzyme patterns.

In Fig. 1, a comparison of isoenzyme patterns of superoxide dismutase, cathodic and anodic peroxidases from different root segments is presented. Since catalase was always represented only by two bands the ratio of which was the same in all root segments studied, the pattern of catalase is not presented.

It may be noticed that the activation of SOD observed during differentiation of the root tissues is mostly due to the activation of already existing isoenzymes. In the case of peroxidases the differences are both qualitative and quantitative. The isoenzyme patterns of the anodic and especially the cathodic peroxidases have proved that peroxidases are tissue specific enzymes. The lowest peroxidase activity as well as the lowest number of isoenzymes was observed in the root meristem and elongation zone. With termination of growth and with gradual maturation of root cells enzyme and isoenzyme activities increase and the number of isoenzymes rises. The specific cathodic isoperoxidases were found for the peripheral tissues and others for the stele. Thus, the isoenzymes with slower relative mobility are characteristic of the stele, while in the cortex and epidermis isoenzymes with relative higher mobility are present. These differences among two tissue complexes confirm the role of peroxidase as indicators of differentiation processes (Del Grosso and Alicchio, 1981). Probably some of these isoperoxidases specific for the stele are involved in the lignification of the cell wall of xylem elements. A trend has emerged in the last few years, where lignin synthesis has been associated with the anionic peroxidases (Boyer *et al.*, 1983; Cottle and Kolattukuddy, 1982) and the oxidation of IAA with the cationic peroxidase isoenzymes (Boyer *et al.*, 1983; Hazel and Murray, 1982). However, the data of Chibbar and Van Huystee (1984) have shown that this fact is not valid for all peroxidase species. In contrast to horseradish peroxidase, the anionic and cationic peroxidases isolated from suspension culture of *Arachis hypogea* cells showed no differences in their capability for IAA oxidation nor in their peroxidative activity. Further experiments are necessary to acknowledge the role of these specific isoperoxidases in the maturing processes of the vascular elements

Acknowledgement

The author would like to thank Mrs Zuzana Vyhnánková for her assistance.

References

Beneš K, Hadačová V and Hlaváček J 1988 The heterogenity of root tip cells and tissues as revealed by enzyme localization *in situ* and isoenzyme patterns: The comparison of proteo- and esterolytic activities. *In* Structural and Functional Aspects of Transport in Roots. Eds. B C Longhman *et al.* Kluwer Academic Publishers, Dordrecht, The Netherlands.

Beneš K, Ivanov V B and Hadačová V 1981 Glycosidases in the root tip. *In* Structure and Function of Plant Roots. Eds. Brouwer *et al.* pp 137–139. Nijhoff/Junk, The Hague, Boston, London.

Boyer N, Desbiez M O, Hofinger M and Gaspar Th 1983 Effect of lithium on thigmomorphogenesis in Bryonia diocia ethylene production and sensitivity. Plant Physiol. 72, 522–524.

Bradford M 1976 A rapid and sensitive method for the quantitation of microgram quantities of protein utilizing the principle of protein-dye binding. Anal. Biochem. 72, 248.

Chibbar R N and van Huystee R B 1984 Characterization of peroxidase in plant cells. Plant Physiol. 75, 956–958.

Cottle N and Kolattukuddy P E 1982 Abscisic acid stimulation of suberization: Induction of enzymes and deposition of polymeric components and associated waxes in tissue cultures of potato tuber. Plant Physiol. 70, 775–780.

Del Grosso E and Alicchio R 1981 Analysis in isozymatic patterns of Solanum melongena: Differences between organized and unorganized tissues. Z. Pflanzenphysiol. 102, 467–470.

Frič F and Fuchs W H 1970 Veränderungen der Aktivität einiger Enzyme in Weizenblatt in Abhängigkeit von der temperaturlabilen Verträglichkeit für Puccinia graminis tritici. Phytopath. Z. 67, 161–174.

Gašparíková O 1985 Protein and isoenzyme patterns in different parts of maize plants (*Zea mays* L.). Acta Univ. Agric. (Brno). 33, 603–606.

Hadačová V and Beneš K 1977 Investigation of isoenzyme patterns of some oxidoreductases, hydrolases and transferases in different growth zones of broad bean (*Vicia faba* L.) roots. Physiol. Veget. 15, 735–745.

Hazell P and Murray D R 1982 Peroxidase isoenzymes and leaf senescence in sunflower, *Helianthus annus* L. Z. Pflanzenphysiol. 108, 87–92.

Hyland K, Voisin E, Banoun H and Auclair C 1983 Superoxide dismutase assay using alkaline dimethylsulfoxide as superoxide anion-generating system. Analytical Biochemistry 135, 280–287.

Khavkin E E 1977 Development of metabolic patterns in growing plant cells (russ.) Nauka, Novosibirsk.

Leshem Y Y 1987 Membrane phospholipid catabolism and Ca^{2+} activity in control of sensecence. Physiol. Plantarum 69, 551–559.

Luxová M 1981 Growth region of the primary root of maize (*Zea mays* L.) *In* Structure and Function of Plant Roots. Eds. R Brouwer *et al.*, pp. 9–14. Nijhoff/Junk Publ. Hague, Boston, London.

Obrucheva N V 1975 Physiology of growing root cells. *In* The Development and Function of Roots. Eds. J G Torrey, D T Clarkson pp 279–298. Academic Press London, New York, San Francisco.

Polter R C and Müller-Stoll W R 1969 The question of identity and homology of the proteins extracted from legume roots separated by means of disc electrophoresis in polyacrylamide gel. Z. Naturforsch. 24, 333–341.

Shannon L M 1968 Plant isoenzymes. Ann. Rev. Plant Physiol. 19, 187–210.

Steward F C, Lyndon R F and Barber J T 1965 Acrylamide gel electrophoresis of soluble plant proteins: A study on pea seedlings in relation to development. Am. J. Bot. 52, 155–164.

Surrey K 1964 Spectrophotometric method for determination of lipoxidase activity. Plant Physiol. 39, 65–70.

Sutcliffe J F and Sexton R 1974 Enzymatic changes during the differentiation of tissues in young pea roots. *In* Structure and Function of Primary Root Tissues. Ed. J Kolek. pp 203–219. Veda, Bratislava.

B. C. Loughman et al. (Eds.), Structural and functional aspects of transport in roots, 57–59.
© 1989 by Kluwer Academic Publishers.

Three-dimensional distribution of plasmodesmata in the rhizodermis of *Trianea bogotensis Karst.*

ELENA B. KURKOVA
K A Timiriazev Institute of Plant Physiology, Academy of Sciences of USSR, Botanical St. 35, SU-127 276 Moscow, USSR

Key words: plasmodesmata, rhizodermis – root hair – hairless cell

Abbreviations: RH, root hair; HLC, hairless cell

Introduction

Rhizodermis (root epidermis) in many plant species consists of two cell types: root hairs (RH) and hairless cells (HLC). Structural differences between these cells appear in the early stages of root development. Trichoblasts (which subsequently form RH) differ from atrichoblasts (which form HLC) by a higher optical and electron density of cytoplasm, higher activity of redox enzymes, and a greater content of proteins and nucleic acids (Cutter, 1970; Danilova, 1974).

We found ultrastructural differences between RH and HLC of the aquatic plant *Trianea bogotensis* in the differentiation zone (Vakhmistrov and Kurkova, 1979). A typical feature of RH is their polyploid nuclei and a cytoplasm with more ribosomes and organelles in the active state than the cytoplasm of HLC cells. Another characteristic feature of *T. bogotensis* was the presence of chloroplasts in all root cells except the RH. In addition, it was shown electrophysiologically that the inflow of potassium from the external environment into RH was at least 10 times higher than in HLC. (Vakhmistrov, 1974).

On the basis of these data, and according to the classical work by Arizs (1969) who demonstrated that parenchymal transport of substances in aquatic plants occurred predominantly through the symplast, *i.e.*, through plasmodesmata, one can speculate that the density of plasmodesmata at the boundary between RH and cortex will be significantly higher than in the base of HLC (on the boundary between HLC and cortex).

Materials and methods

Roots of the freshwater plant *Trianea bogotensis Karst.* (Hydrocharitaceae) served as the object of our study.

Segments of 3–4 mm long roots with RH were taken for fixation by the standard method (3% glutaraldehyde and 1% OsO_4) and the material was embedded in Epon-812. Ultrathin sections were prepared on an LKB-4800 ultratome and examined in the JEM-100 B electron microscope. The number of plasmodesmata were counted during electron microscopic examination of transverse and longitudinal root sections (average thickness 70 nm) in the young RH zone. Several thousands of sections were examined, more than 800 of which were used to estimate plasmadesmatal densities. The presence of chloroplasts in all cells of the root except RH helped us to find the required cell walls in the electron microscope (Kurkova and Vakhmistrov, 1984; Vakhmistrov and Kurkova, 1979). The results obtained were expressed per 1 μm^2 of cell wall and per the entire intercellular boundary.

Cell wall areas were estimated on the basis of

measuring the linear wall dimensions on electron micrographs prepared at a magnification of 4,000–5,000 ×.

Results and discussion

The RH were mosaically disposed in the *T. bogotensis* rhizodermis (one RH falls on every five HLC on the average). We selected areas between two RH separated from each other by three to four HLC. Plasmodesmata were usually unbranched and single, although sometimes they occurred in groups of two to five, more often in the boundary between RH and cortex cells.

The frequency of plasmodesmata per 1 μm^2 and per total surface of the cell wall is significantly higher in RH than that in HLC (Table 1). Plasmodesmata are ten times more dense in the boundary between RH and the cortex than in the boundary between HLC and the cortex. As for radial and transverse walls, the density of plasmodesmata in them decreases in proportion to the distance of the cells from the RH.

The results of Table 1 are shown in Fig. 1, indicating three-dimensional distribution of plasmodesmata in a single RH and four adjacent cells. The average density of plasmodesmata is obtained on the basis of values of two HLC sequentially adjacent to the RH (see Table 1).

We are confident that there should be no exact

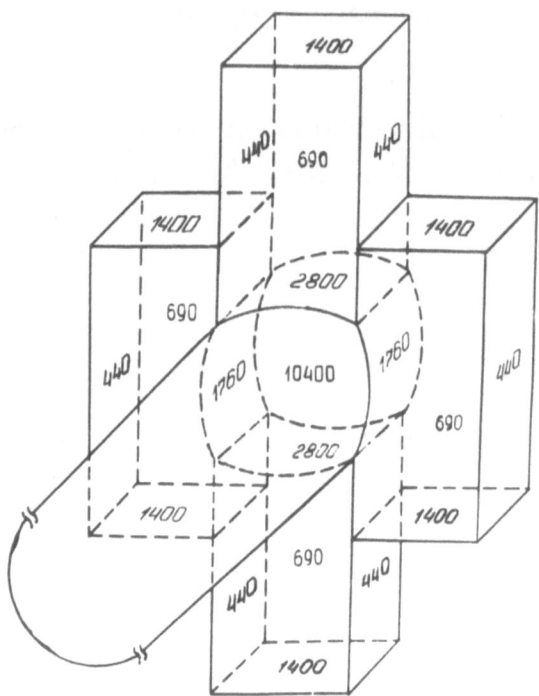

Fig. 1. Three-dimensional reconstruction of plasmodesmata distribution in cell walls of root hair and four adjacent hairless cells in the rhizodermis of *Trianea bogotensis*.

Table 1. Frequency of plasmodesmata in rhizodermis cell walls of *Trianea bogotensis*

Intercellular boundary	Number of plasmodesmata	
	per 1 μm^2 of cell wall	per intercellular boundary
Tangential walls		
RH – cortex	2.06	10419
HLC_1 – cortex	0.10	630
HLC_2 – cortex	0.12	756
Radial walls		
RH–HLC_1	0.60	1764
HLC_1–HLC_2	0.21	529
HLC_2–HLC_3	0.14	353
Transverse walls		
RH–HLC_1	1.02	2860
HLC_1–HLC_2	0.70	1470
HLC_2–HLC_3	0.63	1340

parallel between the number of plasmodesmata and their transport function since the latter may depend on both plasmodesmata ultrastructure and the molecular radius of the mobile substances (Gamalei, 1985; Goodwin, 1983; Olesen, 1979; Terry and Robards, 1987). However, in certain cases the frequency of plasmodesmata and their distribution may play a significant rôle in the direction and in the rate of the symplastic fluxes as in ionic radial transport in the root (Robards, 1976), and the outflow of assimilates from the photosynthetic cells (Kursanov, 1976).

The three-dimensional scheme (Fig. 1) gives a picture of the density of symplastic connections directed along either the root axis or the radial direction through the rhizodermis to internal tissues. In addition, it permits us to compare RH and HLC in order to elicit their specifticity.

The RH are characterized by a high density of plasmodesmata (53% of their total number) at their base on the boundary with the cortex. This provides the possibility either for the passage of absorbed ions to, or the removal of metabolites from the neighbouring photosynthetic cells during the period of RH growth. It is worthy of note that the RH of *T. bogotensis* are on average 2–3 mm long (and may grow up 8 mm). At the same time their volume exceeds that of HLC by 40–60-fold (Kurkova, 1984).

The HLC plasmodesmata are distributed in such a way that over 60 per cent of their number are concentrated in transverse cell walls. This enables us to assume that the symplastic flux of substances in vertical rows of HLC are mainly directed along the root axis. Juniper and Barlow (1969) associate such an asymmetric distribution of plasmodesmata in the tip of corn root with the direction of movement of metabolites and signals necessary for differentiation of linear cell rows, expanding from the quiescent centre.

References

Arizs W H 1969 Intercellular polar transport and the role of the plasmodesmata in coleoptiles and *Vallisneria* leaves. Acta Bot. Neerl. 18, 14–38.

Cutter E G and Feldmann I J 1970 Trichoblasts in Hydrocharis. II. Nucleic acids, proteins and a consideration of cell growth in relation to endopolyploidy. Am. J. Bot. 57, 202–211.

Danilova M F 1974 Structural bases of absorption of substances by the root (*In Russian*). Nauka, Leningrad.

Gamalei Ju V 1985 Plasmodesmata: Intercellular Communication in plants (*In Russian*). Fiziol. rast. 32, 176–190.

Goodwin P B 1983 Molecular size limit for movement in the symplasm of the *Elodea* leaf. Planta. 157, 124–130.

Juniper B E and Barlow P W 1969 The distribution of plasmodesmata in the root tip of maize. Planta. 89, 352–360.

Kurkova E B and Vakhmistrov D B 1984 Distribution of plasmodesmata in the rhizodermis along the root axis in *Trianea bogotensis* (*In Russian*). Fiziol. rast. 31, 141–145.

Kursanov A L 1976 Transport of assimilates in plants (*In Russian*). Nauka. Moscow.

Olesen P 1979 The neck constriction in plasmodesmata: Evidence for a peripheral sphinkter-like structure revealing by fixation with tannic acid. Planta. 144, 349–358.

Robards A W and Clarkson D T 1976 The role of plasmodesmata in the transport of water and nutrients across roots. *In* Intercellular communication in Plants: Studies on Plasmodesmata. Eds. B E S Gunning and A W Robards pp 181–201. Springer-Verlag, Berlin.

Terry B R and Robards A W 1987 Hydrodynamic radius alone governs the mobility of molecules through plasmodesmata. Planta 171, 145–157.

Vakhmistrov D B and Kurkova E B 1979 Symplastic connections in the Rhizodermis of *Trianea bogotensis* Karst (*In Russian*). Fiziol. rast. 26, 763–771.

Vakhmistrov D B, Melnikov P V and Vorob'ev L N 1974 Differences of potassium absorption by root hairs and hairless cells of root epidermis in *Trianbea bogotensis* (*In Russian*). Fiziol. rast. 21, 554–562.

Session 2

Absorption, transport and
utilization of ions

B. C. Loughman et al. (Eds.), Structural and functional aspects of transport in roots, 63–67.
© 1989 by Kluwer Academic Publishers.

Effect of auxin on growth, proton secretion and transmembrane electron transfer in intact maize roots

H. LÜTHEN, F. HILGENDORF and M. BÖTTGER
Institut für Allgemeine Botanik der Universität Hamburg, Ohnhorststrasse 18, D-2000 Hamburg 52, FRG

Key words: auxin, fusicoccin, growth control, proton pump, redox chain, roots

Abbreviations: IAA, indole-3-acetic acid; FC, fusicoccin; HCF, hexacyanoferrate; ADH, alcohol dehydrogenase; AVG, aminoethoxyvinylglycine

Introduction

Auxin is a phytohormone known to inhibit root growth. The scheme of the acid growth hypothesis, originally developed for the stimulation of growth of shoots and coleoptiles (Hager *et al.*, 1971), has also been applied on this inhibitory effect in roots (Edwards and Scott, 1977; Evans *et al.*, 1980). It was found that IAA inhibits proton secretion (Evans *et al.*, 1980). Furthermore acid-induced growth has been detected in roots (Evans, 1976). Fusicoccin, a fungal toxin stimulating proton secretion has been shown to induce a growth response (Lado *et al.*, 1976). However, doubts have been raised whether the acid growth theory quantitatively explains the induction of growth in the maize coleoptile (Kutschera and Schopfer, 1985a). A quantitative study on the relations between the auxin effects on H^+-efflux and on growth has not yet been achieved.

Especially in the early years of hormone research there were reports that IAA at very low concentrations ($10^{-10} M$) enhances growth (Åberg, 1957). More recent researchers failed to confirm any promotion of growth by auxin. Mulkey *et al.* (1983) proposed that auxin induces formation of ethylene that suppresses the growth rate thus overcoming the IAA-effect. They reported that pretreatment of roots with AVG and Co^{2+} sensitizes the roots for auxin so that they can be stimulated by IAA.

Proton secretion is largely accomplished by an ATPase, but it is possible that an additional redox system is involved in proton pumping (Böttger *et al.*, 1985, 1986; Craig and Craine, 1981). Reduction of external electron acceptors is tightly coupled to proton secretion (Rubinstein *et al.*, 1986) or the redox system may be a proton pump by itself (Böttger and Lüthen, 1986). Redox activity has been proposed to be linked to growth control in animal and plant systems (Crane *et al.* 1985).

The aim of the present study is the quantitative reinvestigation of hormonal control of proton efflux and growth and the evaluation of the role of redox processes on growth regulation.

Materials and methods

To avoid the effects of the strong pH-dependence of proton secretion (Böttger, 1986) and of the buffer capacity of substances in the bathing medium or in the cell wall, we decided to measure proton secretion by continuous titration at a constant pH (Böttger *et al.*, 1985), thus departing from the common practice of simply measuring the fall in pH. HCFIII concentration in the bathing medium was measured photometrically and also automatically held constant. The amount of titrators used was registrated by a computer (Apple IIe) and net flux rates could be computed and plotted.

We used intact roots rather than cuttings, since these conditions are more physiological and since proton pumping rates are much higher in intact material. Growth rates were measured using a horizontal microscope.

Results and discussion

Fusicoccin

Figure 1 shows the effect of $10^{-5} M$ FC on growth and proton secretion at pH 5.5. As could be expected, both parameters were enhanced by the toxin and obviously FC increased growth and proton secretion in a proportional manner. Since there is little doubt that FC acts on growth *via* acidification of the cell wall, we can conclude that at pH 5.5 there is a proportional relation between the change of proton efflux and growth. This relation should also be revealed using other substances regulating growth *via* the mechanism of the acid-growth theory.

Auxin at high concentrations

IAA reduced both growth and proton secretion within minutes if applied at concentrations higher than $10^{-8} M$. The effects on growth and proton secretion were far from being proportional, and this might be the most striking apparent difference between FC and IAA. Auxin reduced H^+-secretion only slightly but there was a nearly complete inhibition of straight growth. These discrepancies do not contradict the acid-growth theory but are caused by a high specificity of auxin on the proton secretion of cells in the elongation zone. We demonstrated this by measuring net H^+ efflux of the 10^- mm zone behind the root tip. This was achieved by dipping only the root tips into the bathing medium.

With the secreting surface reduced, net proton efflux dramatically decreased. But the absolute IAA-induced effect on H^+ extrusion remained virtually unchanged. The relative fall, however, increased from 10% to nearly 80% at higher auxin levels (Fig. 2). Recently, studies by Pilet *et al.* (1983) indicated that the zone of maximal elongation and of maximal acidification coincide in maize roots. Thus, there seems to be no evidence for basic differences between the growth effects induced by IAA and FC in maize roots. The acid-growth theory of FC action has never been questioned so far (Kutschera and Schopfer, 1985b).

Auxin at very low concentrations

Very low auxin concentrations have been often suspected to increase root growth. It was argued that the optimum of the bell-shaped dose response curve is shifted to lower concentrations in roots. In our experiments, however, no effect of $10^{-10} M$ IAA could be detected. These results are in agreement with those of Edwards and Scott and Evans *et al.*, who also failed to confirm any stimulation of growth or acidification by such low auxin levels. Mulkey *et al.* (1982) stated that this situation is changed after pretreatment with inhibitors of ehtylene biosynthesis. We tried to confirm their results without success. Cobaltous ions turned out to inhibit proton secretion dramatically, and AVG-pretreatment did not reveal any stimulatory activity of auxin at low concentrations (data not shown).

Transmembrane electron transfer: some facts and a model

If impermeable electron acceptors are applied to the plants, the following effects can be demonstrated:
a) The electron acceptors are reduced by the plants. (Craig and Crane, 1981).
b) Reduction of the electron acceptor triggers a rise in proton secretion rate.
c) These effects can be stimulated by small concentrations of alcohols that increase internal

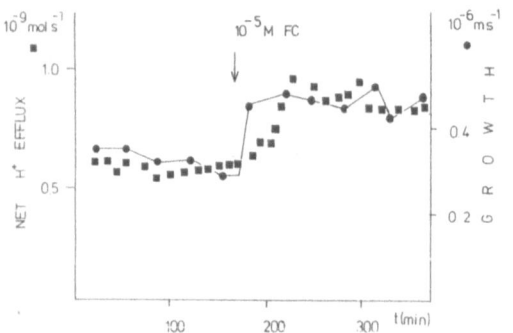

Fig. 1. Effect of fusicoccin on proton secretion and growth. Note that FC enhances both parameters by the same factor.

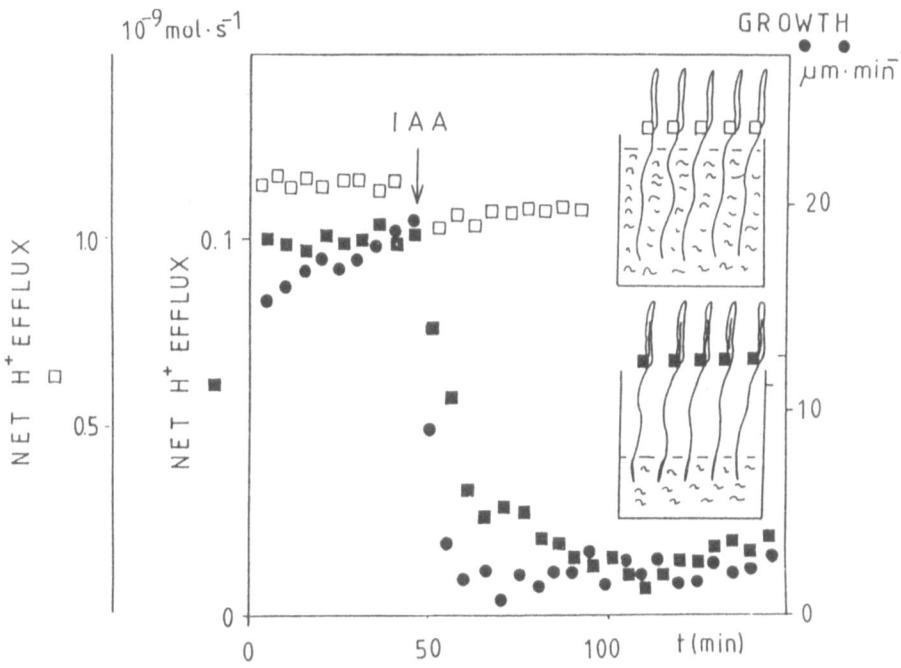

Fig. 2. Effect of $2*10^{-6}$ M IAA on proton secretion and growth. IAA nearly completely inhibits growth (●) but only slightly suppresses proton secretion (□). If the plants are dipped into the solution with their root tips only, proton secretion is massively inhibited (■). This indicates a high sensitivity of cells in the elongation zone for auxin.

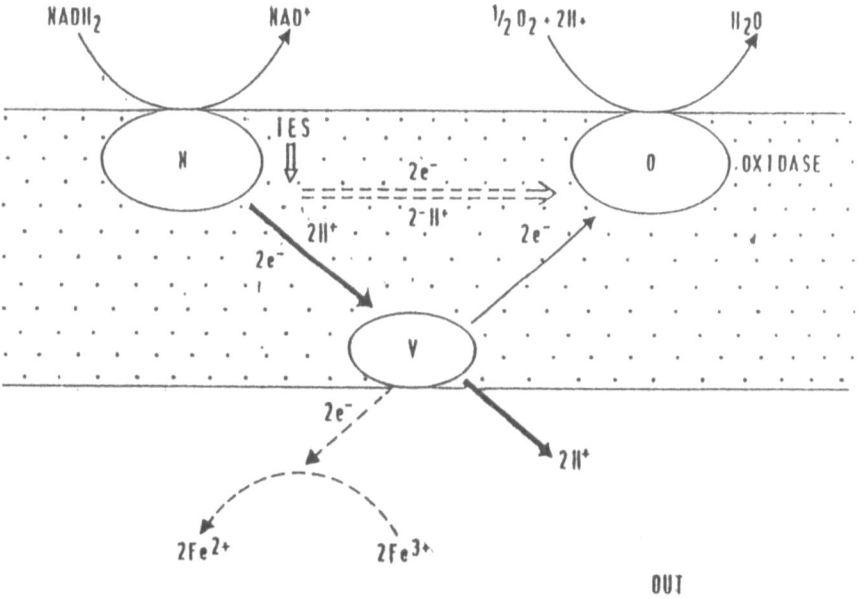

Fig. 3. Hypothetical model of a redox driven proton pump. HFCIII competes with oxygen for electrons. IAA transfers electrons and protons in a kind of short-circuit to the terminal oxidase.

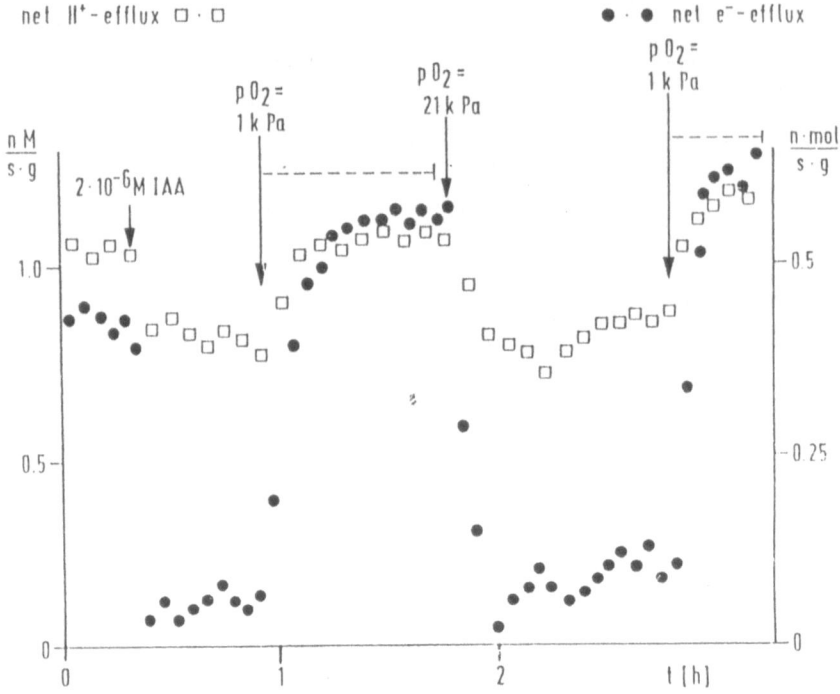

Fig. 4. Effect of auxin on proton secretion and transplasmamembrane electron transfer. Both parameters are inhibited. The inhibition is very dependent on a sufficient supply of oxygen and can be switched off and on by reducing or enhancing oxygen concentration in the incubation medium.

NADH level (Böttger and Hilgendorf, 1988; Craig and Crane, 1981) *via* the action of ADH. It is concluded that NADH can be a substrate for the redox process. Furthermore, electron acceptors reduce the internal NADPH and NADH-concentrations (S Krüger, personal communication).

d) The proton secretion loses its high requirement for oxygen (Böttger and Hilgendorf, 1987).

e) Low oxygen supply induces an increase of transmembrane electron transfer (Böttger and Hilgendorf, 1987).

f) Oxygen consumption in reversibly inhibited by HFCIII. The effect can only be measured if mitochondrial respiration is inhibited by Disulfilram and Cyanide (Crane, pers. comm.).

From these data, we derived a model, in which we postulated a direct competition of HCFIII and oxygen. This redox chain was further postulated to be directly involved in proton pumping (Fig. 3).

We further investigated whether electron transfer is influenced by hormones that have been shown to regulate proton secretion.

Auxins and electron transfer

HCFIII reduction is inhibited by auxin as shown in Fig. 4. This inhibition is very dependent on oxygen concentration and can be switched off and on by varying oxygen partial pressure. The K_M of this process is about $40 \, \mu M$ Oxygen.

High amounts of HCFIII, especially at high calcium levels, induce a very high reduction rate that cannot be regulated by the hormone.

Plasmalemma redox and growth

HCFIII also influences other physiological parameters. It inhibits growth (data not shown, see Böttger and Lüthen, 1985, Fig. 1). This is surprising, since the acid growth theory would predict a stimulation of growth induced by the massive increase in proton efflux. The effect seems not to be due to osmotic shock because Ferrocyanide suppresses proton efflux but does not change growth rate. It seems that redox potential is a crucial factor in growth control. Griesebach (1981) has shown that lignification depends on reducing substances in the cell wall. It will be a target for further research to evaluate to what extent redox processes mediate the hormone-induced changes of root growth.

Acknowledgements

We thank DFG for funding and E Marre for the kind gift of Fusicoccin.

References

Åberg B 1957 Auxin relations in roots. Ann. Rev. Plant Physiol. 8, 153–180.

Böttger M, Bigdon M and Soll H J 1985 Proton translocation in corn coleoptiles: ATPase or redox chain? Planta 163, 376–380.

Böttger M and Lüthen H 1986 Possible linkage between NADPH oxidation and proton secretion in *Zea mays* L. roots. J. Exp.Bot. 37, 666–675.

Böttger M 1986 Proton translocation systems at the plasmalemma and its possible regulation by auxin. Acta Horticult. 179, 83–93.

Böttger M and Hilgendorf F 1988 Hormone action on transmembrane electron and H$^+$-transports. Plant Physiol. 86, 1038–1048.

Craig T and Crane F L 1981 Evidence for a transplasmamembrane electron transport system in plant cells. Proc. Indiana Acad. Sci. 90, 150–155.

Crane F L, Sun I L, Clark M G, Grebing C and Löw H 1985 Transplasmamembrane redox systems in growth and development. Biochim. Biophys. Acta 811, 233–64.

Edwards K L and Scott T K 1974 Rapid growth response of corn root segments: Effect of pH on elongation. Planta 119 27–37.

Edwards K L and Scott T K 1977 Rapid growth responses of corn root segments: Effect of auxin on elongation. Planta 135, 1–5.

Evans M L 1976 A new sensitive root auxanometer. Plant Physiol 58, 599–601.

Evans M L, Mulkey T J and Vesper M 1980 Auxin action on proton influx in corn roots and its correlation with growth. Planta 148, 510–512.

Griesebach H 1981 Lignins. *In* The Biochemistry of Plants, Vol. 7. pp 457–458. Academic Press.

Hager A, Menzel H and Krauss A 1971 Versuche und Hypothese zur Primärwirkung des Auxins beim Streckungswachstum. Planta 100, 47–75.

Kutschera U and Schopfer P 1985a Evidence against the acidgrowth theory of auxin action. Planta 163, 483–493.

Kutschera U and Schopfer P 1985b Evidence for the acidgrowth theory of fusicoccin action. Planta 163, 494–499.

Lado P, Michaelis M I, Cerena R and Marre 1976 Fusicoccininduced K$^+$-stimulated proton secretion and acid-induced growth of apical root segments. Plant Sci. Lett. 6, 5–20.

Mulkey T J, Kuzmanoff and Evans M L 1982 Promotion of growth and hydrogen ion efflux by auxin in roots of maize pretreated with ethylene biosynthesis inhibitors. Plant Physiol. 70, 186–188.

Pilet P E, Versel J M and Mayor G 1983 Growth-distribution and surface pH along maize roots. Planta 158, 398–402.

Rubinstein B and Stern A I 1986 Relationship of transplasma membrane redox activity to proton secretion. Plant Physiol. 80, 805–811.

B. C. Loughman et al. (Eds.), Structural and functional aspects of transport in roots, 69–72.
© 1989 by Kluwer Academic Publishers.

Separation of wheat root microsomal membranes by countercurrent distribution
An evaluation of plasma membrane markers

ALAJOS BÉRCZI[1], CHRISTER LARSSON[2], SUSANNE WIDELL[2] and IAN M. MØLLER[2,3]
[1]*Institute of Biophysics, Biological Research Center, Hungarian Academy of Sciences, P.O. Box 521, H-6701 Szeged, Hungary.* [2]*Department of Plant Physiology, University of Lund, Box 7007, S-220 07 Lund, Sweden and* [3]*Corresponding author*

Key words: ATPase, calcium, glucan synthase II, plasma membrane, redox activities, vesicle orientation

Abbreviations: CCD, countercurrent distribution; CCO, cytochrome c oxidase; ER, endoplasmic reticulum; GS II, glucan synthase II; IO, inside-out; PM, plasma membrane; RO, right side-out; STA, silico-tungstic acid; ΔVO_4-ATPase, vanadate inhibition of K^+, Mg^{2+}-ATPase

Introduction

Leonard and Hodges in 1972 were the first to describe a method of purifying plasma membrane (PM) vesicles from microsomal fractions of plant tissues, specifically roots, using sucrose gradient centrifugation (see Hodges and Leonard, 1974). In these membranes, a vanadate-sensitive K^+, Mg^{2+}-ATPase is localized (Hodges and Leonard, 1974; Sze, 1985), and such relatively crude PM fractions, or even microsomal fractions, have been used extensively to study transport phenomena across the PM (for a review see Sze, 1985). Thus transport which is ATP-dependent and vanadate-inhibited has often been assumed to be associated with PM vesicles (Sze, 1985).

Using aqueous polymer two-phase partitioning for the isolation of PM vesicles (Widell and Larsson, 1981) a substantially higher purity is obtained: ca 95% PM (for reviews see Larsson, 1985; Larsson et al., 1987) compared to the ca 50% PM achieved with sucrose gradient centrifugation (Berczi and Moller, 1986; Hodges and Mills, 1986). These PM vesicles are almost entirely sealed and

right side-out (RO) with the binding site for MgATP on the inner, cytoplasmic surface (Larsson et al., 1984). Thus, RO–PM vesicles cannot be responsible for the ATP-dependent transport phenomena (usually proton uptake) observed with microsomal fractions (Sze, 1985). Thus, the question is: Does the microsomal fraction also contain inside-out (IO) PM vesicles, or is the observed ATP-dependent transport due to vesicles of a different origin?

We have investigated this problem by separating vesicles in the microsomal fraction from wheat roots using repeated aqueous polymer two-phase partitioning — countercurrent distribution (CCD; see Larsson, 1985). In addition to measuring the distribution and latency of vanadate-sensitive (ΔVO_4) K^+, Mg^{2+}-ATPase in the different fractions we have followed the activity and latency of marker enzymes like cytochrome c oxidase (CCO; mitochondria), glucan synthase II (GS II; plasma membrane) and antimycin-insensitive NADH-cytochrome c reductase (usually regarded as an ER marker) as well as NADH-ferricyanide reductase, and STA staining (a PM-specific stain for electron microscopy).

Materials and methods

Seven-day-old spring wheat (*Triticum aestivum.*
cv. Drabant) seedlings were grown in darkness. A
microsomal membrane fraction was prepared from
the roots and subjected to phase partitioning using
a 5-step CCD procedure (see legend to Fig. 1).

ATPases were measured in the presence of
0.1 mM EDTA, 0.1 mM molybdate, 1 mM azide,
3 mM ATP, 25 mM Tris-MES (pH 6.0), 250 mM
sucrose and other compounds specified in the figure
legends. The P_i liberated by ATP hydrolysis was
measured according to Hodges and Leonard
(1974).

GS II activity was assayed essentially as described by Kauss and Jeblick (1986). CCO and
NADH-cyt. c reductase were determined essentially according to Hodges and Leonard (1974) using
0.025% (w/v) Triton X-100. NADH-ferricyanide
reductase was determined by measuring the reduction of ferricyanide at 420 nm under similar conditions to the NADH-cyt. c reductase assay.
Protein was measured according to Lowry *et al.*

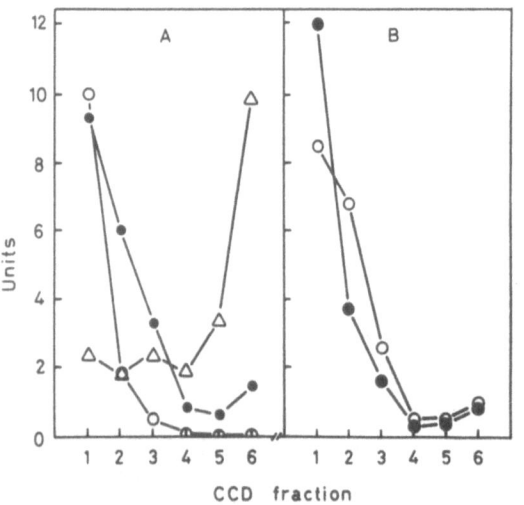

Fig. 1. Distribution of total protein and total marker enzyme
activities in the presence of detergent (see Fig. 3) after countercurrent distribution of a microsomal fraction from wheat roots.
The microsomal fraction was loaded in tube 1, and 5 transfers
were made of the upper phase. Thus, membrane vesicles partitioning to the lower phase will end up in early fractions and
vesicles partitioning to the upper phase in late fractions. A: ○,
CCO (1 unit = 1 μmol min^{-1}); △, GSII (20 nmol min^{-1}); ●,
protein (3 mg). B: ○, NADH-cyt. c reductase (0.2 μmol min^{-1});
●, NADH-ferricyanide reductase (5 μmol min^{-1}).

(1951) after solubilization of membrane proteins
with 1% (w/v) deoxycholic acid.

All data are from one preparation, but similar
results were obtained with two other preparations.

Results and discussion

Protein distributed mainly to the lower phase
and about 43% of total protein was found in fraction 1 (F1). However, some protein distributed
preferentially to the upper phase and 7% was recovered in F6 (Fig. 1A). CCO is a mitochondrial
marker and its activity was used to follow the distribution of intracellular membranes (Larsson, 1985).
These had a high affinity for the lower phase (80%
in F1) and only 0.4% of CCO activity was recovered in F6 (Fig. 1A). In contrast, GS II (Fig.
1A) as well as STA staining (data not shown), both
markers for the PM (Hall, 1983), distributed
mainly to the upper phase.

Thus, ca 50% of total GS II activity was recovered in F6 compared to only ca 10% in F1 (Fig.
1A). An even more extreme distribution of PM
vesicles to the upper phase was suggested by the
PM-specific staining of vesicles in F1 and F6 and by
the distribution of GS II activity with the other two
microsomal preparations (data not shown).

Total activity of antimycin A-insensitive
NADH-cyt. c reductase distributed in the same way
as total protein (Fig. 1B). Consequently, its specific
activity was about the same in all six fractions (data
not shown). This activity is often referred to as 'a
well-known ER marker' although it is known to be
found also in the outer mitochondrial membrane
(Møller and Lin, 1986). It is clearly also present in
the PM (Lundborg *et al.*, 1981; Fig. 1B) and in
some cases actually enriched several-fold in this
membrane (Larsson *et al.*, 1987). This is not surprising since the PM contains flavoprotein and
b-type cytochrome just like the outer mitochondrial membrane and the ER (Crane *et al.*, 1985;
Møller and Lin, 1986). Antimycin-insensitive
NADH-cyt. c reductase is therefore not always a
useful marker for ER, particularly not in PM
fractions.

NADH-ferricyanide reductase distributed similarly to NADH-cyt. c reductase (Fig. 1B), but the
former showed a slight enrichment (higher specific
activity) in F1 compared to F6 (results not shown).

K^+, Mg^{2+}-ATPase and ΔVO_4-ATPase both showed two peaks of total activity, one in F1 and one in F6 (Fig. 2B). The specific activities in F1 were about half those in F6 (Fig. 2A). Both K^+, Mg^{2+}-ATPase and ΔVO_4-ATPase activities are often used as PM markers. However, considering the discrepancies between the distribution of GS II and STA staining on the one hand and the ATPase activities on the other hand (compare Figs. 1 and 2), there is clearly also a non-PM membrane fraction in wheat roots containing these ATPase activities.

The Mg^{2+}-ATPase activity in F6 was strongly inhibited by Ca^{2+}; only a small inhibition was observed in F1 (Fig. 2). Ca^{2+}-inhibited Mg^{2+}-ATPase has earlier been reported in PM of, *e.g.*,

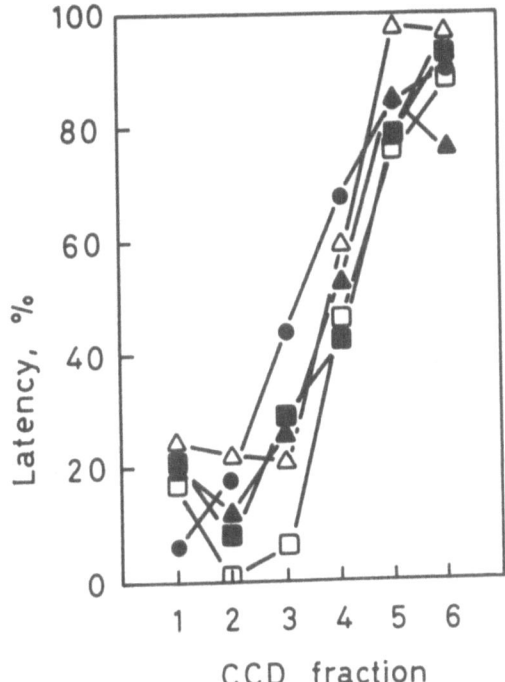

Fig. 3. The latencies of marker enzymes after CCD of a microsomal fraction from wheat roots. The latencies were calculated according to Larsson *et al.* (1984) with 0.025% (w/v) Triton X-100 for the ATPases (□, K^+, Mg^{2+}-ATPase; ■, ΔVO_4-ATPase), NADH-cyt. c reductase (▲) and NADH-ferricyanide reductase (△) and with 0.015% (w/v) digitonin for GS II (●).

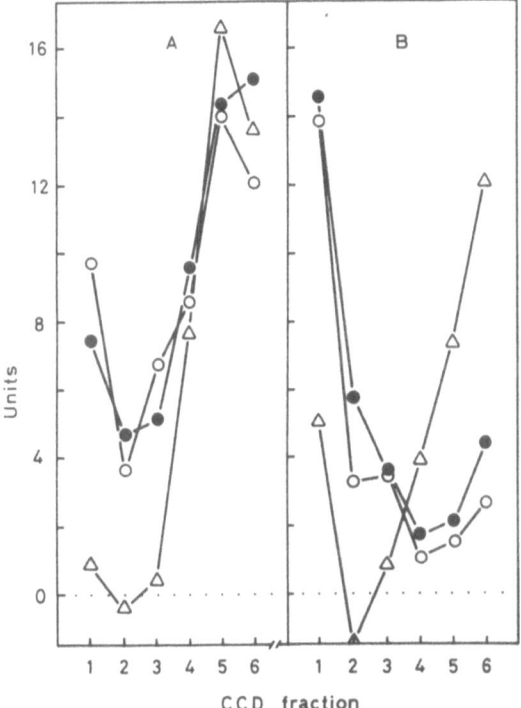

Fig. 2. The distribution of specific (**A**) and total (**B**) ATPase activities in the presence of detergent (see Fig. 3) in fractions from CCD of a microsomal fraction from wheat roots. K^+, Mg^{2+}-ATPase activity was measured as the increase in activity upon addition of $3\,mM$ $MgSO_4$ + $25\,mM$ KNO_3. Vanadate-inhibition of the K^+, Mg^{2+}-ATPase activity was achieved with $0.1\,mM$ Na_3VO_4. Ca^{2+}-inhibition of the Mg^{2+}-ATPase activity ($3\,mM$ $MgSO_4$) was achieved with $0.3\,mM$ $Ca(NO_3)_2$. ●, K^+, Mg^{2+} − ATPase [A, 1 unit = $20\,nmol$ (mg protein)$^{-1}$ min^{-1}; B, 1 unit = $300\,nmol$ min^{-1}]; ○, ΔVO_4-ATPase [$10\,nmol$ (mg protein)$^{-1}$ min^{-1}; $200\,nmol$ min^{-1}]; △, ΔCa^{2+}-ATPase [$10\,nmol$ (mg protein)$^{-1}$ min^{-1}; $50\,nmol$ min^{-1}].

cucumber seedlings (Memon *et al.*, 1987). In the microsomal fraction from wheat roots there is more Ca^{2+}-ATPase than Mg^{2+}-ATPase activity (Kähr and Møller, 1976) and in the present study Ca^{2+} often stimulated Mg^{2+}-ATPase activity slightly in the microsomal fraction and sometimes also in F1 (results not shown).

In Fig. 3 are shown the latencies of the various enzyme activities in the CCD fractions. In all cases, the latencies were low in F1 ($< 25\%$) and high in F5 and F6 ($> 75\%$). Since the active site for the ATPase is assumed to be located on the inner, cytoplasmic surface of the PM (*e.g.*, Larsson *et al.*, 1984) this means that active site(s) of GS II, NADH-cyt. c reductase and NADH-ferricyanide reductase are also located on this surface. For GS II and NADH-ferricyanide reductase this confirms the results of Larsson *et al.* (1984) and Buckhout and Hrubec (1986), respectively. However, in general, the two reductase activities are thought to be found on both sides of the PM (Crane *et al.*, 1985; Møller and Lin, 1986).

Conclusions

1) CCD of membrane vesicles in the microsomal fraction of wheat roots lead to the isolation of predominantly sealed, pure RO-PM vesicles in F5 and F6 as shown by GS II activity (Fig. 1A), enzyme latencies (Fig. 3) and by STA staining.
2) Antimycin A-insensitive NADH-cyt. c reductase, NADH-ferricyanide reductase (Fig. 1B), K^+, Mg^{2+}-ATPase and ΔVO_4-ATPase (Fig. 2) are all located both on the inner surface of the RO-PM vesicles (high latency) and on the outer surface (low latency) of some intracellular membrane vesicles. We therefore advise against uncritical use of NADH-cyt. c reductase as a marker for ER, and the K^+, Mg^{2+}-ATPase and ΔVO_4-ATPase as markers for the PM.
3) Ca^{2+}-inhibited Mg^{2+}-ATPase activity is a PM property (Fig. 2) and at least in wheat root it is a much better PM marker than vanadate-inhibited K^+, Mg^{2+}-ATPase.
4) The presence of ATP-dependent, vanadate-inhibited transport in microsomal fractions should not be taken as firm evidence that IO-PM vesicles are present.
5) There remains the possibility that IO-PM vesicles are copurified with the RO-PM vesicles in F5 and F6 and that these are the cause of the ca 10% non-latent activity (Fig. 3). This is indicated by a recent study on phase partitioned PM vesicles from soybean hypocotyls (Canut et al., 1988).

Acknowledgements

We are grateful to Ms Ann-Christine Holmström for excellent technical assistance and to the Swedish Institute for a grant to IMM to allow AB to visit Lund for 3 months during late 1986. This study was supported by grants from the Swedish Natural Science Research Council to CL and IMM and from the Carl Tesdorpf Foundation.

References

Bérczi A and Møller I M 1986 Comparison of the properties of plasmalemma vesicles purified from wheat roots by phase partitioning and by discontinuous sucrose gradient centrifugation. Physiol. Plant. 68, 59–66.

Buckhout T J and Hrubec T C 1986 Pyridine nucleotide-dependent ferricyanide reduction associated with isolated plasma membranes of maize (*Zea mays* L.) roots. Protoplasma 135, 144–154.

Canut H, Brightman A, Boudet A M and Morré D J 1988 Plasma membrane vesicles of opposite sidedness from soybean hypocotyls by preparative free-flow electrophoresis. Plant Physiol. 86, 631–637.

Crane F L, Sun I L, Clark M G, Grebing C and Löw H 1985 Transplasmamembrane redox systems in growth and development. Biochim. Biophys. Acta 811, 233–264.

Hall J L 1983 Plasma membranes. *In* Isolation of Membranes and Organelles from Plant Cells. Eds. J R Hall and A L Moore pp 55–81. Academic Press, London.

Hodges T K and Leonard R T 1974 Purification of a plasma membrane-bound adenosine triphosphatase from plant roots. Methods Enzymol. 32, 392–406.

Hodges T K and Mills D 1986 Isolation of the plasma membrane. Methods Enzymol. 118, 41–54.

Kauss H and Jeblick W 1986 Synergistic activation of 1.3-β-d-glucan synthase by Ca^{2+} and polyamines. Plant Sci. 43, 103–107.

Kähr M and Møller I M 1976 Temperature response and effect of Ca^{2+} and Mg^{2+} on ATPases from roots of oats and wheat as influenced by growth temperature and nutritional status. Physiol. Plant. 38, 153–158.

Larsson C 1985 Plasma membranes. *In* Modern Methods of Plant Analysis. New Series Vol. 1. Cell Components. Eds. H F Linskens and J F Jackson. pp 85–104. Springer-Verlag, Berlin.

Larsson C, Kjellbom P, Widell S and Lundborg T 1984 Sidedness of plant plasma membrane vesicles purified by partition in aqueous two-phase systems. FEBS Lett. 171, 271–276.

Larsson C, Widell S and Kjellbom P 1987 Preparation of high-purity plasma membranes. Methods Enzymol. 148, 558–568.

Lowry O H, Rosebrough N J, Farr A L and Randall R J 1951 Protein measurements with the Folin phenol reagent. J. Biol. Chem. 193, 265–275.

Lundborg T, Widell S and Larsson C 1981 Distribution of ATPases in wheat root membranes separated by phase partition. Physiol. Plant. 52, 89–95.

Memon A R, Sommarin M and Kylin A 1987 Plasmalemma from the roots of cucumber: Isolation by two-phase partitioning and characterization. Physiol. Plant. 69, 237–243.

Møller I M and Lin W 1986 Membrane-bound NAD(P)H dehydrogenases in higher plant cells. Annu. Rev. Plant Physiol. 37, 309–334.

Sze H 1985 H^+-translocating ATPases: Advances using membrane vesicles. Annu. Rev. Plant Physiol. 36, 175–208.

B. C. Loughman et al. (Eds.), Structural and functional aspects of transport in roots, 73–77.
© 1989 by Kluwer Academic Publishers.

Effects of K^+ status and phytohormones on K^+ transport in wheat

L. ERDEI and M. R. DHAKAL[1]
Institute of Biophysics, Biological Research Center, Hungarian Academy of Sciences, P.O.Box 521, H-6701 Szeged, Hungary. [1]Permanent address: Mahendra Morang Campus, Biratnagar, Nepal

Key words: hormones, potassium status, potassium transport, wheat

Abbreviations: ABA, abscisic acid; BA, N^6-benzylaminopurine; GA_3, gibberellic acid; NAA, naphtaleneacetic acid

Introduction

Among the many effects of plant growth substances they are known to have a profound influence on ion transport in plants (Van Steveninck, 1976). The appearance of their effects, however, depends on the experimental conditions, *e.g.* i) the concentration of the hormone applied; ii) the duration of the treatment (short-term or long-term effects); iii) the ion status of the plant; iv) the transport process studied (ion species, short or long distance transport).

To interpret unequivocally responses of plants to hormonal treatments each of these factors should clearly be taken into account. For instance, in the short term, changes in ion transport can be observed immediately (De Boer *et al.*, 1985), while long term treatments may act through gene expression and lead to changes in the protein components of the transport mechanism. Besides transport kinetics, selectivity can also be influenced by hormones (Ilan, 1971; Oláh *et al.*, 1983). The interaction between transport processes and hormones, however, can be strongly modified by the prevailing nutrient status of the plant. For example, the effect of ABA on stomatal responses was modified by nutrient status (Radin, 1984). In addition, differential influences on uptake and translocation have also been reported (Pitman *et al.*, 1984; Behl and Jeschke, 1979).

In the present work the effects of long-term (7-days) pretreatments with different concentrations of the main plant hormones were investigated on K^+($^{86}Rb^+$) uptake and translocation and on Na^+/K^+ selectivity in young wheat plants of different K^+-status.

Materials and methods

Experiments were carried out as described by Erdei *et al.* (1984) and Dhakal and Erdei (1986a; b). Briefly, seedlings of winter wheat (*Triticum aestivum* L. cv. MV-8) were grown in plastic beakers (50 seedlings in $3.5\,dm^3$ nutrient solution) in a climate chamber (Conviron PGW 36) for 14 days (11 h light periods, $60\,W\,m^{-2}$). Complete nutrient solution was used in all beakers but with different K^+ supplies; K^+ (supplied as KCl) concentration ranged from 0.01 to 10 mM.

For each hormone treatment, which started from the 8th day of growth in the phytotron, a series of beakers of seedlings of different K^+ supplies was treated with hormones added to the nutrient solution in final concentrations of 10^{-8} to $10^{-5}\,M$. NAA (Fluka), BA (Sigma), GA_3 (Calbiochem), ABA (Serva) and ethrel (Union Carbide) for ethylene were used in these experiments.

K^+($^{86}Rb^+$) uptake and translocation experiments were carried out using freshly prepared growth solutions (37 kBq ^{86}Rb $[100\,\mu mol\ K^+]^{-1}$). After a 5 h uptake period roots and shoots were rinsed and separated and the $^{86}Rb^+$ taken up was assayed in a γ-spectrometer.

Na^+/K^+ selectivity is expressed as the Na^+/K^+ ratio in the root related to their ratio in the growth solution (distribution factor, $D_{Na\,+\,K}\,+$).

Results and discussion

In order to illustrate the interaction between K^+-status and hormonal action, the effects of hormones are compared on low-K^+ and high-K^+ plants. Low-K^+ plants were grown in a complete nutrient solution containing $0.1\,M\,K^+$ and are characterized by an internal K^+ concentration of $0.2\,$mmol $(g\,dry\,weight)^{-1}$; high-K^+ plants were grown in the presence of $10\,mM\,K^+$ resulting in an internal K^+ concentration of $2.0\,$mmol $(g\,dry\,weight)^{-1}$ in the root (Erdei *et al.*, 1984).

The effects of representatives of the main hormone groups on the $K^+(^{86}Rb^+)$ uptake in the roots are shown in Fig. 1. In low-K^+ plants GA_3, ethylene and ABA stimulated $K^+(^{86}Rb^+)$ uptake in the concentration region of 10^{-8} to $10^{-6}\,M$, whereas at $10^{-5}\,M$ they had no effect (GA_3 and ethylene) or brought about inhibition (ABA). NAA and BA inhibited K^+ uptake at all concentrations, more effectively with increasing concentration. In high-K^+ plants the effects of GA_3, ethylene and ABA were similar to those in low-K^+ plants. However, BA and NAA had opposite effects, stimulating the

K^+ uptake process. The influence of hormones on the translocation of $K^+(^{86}Rb^+)$ from roots to shoots in low-K^+ plants differed quantitatively but not qualitatively from their action on the uptake process (Fig. 2). NAA, BA and ABA at $10^{-5}\,M$ caused a greater than 80% reduction. In high-K^+ plants BA stimulated (60%) and ethylene inhibited (65%) $K^+(^{86}Rb^+)$ translocation. Ethylene had an inhibitory effect only in this case. A qualitative overview of these hormone effects is given in Table 1.

Selectivity data (Fig. 3) show that BA and NAA increases selectivity in favour of K^+ (*i.e.* D_{Na^+, K^+} decreases) in low-K^+ plants. In high-K^+ plants, however, BA has no effect on D_{Na^+, K^+}, while NAA brings about a dramatic increase, *i.e.* selectivity for K^+ decreases. GA_3 exerted slight effects only both in low- and high-K^+ plants, increasing or decreasing D_{Na^+, K^+} values, respectively. ABA seems to increase K^+ selectivity in low-K^+ plants, and ethylene is without effect.

It has previously been shown that in young wheat plants NAA, BA and ABA decreased the internal K^+ (and also Na^+) levels with their increasing concentrations, while GA_3 and ethylene

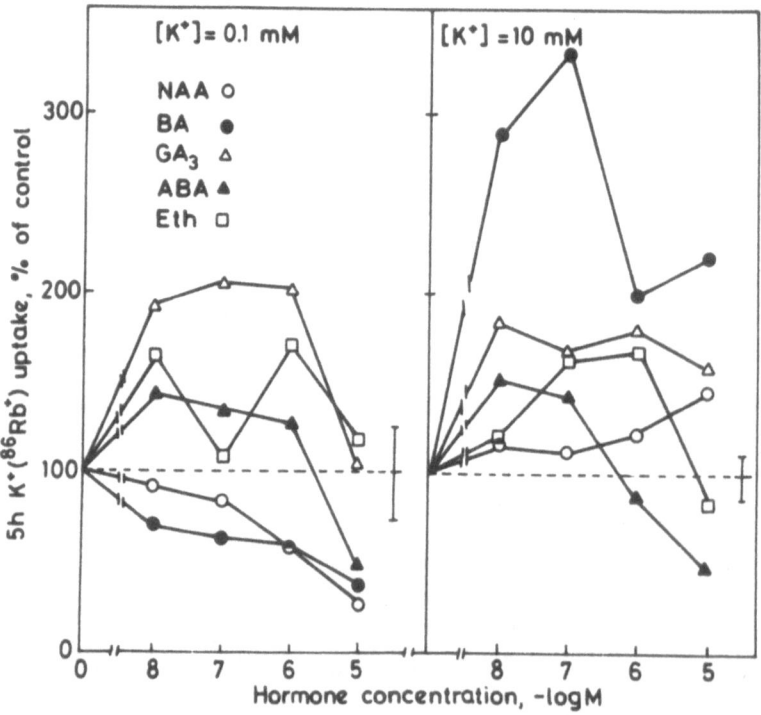

Fig. 1. The effects of hormones on $K^+(^{86}Rb^+)$ uptake in roots of wheat at different K^+ status. Control (100%) values: 178 ± 46 and $130 \pm 18\,\mu$mol $(g\,dry\,weight)^{-1}$ for low-K^+ and high-K^+ plants, respectively. Eth, ethylene.

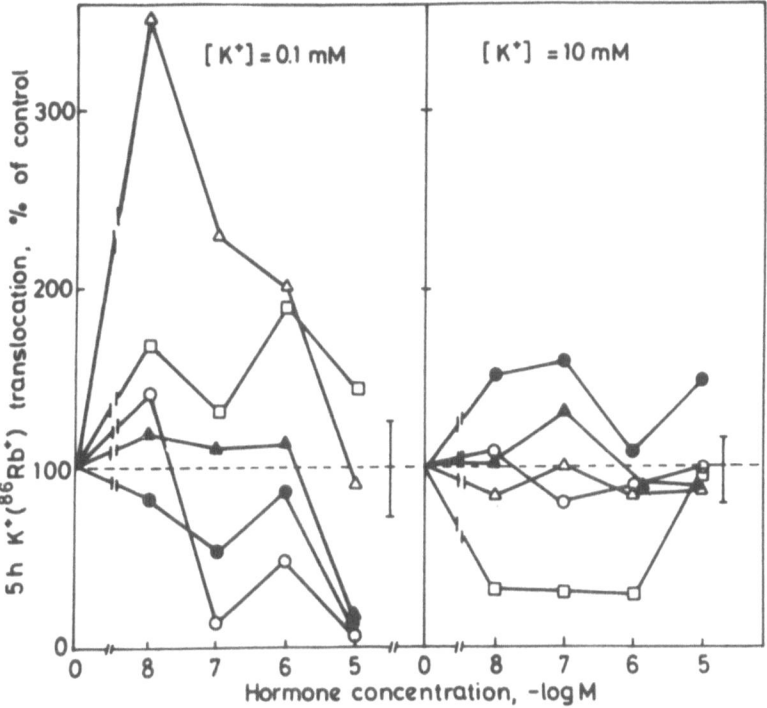

Fig. 2. The effects of hormones on $K^+ (^{86}Rb^+)$ translocation from roots to shoots. Control (100%) values: 29 ± 8 and $22 \pm 1 \, \mu$mol (g dry weight)$^{-1}$ for low-K^+ and high-K^+ plants, respectively. *Symbols* as in Fig. 1.

had no effect (Dhakal and Erdei 1986a, b). The present uptake (influx) rate measurements are in agreement with the data for internal ion levels only in the cases of NAA and BA in low-K^+ plants; for the remaining cases we have to suppose that efflux rates were highly accelerated resulting in little (ABA) or no (GA$_3$ and ethylene) change in net K^+ levels. In high-K^+ plants, in all cases, an exaggerated efflux must be supposed in order to explain zero changes or net loss in internal K^+ concentration.

From Table 1 it can be seen that the effects of most of the hormones on the K^+ transport process are different depending on the K^+ status of the plants. In this respect BA is the most outstanding subtance: K^+ uptake and translocation are both inhibited in low-K^+ plants, yet stimulated in high-K^+ plants. NAA showed similar, but less pronounced, effects ot those of BA. GA$_3$ in low-K^+ plants stimulated translocation but did not influence it in high-K^+ plants. ABA acted in a concentration dependent manner. Ethylene exerted stimulation in general, but at $10^{-7} M$ it reduced

translocation in high-K^+ plants. Differential effects on uptake and translocation was observed only in high-K^+ plants; NAA slightly stimulated uptake

Table 1. Qualitative effects of hormones on $K^+ (^{86}Rb^+)$ uptake in roots (U) and translocation to shoots (T) in low-K^+ (0.1 mM-grown) and high-K^+ (10 mM-grown) plants. Symbols: 0, no effect; +, stimulation, −, inhibition (single symbols are within the $\pm$ SE of the control value)

Hormones	M	Low-K^+ plants		High-K^+ plants	
		U	T	U	T
NAA	10^{-7}	−	− −	+	−
	10^{-5}	− −	− −	+ +	0
BA	10^{-7}	− −	− −	+ +	+ +
	10^{-5}	− −	− −	+ +	+ +
GA$_3$	10^{-7}	+ +	+ +	+ +	0
	10^{-5}	0	0	+ +	0
ABA	10^{-7}	+ +	+	+ +	+ +
	10^{-5}	− −	− −	− −	0
Ethylene	10^{-7}	+	+ +	+	− −
	10^{-5}	+ +	+ +	0	0

76   *Erdei and Dhakal*

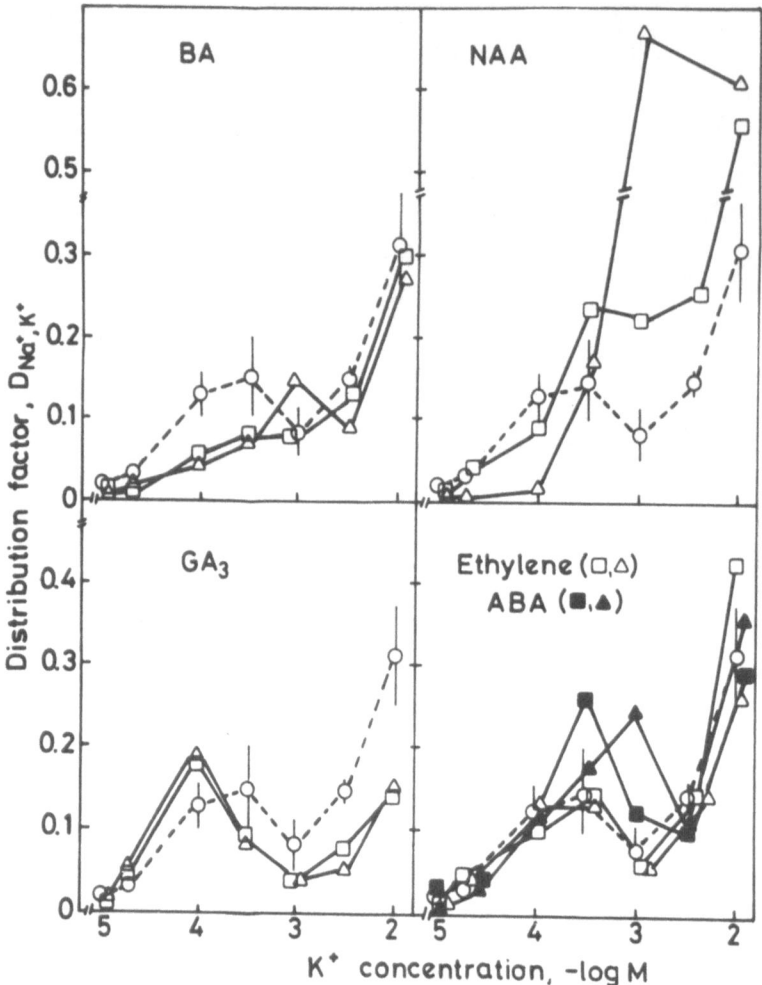

Fig. 3. The effects of hormones on NA$^+$/K$^+$ selectivity expressed as the changes in distribution factor (D$_{Na}$ +, $_K$ +) in roots of wheat grown at different K$^+$ supplies.

$$D^{Na} +, _K + = \frac{Na^+/K^+ \text{ in the plant}}{Na^+/K^+ \text{ in the growth solution}} \cdot$$ Hormones, with the exception of ABA, were applied at 10^{-8} (□) or 10^{-6} (△) M

concentrations. ABA was present at 10^{-7} (■) or 10^{-5} (▲) M. Average $\pm$ SE (n = 4) of control values are shown by (○).

and slightly inhibited translocation; ABA at 10^{-5} M strongly inhibited uptake but was without effect on translocation.

Selectivity (Fig. 3) was affected in different ways by the different hormones, but in general, only in plants of low-K$^+$ status (exception: NAA).

In our view these results can be interpreted using the model of the regulation of K$^+$ transport by feedback control. It is well known, and recently reviewed by Jensén *et al.* (1987), that the cytoplasmic concentration of K$^+$ is one of the most powerful factors in the regulation of K$^+$ uptake. In

low-K$^+$ plants the transport mechanism is responsive to changes in external and internal K$^+$ concentrations and is tightly coupled to metabolic energy. In high-K$^+$ plants, on the other hand, the transport process is predominantly passive, possibly a K$^+$/K$^+$ exchange reaction (Erdei *et al.*, 1984). We suggest therefore, that the long-term effects of most of the hormones in low-K$^+$ plants are due to interference with the transport system caused by changes in the molecular background, while in high-K$^+$ plants, where the K$^+$ transport system is inhibited by feedback control, the interaction between the hormones and the transport mechanism cannot lead to dramatic changes (inhibition) in the K^+ transport process. However, the exceptions (ABA, ethylene) indicate that this explanation for the interaction between hormones and the K^+ transport mechanism does not provide the answer in all cases.

Acknowledgement

This work was supported by a grant to MRD from the Biological Research Center, Hungarian Academy of Sciences, Szeged, Hungary. Crop Nutrition Research Report publication No. 17.

References

Behl R and Jeschke W D 1979 On the action of abscisic acid on transport, accumulation and uptake of K^+ and Na^+ in excised barley roots: Effects of the accompanying anions. Z. Pflanzenphysiol. 95, 335–353.

De Boer A H, Katou K, Mizuno A, Kojima H and Okamoto H 1985 The role of electrogenic xylem pumps in K^+ absorption from the xylem of *Vigna unguiculata*: The effects of auxin and fusicoccin. Plant Cell Environ. 8, 579–586.

Dhakal M R and Erdei L 1986a Long-term effects of plant hormones on K^+ levels and transport in young wheat plants of different K^+ status. Physiol. Plant. 68, 632–636.

Dhakal M R and Erdei L 1986b Long-term effects of abscisic acid in K^+ transport in young wheat plants of different K^+ status. Physiol. Plant. 68, 637–640.

Erdei L, Oláh Z and Bérczi A 1984 Phases in potassium transport and their regulation under near-equilibrium conditions in wheat seedlings. Physiol. Plant. 60, 81–85.

Ilan I 1971 Evidence for hormonal regulation of the selectivity of ion uptake by plant cells. Physiol. Plant. 25, 230–233.

Jensén P, Erdei L and Møller I M 1987 K^+ uptake in plant roots: Experimental approach and influx models. Physiol. Plant. 70, 743–748.

Oláh Z, Bérczi A and Erdei L 1983 Benzylaminopurine-induced coupling between calmodulin and Ca-ATPase in wheat root microsomal membranes. FEBS Lett. 154, 395–399.

Radin J W 1984 Stomatal responses to water stress and abscisic acid in phosphorus-deficient cotton plants. Plant Physiol. 76, 392–394.

Van Steveninck R F M 1976 Effects of hormones and related substances on ion transport. *In* Encyclopedia of Plant Physiology, New Series. Vol. 2B, Eds. U Lüttge and M G Pitman. pp 307–342. Springer Verlag, Berlin. ISBN 3-340-07453 8.

B. C. Loughman et al. (Eds.), Structural and functional aspects of transport in roots, 79–84.
© 1989 by Kluwer Academic Publishers.

Transport and distribution of solutes in sugar beet roots

V. P. KHOLODOVA, Yu. P. BOLYAKINA, A. B. MESHCHERYAKOV, E. RICHTER[1],
R. EHWALD[1], A. Kh. MASHKOVA and T. V. PECHENOVA
KA Timiriazev Institute of Plant Physiology, Academy of Science of USSR, SU-127 276 Moscow, USSR.
[1]*Section of Biology, Humboldt-University of Berlin, DDR-1040, Berlin, GDR*

Key words: distribution, ions, plasmodesmata, sugars, sugar beet, transport

Inorganic ions and sugars are the main osmotic components of sugar beet root tissues. While the roots function exclusively as nutritive organs, sugar concentrations are low in them, and their osmotic potential is created by inorganic ions. Later in the course of differentiation of the tertiary structure of the roots, they transform into storage organs where sucrose prevails, its concentration being up to $0.6\,M$ or even high.

In sugar beet roots, the pathway is rather long for lateral translocation of solutes out of the conductive bundles into the storage parenchyma. It consists of approximately 10 parenchyma cells in young roots and increases up to 30–40 cells in

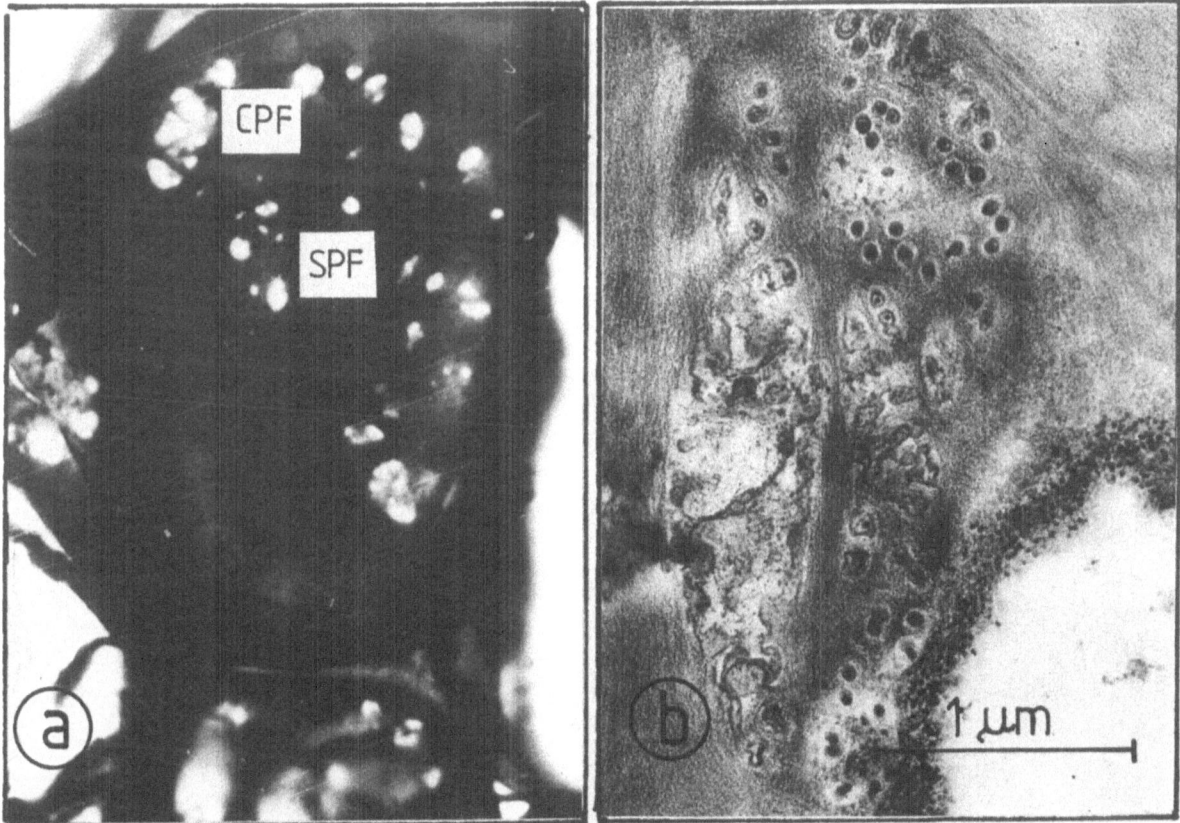

Fig. 1. **a**) The plasmodesmatal fields (SPF-simple; CPF-composite) × 2000. **b**) Transverse section of plasmodesmata in mature root parenchyma cell with a common large median cavity.

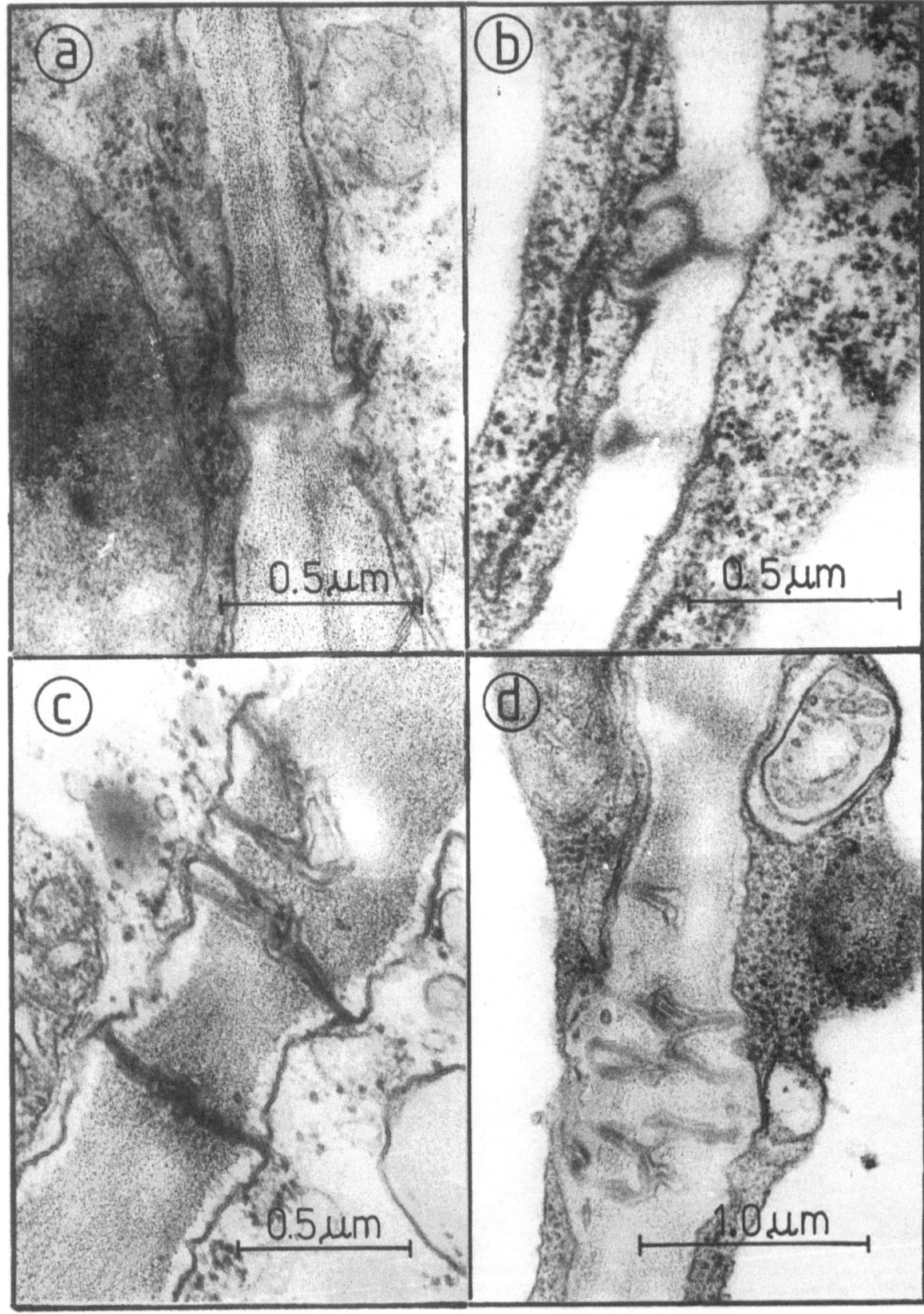

mature ones. In the inner, well developed zones of storage roots, the length of the pathway may be up to 5 mm.

Apoplast may be used for lateral movement of the solutes because no structures capable of preventing it are found in the cell walls and intercellular spaces of the roots. At the same time, a symplastic pathway exists through plasmodesmata connecting the storage parenchyma cells (Boljakina and Kholodova, 1977; Kholodova *et al.* 1980; 1981).

Plasmodesmata are not uniformly distributed throughout the cell wall surface, most of them being arranged in large groups. In the locations of numerous plasmodesmata, the thickness of the cell wall increases more slowly during cell growth and it does not reach such a thickness as walls found outside these zones. As a result, thin-walled plasmodesmal fields (pit fields) are formed (Fig. 1a). Pit fields are simple (individual) or mostly they consist of a few simple pit fields separated by thick-walled strips without plasmodesmata (composite pit fields).

The pit fields vary greatly in size, the average value being equal to $27 \pm 18 \, \mu m^2$. They occupy $148{-}161 \, \mu m^2$ per $10^3 \, \mu m^2$ of cell walls, or appr. 15% of the cell wall surface. No significant difference is found between the transversal and longitudinal walls of root parenchyma cells.

The density of the plasmodesmata is rather high in the pit fields – up to 35 plasmodesmata per μm^2; this is partly due to the manner of their development in storage parenchyma cells.

In dividing cells plasmodesmata are initiated by strands of the endoplasmic reticulum which pass through newly formed cell walls. During the short initial stage of cell development plasmodesmata appear as simple cylindrical channels (Fig. 2a) and their further reorganization depends on the growth of the cell wall. When cell walls become thicker branching of the plasmodesmata take place due to cell wall material trapping the branched channels of the endoplasmic reticulum. Different stages of this process which was first shown by Danilova and Telepova (1978) are demonstrated in Fig. 2a–d.

In fully expanded cells, most plasmodesmata are branched in two directions (Kholodova *et al.*, 1981)

and open into adjacent cells with 3–4 channels (Fig. 1b, 2d). When cell expansion is complete, large median cavities of plasmodesmata are formed in the region of central lamella of the cell walls (Kholodova *et al.*, 1980; 1981). An increase in the number of plasmodesmal channels together with formation of median cavities uniting all the branches of the plasmodesma may favour symplastic solute movement.

The diameter of plasmodesmal channels (lumens) is in the range of 35–50 nm increasing slightly in the cells of mature roots. The cross section of the channels occupies up to 0.5% of the cell surface but the value is reduced to 0.1% when neck constrictions of the plasmodesmata are considered. These calculations permit us to evaluate the capacity of symplastic translocation of solutes to storage cells of sugar beet roots.

In sugar beet root cells, sucrose is predominantly localized in vacuoles and therefore membrane transport seems to be an important step in sucrose movement into the storage compartment. Membrane transport can also return some substances, especially monosaccharides, which leaves the cells as a result of membrane leakage (Meshcheryakov *et al.*, 1983).

Membrane transport of sugars has been studied in detail in sugar beet root suspension cells which are a more suitable model than the root tissues, in particular due to the higher rates of membrane transfer – 10–15 nmol . min^{-1} . ml^{-1} sedimented volume for $100 \, \mu M$ D-glucose or 3-0-methyl-D-glucose (Kholodova *et al.*, 1982). A kinetic analysis reveals an active system of monosaccharide transport. Some evidence in favour of H$^+$-ATPase energization of active monosaccharide transport was obtained with the help of an inhibitory analysis (Meshcheryakov and Kholodova, 1984). However, H$^+$-fluxes coupled with monosaccharide transport could not be observed by the conventional technique. The high buffering capacity of the culture which reached 2.5–3.0 mmol H$^+$ per pH unit per ml sedimented cells seems to be the main reason for the failure to observe the H$^+$-influx (Meshcheryakov and Kholodova, 1985).

The use of a flow-through system (Fig. 3) with cultured cell aggregates packed in a column is

Fig. 2. The stages of the development of branching plasmodesmata.

82   *Kholodova* et al.

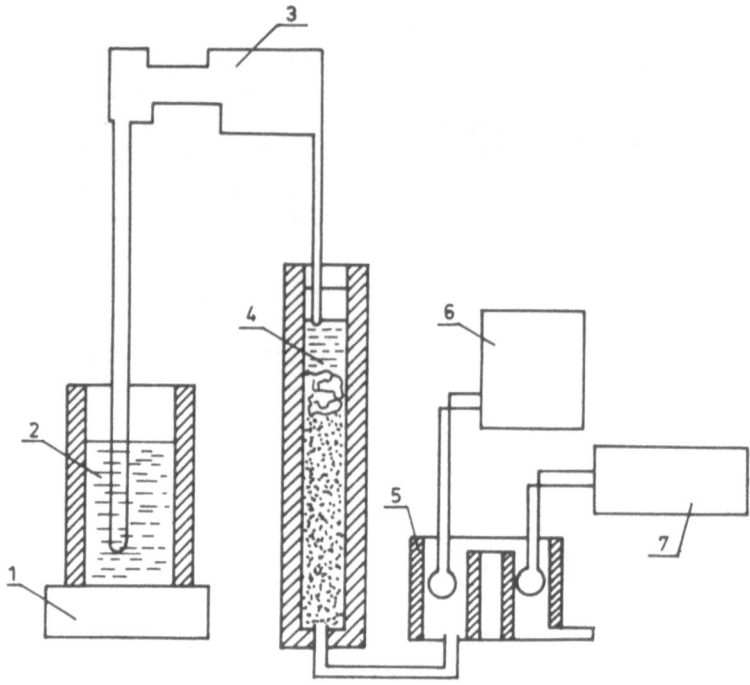

Fig. 3. Scheme of device for H^+-fluxes registration in suspension-cultured cells packed into column with flowing medium. **1** – magnetic stirrer, **2** – thermoequilibrated vessel with aerated cultural medium, **3** – pump, **4** – thermoequilibrated column with cells, **5** – thermoequilibrated cell with pH-electrode, **6** – pH-meter, **7** – titrimeter.

found to reduce the buffering capacity (β) by a factor of 20–50 since the concentration of the buffering substances of the medium decreases sharply; on the other hand, the sealing of the system helps to avoid uncontrolled CO_2 accumulation. High velocities of the solution flow – 3.0–3.5 ml . min^{-1} are necessary for registration of pH-changes at the exit of the thermoequilibrated column (Figs. 3, **4**) followed by potentiometric titration for β measurement (Figs. 3, **4, 5**).

The H^+-influxes induced by addition of sugars (Fig. 4) are equal to 12.4 $\pm$ 1.30 nmol . min^{-1}.ml^{-1} for glucose and 11.6 $\pm$ 1.21 for methylglucose, the stoichiometry of H^+/sugar transport being 0.96 and 0.92 for the monosaccharides. Along with H^+/sugar symport, K^+-efflux can be simultaneously measured by routine techniques due to the relatively low buffering capacity of the culture for K^+. Large fluctuations of the ratio of H^+ to K^+ fluxes (in the range of 0.65–1.2) seems to indicate their electrical but not mechanical interaction. The data obtained for sugar beet suspension cells favour H^+/monosaccharide symport which was shown to function in the plasmalemma

of many other plants (Reinhold and Kaplan, 1984).

Sucrose seems to be transported by another mechanism. H^+-influx cannot be registered after sucrose addition but the reason may be in the 5 to 10 times lower rate of sucrose uptake compared to monosaccharides. Sucrose administration results in K^+-influx (but not efflux as for monosaccharides) which starts after a significant lagphase and proceeds over a long time. It is suggested that active sucrose transport may be located in the tonoplast rather than in the plasmalemma. Investigations of sucrose-transporting mechanism in vacuoles isolated from sugar beet root cells are in progress.

Besides sucrose, some mobile inorganic ions (MII) make a significant contribution to the osmotic potential of sugar beet root tissues. Measured with the aid of ORION ion-seletive electrodes, their concentrations decrease in the row: $K^+ \gg Cl^- = NO_3^- > Na^+ > NH_4^+$. High levels of NO_3^- are found to be characteristic for sugar beet roots, and at the same time NO_3^- concentrations vary to a greater extent than other ions investigated depending on administration of fertilizers (Fig. 5), on stages of root development *etc.*

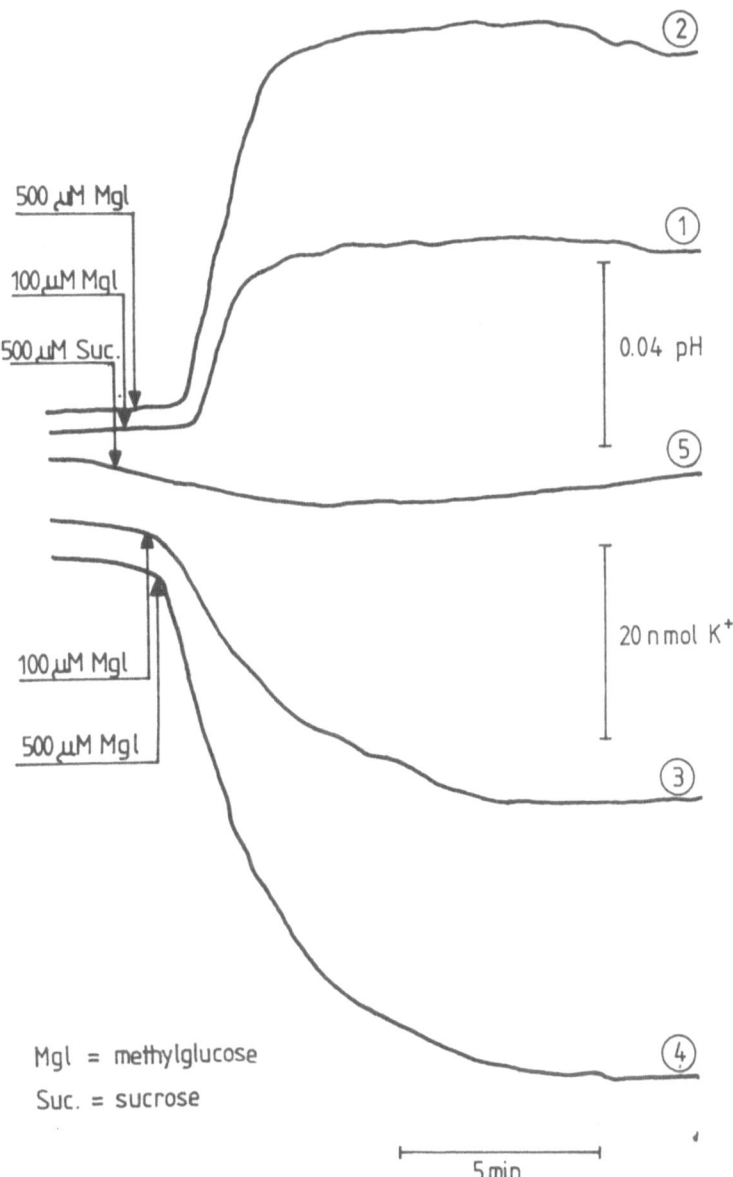

Fig. 4. Fluxes of H$^+$ ions registered in flow system (*curves* **1, 2**) and K$^+$-ions registered by routine method (*curves* **3, 4, 5**), which are induced by addition 100 μM methylglucose (*curves* **1, 3**), 500 μM methylglucose (*curves* **2, 4**) and 500 μM sucrose (*curve* **5**) in suspension culture.

To investigate the interrelation between the main osmotic components – sucrose and MII – more than 15 zones have been separately analyzed in storage roots with differentiated tertiary anatomical structure. The average values of the ratio of sucrose to MII concentrations (S/MII) exceed 5–8 in the roots of mature plants. Osmotic potentials of separate root tissues or zones differed moderately but the individual components are distributed unevenly

which is reflected in the S/MII index variability (Table 1). Small variations occur between the adjacent conducting and storage parenchyma tissues or between the phloem and xylem zones of the roots. Much greater differences are found for periderma and underlying layers or for the central parenchyma zone of the crown where the S/MII index decreases down to 0.2–0.4 as compared to 2–6 in the neighbouring tissues. The strong negative

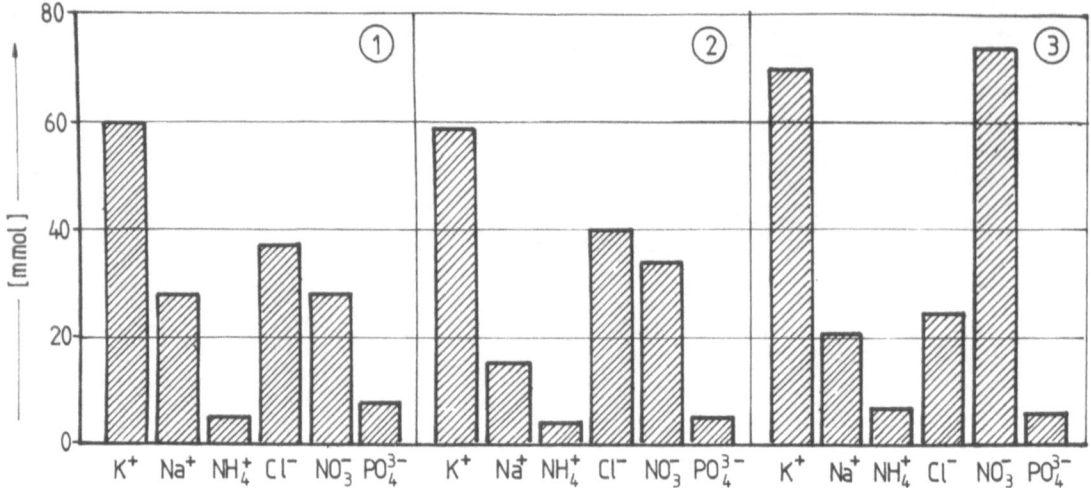

Fig. 5. Composition of mobile inorganic ions in sugar beet roots. **1** – without fertilizer, **2** – NPK-fertilizer, **3** – N-fertilizer.

Table 1. Distribution of sucrose and potassium within sugar beet root tissues

Tissue	Sucrose, mM	Potassium, meq	S/MII
Central core	360	33	6.3
Vascular, the 1st ring, xylem	400	50	4.5
phloem	400	30	5.8
Parenchyma between the 1st and 2nd vasculars			
near phloem	320	33	5.3
middle	300	40	4.5
near xylem	390	48	4.4
Vascular, the 2nd ring, xylem	377	48	5.4
phloem	412	36	7.8
Parenchyma, between the 2nd and 3rd vasculars			
near phloem	402	36	6.7
middle	418	30	6.5
near xylem	388	42	5.7
Vascular, the 3rd ring, xylem	363	38	5.7
phloem	415	35	6.7
Outer rings	295	66	3.4
Periderma	114	207	0.22
Crown pith	196	321	0.35
Crown vascular	339	129	1.6

correlation between MII and sucrose concentrations allows one to suggest a turgo- or osmoregulatory mechanism of the distribution of ions and sucrose in sugar beet roots.

Acknowledgement

The authors heartily acknowledge E N Ushakova for her skillful technical assistance.

References

Boljakina Ju P and Kholodova V P 1977 Destructive changes in storage parenchyma cells of sugar beet roots during the second vegetative period. Soviet Plant Physiol. 24, 933–940.

Danilova M Ph and Telepova M N 1978 Differentiation of protophloem sieve elements in seedling roots in *Hordeum vulgare*. Phytomorphology. 28, 418–425.

Kholodova V P, Boljakina Ju P and Buzulukova N P 1980 Storage tissues. *In* Atlas of plant tissue ultrastructure (in Russian). Eds M Ph Danilova and G M Kozubov. pp 347–384. Publishing House 'Karelia', Petrozavodsk.

Kholodova V P, Boljakina Ju P, Meshcheryakov A B and Orlova M S 1981 Sugar beet root as an organ for sucrose accumulation. *In* Structure and function of plant roots. Eds R Brouwer *et al..* pp 209–213. Nijhoff, The Hague/Boston/London.

Meshcheryakov A B and Kholodova V P 1984 Role of the electrochemical proton gradient in membrane transport of monosaccharides into cells of a sugar beet suspension culture. Soviet Plant Physiol. 30, 751–757.

Meshcheryakov A and Kholodova V 1985 Die Puffereigenschaften der Zellsuspensionskultur der Zuckerrübenwurzel. Colloquia Pflanzenphysiologie der Humboldt-Universität zu Berlin. 8, 85–93.

Kholodova V P, Meshcheryakov A B and Chernyavskaya T N 1982 Transport of 3-0-methyl-D-glucose into cells of sugar beet suspension culture. Soviet Plant Physiol. 29, 667–675.

Meshcheryakov A B, Kholodova V P and Ehwald R 1983 Hexose uptake in Beta root parenchyma: stimulation by wounding, washing and reduced water potential, inhibition by different electrolytes. Biochem. Physiol. Pflanzen. 178, 273–377.

Reinhold L and Kaplan A 1984 Membrane transport of sugars and amino acids. Annu. Rev. Plant Physiol. 35, 45–83.

B. C. Loughman et al. (Eds.), Structural and functional aspects of transport in roots, 85–87.
© 1989 by Kluwer Academic Publishers.

Ecto Ca-ATPase of plasma membrane fraction from barley roots

NATALIA I. TIKHAYA and DMITRIY B. VAKHMISTROV
K A Timiriazev Institute of Plant Physiology, Academy of Sciences of USSR, SU-127 276 Moscow, USSR

Key words: ATPase, barley, calcium, plasmalemma, roots

Abbreviations: DCCD, dicyclohexylcarbodiimide; DES, diethylstilbestrol; EDTA, ethylenediaminetetra-acetic acid; MES, N-morpholino ethanesulphonic acid

Calcium is thought to be a cellular messenger in many physiological processes in plants (Deiter, 1984). Therefore, the regulation of cellular calcium homeostasis is important because the excess intracellular Ca leads to cell dysfunction and death. It is supposed that this is mainly achieved by transport systems which are responsible for the control of Ca entry and extrusion across the plasma membrane. We suggested that in addition to the Mg, Ca-ATPase, discovered by Deiter and Marmé (1981), the similar enzyme, ecto-Ca-ATPase of animal cells (Tuana and Dhalla, 1982) may function in the plasma membrane of plant cells. The results are discussed in terms of the possibility of a ecto-Ca-dependent ATPase in microsomal fractions enriched in plasma membranes and on the surface of intact barley roots.

The ATPase with optimum activity at pH 6.0 was found in the purified microsomal fraction (Fig. 1). The enzyme was specific to ATP (ATP > UTP > CTP > GTP; in the presence of AMP, β-glycerophosphate or p-nitrophenylphosphate hydrolysis did not occur) to Ca (Ca > Mg = Mn $\gg$ Co > Ba). It was activated by $1\,mM$ Ca with an apparent Km of $9.5 \times 10^{-5}\,M$ (Fig. 2).

The distinction between this enzyme and Ca, Mg-ATPase was obvious in the experiments with calmodulin. It is known that calmodulin activates the Ca, Mg-ATPase of root cells in corn (Deiter and Marme, 1981) and wheat (Olah *et al.*, 1983). Addition of calmodulin ($0.3–30\,\mu g\,ml^{-1}$) did not stimulate the Ca-ATPase of barley root cells.

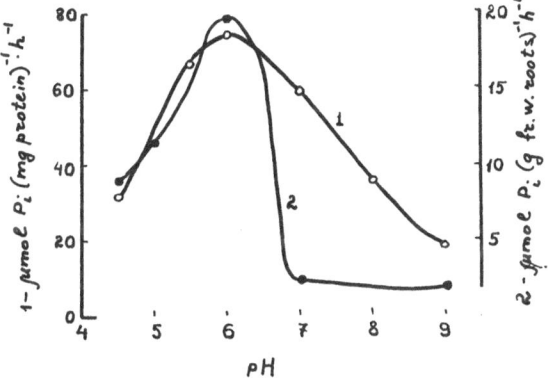

Fig. 1. Effect of pH on the Ca-ATPase activity of a microsomal fraction (**1**) and intact roots (**2**). Reaction mixture (1 ml) for microsomal fraction: $33\,mM$ Tris-MES, $3\,mM$ ATP, $1\,mM$ EDTA, $3\,mM$ $CaCl_2$; for intact roots (6 ml): $33\,mM$ Tris-MES, $3\,mM$ ATP, $1\,mM$ ammonium molybdate, $3\,mM$ $CaCl_2$. The average is of three independent experiments.

On the other hand, it is possibly that H-ATPase is responsible for ATP-hydrolysis. In this case it seems reasonable to expect that the inhibitors of H-ATPase – DCCD, DES and sodium orthovanadate will decrease the stimulation of the enzyme induced by Ca. However, the enzyme was only slightly inhibited by DCCD, DES and was insensitive to sodium orthovanadate (Table 1).

It should be emphasized that an enzyme with similar properties was found on the surface of intact roots (see Fig. 1, Table 1) (Tikhaya *et al.*, 1985). These results permit us to suppose the presence of the ecto-Ca-ATPase possessing its

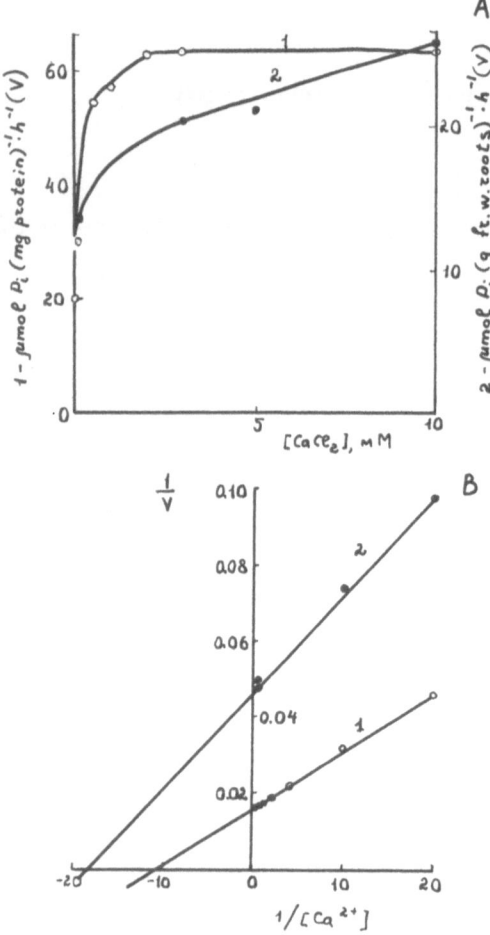

Fig. 2. Ca-ATPase activity as a function of Ca concentration in a microsomal fraction (**1**) and intact roots (**2**) – **A**. Lineweaver–Burke plot of the same data – **B**. The reaction mixtures were the same as for Fig. 1. The average is of three independent experiments.

Table 1. Effect of inhibitors on the Ca-ATPase activity. The reaction mixtures were the same as for Fig. 1. The average is of three independent experiments.

Inhibitor	Microsomal fraction		Intact roots	
	μmoles Pi mg protein $\times$ h	%	μmoles Pi mg protein $\times$ h	%
—	56.10 ± 0.61	100	25.20 ± 3.08	100
Lead nitrate, $3.10^{-3}\,M$	—**		$2.78 + 1.71$	11
Sodium fluoride, $10^{-2}\,M$	11.22 ± 0.76	20	2.52 ± 1.78	10
Ammonium molybdate, $10^{-3}\,M$	56.19 ± 1.38	100	18.90 ± 2.87	75
DES, $5.10^{-5}\,M$	43.68 ± 1.26	78	24.20 ± 3.51	96
DCCD, $5.10^{-5}\,M$	44.89 ± 1.31	80	23.69 ± 3.02	94
Sodium orthovanadate, $5.10^{-5}\,M$	55.59 ± 1.07	100	**	

** – experiments were not carried out

Table 2. Effect of alamethycin on K, Mg- (pH 5.5) and Ca-ATPases (pH 6.0) activities of microsomal fraction, μmoles Pi $\times$ (mg protein)$^{-1}$ $\times$ h^{-1}. Incubation mixture (1 ml) for K, Mg-ATPase activity: $33\,\text{m}M$ Tris-MES, $3\,\text{m}M$ ATP, $3\,\text{m}M$ MgCl$_2$, $50\,\text{m}M$ K$_2$SO$_4$. Alamethycin: protein = $1.2\,\mu$g: $5\,\mu$g (w/w). The average is of three independent experiments

Type of activity	Without alamethycin (a)	In the presence of alamethycin (b)	Ratio b/a
K, Mg-ATPase	4.97 ± 0.10	18.18 ± 0.07	3.66
Ca-ATPase	71.77 ± 0.17	72.10 ± 0.16	1.00

active center *in situ* on the outside of the plasmalemma. The evidence in support of this suggestion has been obtained from the following experiments. First, lead nitrate, an impermeable agent, inhibited the Ca-ATPase activity of intact roots by 89% and did not effect the basal ATPase activity. Second, the activity of endo- K, Mg-ATPase was increased 3–4 fold after the treatment of the microsomal fraction by the channel-type polypeptide alamethycin whereas the Ca-ATPase activity did not change (Table 2). These results indicate the following: i) the active center of the Ca-ATPase is located *in situ* on the outside of the plasma membrane, ii) the microsomal fraction was mainly composed of right side-out sealed vesicles.

Thus, in the plasma membrane enriched fraction we found the ecto-Ca-ATPase which probably plays a major physiological role in the apoplast compartment of plant cells. Since the outside of the plasmalemma is involved in many physiological processes such as transport of substances, hormone regulation, growth of cell and others, we assume that it is important to study not only endo-enzymes, but ecto-enzymes of the plasma membranes.

References

Deiter P 1984 Calmodulin and calmodulin-mediated processes in plants. Plant Cell and Environment 7, 37--380.

Deiter P and Marme D 1981 A calmodulin-dependent microsomal ATPase from corn (*Zea mays* L.) FEBS Letters 125, 245–248.

Olah Z, Berczi A and Erdei L 1983 Benzylaminopurine-induced coupling between calmodulin and Ca-ATPase in wheat root microsomal membranes. FEBS Letters 154, 395–399.

Tikhaya N I, Tazabaeva K A and Vakhmistrov D B 1985 Ca-ATPase of barley root cell plasmalemma: Properties and location. Sov. Plant Physiol. 32, 823–832.

Tuana B S and Dhalla N S 1982 Purification and characterization of a Ca-dependent ATPase from rat heart sarcolemma. Biol. Chem. 257, 14440–14445.

B. C. Loughman et al. (Eds.), Structural and functional aspects of transport in roots, 89–91.
© 1989 by Kluwer Academic Publishers.

ATP-ase localization in the phloem of the Ricinus root

F. DIDEHVAR and D.A. BAKER
Department of Biochemistry and Biological Sciences, Wye College, University of London, Ashford, Kent, TN25 5AH, UK

Key words: ATPase, phloem, *Ricinus communis*, roots

Introduction

Roots function as one of the major sinks for photoassimilates within the plant and thus a knowledge of the unloading mechanism(s) within root tissues is essential to any detailed understanding of root growth. Assimilate unloading is known to involve both symplastic and apoplastic pathways in reproductive sinks such as seeds but it is probable that a symplastic route is followed within root tissues (Thorne, 1985; 1986).

ATPase activity is often associated with membranes involved in organic solute transport (Didehvar and Baker, 1986) and has been reported for the plasmalemma of the sieve elements of aerial roots of *Monstera deliciosa* using microautoradiographical and cytochemical techniques (Eschrich, 1983). The present study is part of a detailed investigation of ATPase activity in *Ricinus communis*, which extends the investigations by Browning *et al.* (1980) and Didehvar and Baker (1986), and examines the localization of ATPase activity in the phloem tissues of Ricinus roots.

Materials and methods

Material was taken from 3 to 8 month old plants of *Ricinus communis*. Samples were taken from 5–10 mm from the tip of the roots. Tissues were prefixed in 1% glutaraldehyde and 4% formaldehyde buffered to pH 7.2 with 100 mM cacodylate for 2 h at 0°C. Tissues were then washed in 50 mM Tris-maleate buffer pH 7.2, at laboratory temperature for 2 h. During this time 150–300 μm thick transverse sections were cut from samples using an

Oxford vibrotome. Tissues in each sample were then divided into three parts. One part was transferred to an incubation solution containing 2 mM $Mg(NO_3)_2$, 2 mM ATP and 100 mM Tris-maleate buffer, pH 7.2 at laboratory temperature for 2 h. The second part was incubated in a control solution which lacked ATP only, for the same period of time. The last part was transferred to incubating solution to which 0.5 mM of sodium orthovanadate was added. Tissues were washed in distilled water for 3 h and then fixed in 1% OsO_4 in 20 mM cacodylate buffer, pH 7.2, overnight at 4°C. After post fixation sections were washed 3 times in 50 mM cacodylate buffer for a total time of 30 minutes. They were then dehydrated in an acetone

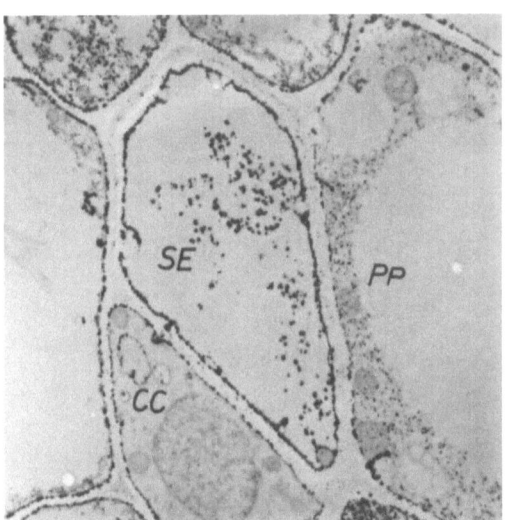

Fig. 1. Transverse section of sieve elements (**SE**), companion cells (**CC**) and phloem parenchyma cells (**PP**) in the root of Ricinus incubated in the presence of ATP. Plasma membrane in all cells is stained with reaction product. × 6000.

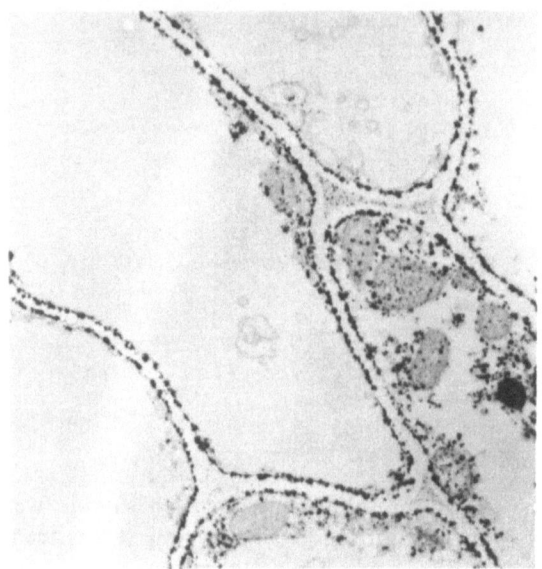

Fig. 2. Transverse section of vascular parenchyma cells in Ricinus root incubated in the presence of ATP. × 10 500.

series and embedded in Epon epoxy resin. Sections were cut using a diamond knife and viewed without staining with a Corinth 500 AEI electron microscope.

Results and discussion

In the phloem tissues the reaction product was localized mainly at the plasmalemma of the sieve

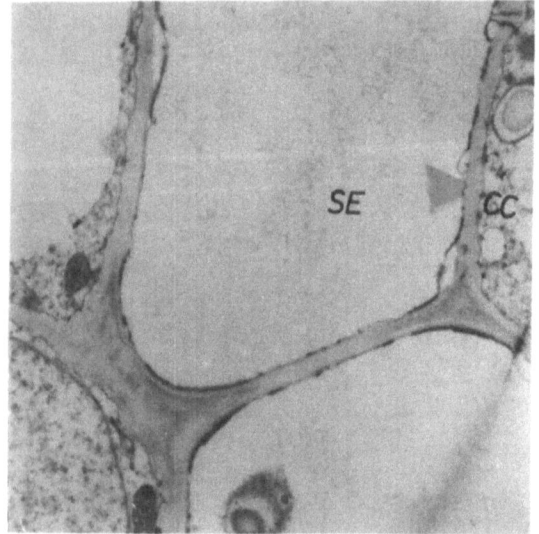

Fig. 3. Transverse section of sieve elements and companion cells from root of Ricinus incubated in the absence of ATP. There is no deposition of reaction product on plasmalemma. × 10 800.

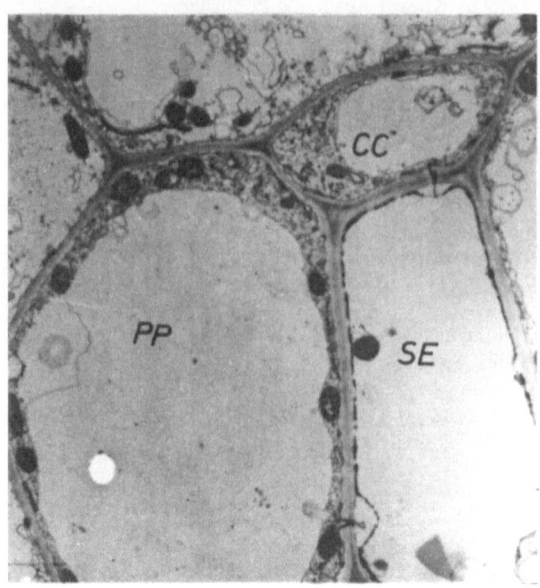

Fig. 4. Transverse section of sieve element, companion cell and phloem parenchyma cell from root of Ricinus incubated in the presence of ATP plus Na-vanadate. × 6000.

elements, companion cells and phloem parenchyma cells (Fig. 1). This activity was also associated with dispersed P-protein in the sieve element lumen. The ATPase activity was also localized on the plasmalemma of vascular parenchymatous cells (Fig. 2). The minus substrate controls showed no reaction as can be seen in Fig. 3. In some of the control preparations some larger granular deposits were present but they were not considered to be associated with enzymic activity. The addition of vanadate considerably reduced the ATPase activity and only residual amounts of reaction product were localized on the plasmalemma of cells in phloem tissues when this inhibitor was used (Fig. 4).

The distribution of ATPase activity observed here in the phloem tissues of root is similar to that previously shown for other plant species (Cronshaw and Gilder, 1972; Eschrich, 1983). The unloading of sucrose in a vegetative sink, such as the root, most probably occurs via the symplastic pathway. Dick and ApRees (1975) have provided convincing evidence for symplastic movement of sucrose between the stele and cortex in pea roots. A symplastic route for sucrose unloading in corn roots has also been indicated by Giaquinta *et al.* (1983).

The presence of ATP-hydrolysing activity in the phloem tissues supports the idea of linkage between phloem loading and ATP-hydrolysis by an ATP-energized proton efflux pump in the plasma membrane (Baker, 1978). The enzyme localization technique employed in the present study provides circumstantial evidence that a similar proton pumping activity might be present within root sink tissues. In vegetative sink tisues, where unloading is predominantly symplastic, a proton gradient may be utilized to recover sugar which has leaked into the apoplast. It is possible that cell wall invertases may hydrolyse any apoplastic sucrose and that this recovery is of hexose monomers. With respect to this latter point it is of relevance that Lin *et al.* (1984) have found only hexose carriers in protoplasts isolated from corn root protoplasts.

References

Baker D A 1978 Proton co-transport of organic solutes by plant cells. New Phytol. 81, 485–497.

Browning A J, Hall J L and Baker D A 1980 Cytochemical localization of ATPase activity in phloem tissues of *Ricinus communis*. Protoplasma 104, 55–65.

Cronshaw J and Gilder J 1972 Localization of adenosine triphosphatase activity in the phloem of *Nicotiana tabacum*. *In* 30th Annual Proceedings of the Electron Microscopy Society of America, Los Angeles. Ed. C J Amceneaux pp 230–231. Claitor's pub. Div. Baton Rouge.

Dick P S and ApRees T 1975 The pathway of sugar tranport in root of *Pisum sativum*. J. Expt. Bot. 26, 305–314.

Didehvar F and Baker D A 1986 Localization of ATPase in sink tissues of Ricinus. Ann. Bot. 57, 823–828.

Eschrich W 1983 Phloem unloading in aerial roots of *Monstera deliciosa*. Planta 157, 540–547.

Giaquinta R T, Lin W, Sadler N L and Franceschi V R 1983 Pathway of phloem unloading of sucrose in corn roots. Plant Physiol. 72, 362–367.

Lin W, Schmitt M R, Hitz W D and Giaquinta R T 1984 Sugar transport in isolated corn root protoplasts. Plant Physiol. 76, 894–897.

Thorne J H 1985 Phloem unloading of C and N assimilates in developing seeds. Annu. Rev. Plant Physiol. 36, 317–343.

Thorne J H 1986 Sieve tube unloading. *In* Phloem Transport. Eds. J Cronshaw, W L Lucas and R T Giaquinta. pp 211–224. Alan R. Liss Inc.

B. C. Loughman et al. (Eds.), Structural and functional aspects of transport in roots, 93–95.
© 1989 by Kluwer Academic Publishers.

Electrical conductivity and capacitance of root tissues in different conditions of energetic metabolism

JANA ČERNOHORSKÁ, MIROSLAV DVOŘÁK and ERNST MANFRED WIEDENROTH[1]
Plant Physiology Department, Charles University, CS-128 44 Praha 2, ČSSR and [1] Department of Biology, Humboldt-University, DDR-1040 Berlin, Invalidenstr. 43, GDR

Key words: anoxia, conductivity, membrane, root

Abbreviations: C-R curve, current-response characteristic

Introduction

This work follows former respirometric and morphological analysis of root tissue during hypoxia (Wiedenroth and Erdmann, 1985; Erdman *et al.*, 1986).

For an analysis of physiological changes in root tips of wheat (*Triticum aestivum* L. cv. Hatri) under the influence of various respiration conditions, *i.e.* aerobiosis, hypoxia and anoxia, the method of the current-response to the applied saw-tooth voltage was used. The C-R curves were employed to characterize the actual functional state of the membrane system of the plant tissue (Dvořák *et al.*, 1981). Pulses 0.5 ms long and the input voltage, $U_{max} = 10 \, V$ were used. The current-response is given by the relationship:

$$I = \frac{U_{max}}{T} \cdot C \cdot [(1 - e^{-a_1 1^t}) + (1 - e 2^t)]$$

$$+ R^{-1} \cdot t;$$

$$I_{max} = {}^i R_{max} + {}^i C_{max}$$

The C-R curves yielded C = the relative capacitance = ${}^i C_{max}/I_{max}$, a_1, a_2 = the velocity of capacitor charging (fast and slow, respectively) = $(R_i \cdot C)^{-1}$.

Cultivation

The plants were grown for 10 days in Knop's nutrient solution, in 31 glass vessels and were continuously aerated (control), or in pots with sand flooded with the same nutrient solution immersed in parallel vessels (hypoxia). For anoxia treatment: 2-cm long root tips (1 to 3 roots) from aerated cultures were cut, immersed in the nutrient solution and flushed with nitrogen (anoxia) or aerated in the controls. Root tips were sampled at time intervals of 4 h, 7.5 h, 17 h and the C-R curves were measured.

Results and discussion

Anoxia

The stability of the membrane system is dependent on the supply of metabolic energy; the consumption of reserve matter is higher in anaerobic conditions than in aerobic ones with heterotrophic organs. Therefore, the dynamics of the changes in the membrane stability were studied using the C-R curves. This short-time treatment with N_2 does not lead to adaptation; the conductivity of the plant tissue increases gradually, indicating a loss of the membrane semipermeability (Fig. 1). This is primarily demonstrated by a decrease in the

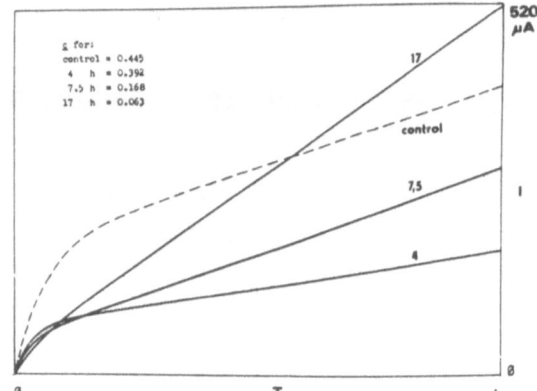

Fig. 1. C-R curves obtained during N_2 treatment. Numbers on the curves = hours of treatment; *x-axis* = relative time of 1 pulse (= 2 ms); *y-axis* = I_{max}, μA.

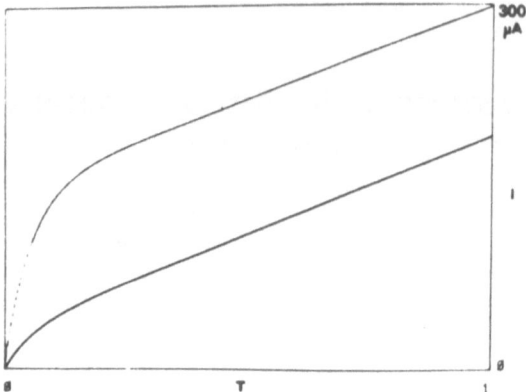

Fig. 2. C-R curves obtained after 10 days of hypoxia treatment. Upper curve is control. *x-axis* = relative time of 1 pulse (= 2 ms); *y-axis* = I_{max}, μA.

capacitance in the relative expression (Table 1). Simultaneously, the velocity of saturation of the capacitors is changed (expressed as parameter a_1) — increases up to 7.5 h and then decreases. This corresponds to the theory that ions from different inner compartments gradually pass through the cytoplasm and the free space and finally enter the ambient solution.

Hypoxia

In the intact plants, after long-lasting cultivation under the conditions of hypoxia, the changes in the C-R curves of the root tips are not so pronounced. Significant symptoms of membrane damage appeared after 10 days of the hypoxia treatment — $c_{relative}$ decreases from 0.492 to 0.164, although the other parameters of the current characteristics indicate that the membrane activity is preserved (Fig. 2). The decrease in the capacitance could result from a lowered energy input. In this respect, the state of the whole plant is different from that of root tips which could deplete their limited energy reserves within a few hours.

With root hypoxia, adaptive changes in the whole plant and especially in the roots must be considered (Fig. 3). The principle of adaptation is based on morphological changes in the root structure. Lysigenically and schizogenically formed aerenchyma — the gas channels — probably enable

Table 1. Characteristic parameters of Current-Response curves

Hours of treatment	$c_{relative}$	a_1	I_{max}, μA
Control	0.445	21.82	406.6
4	0.392	21.18	174.8
7.5	0.168	41.47	289.2
17	0.063	20.43	516.9

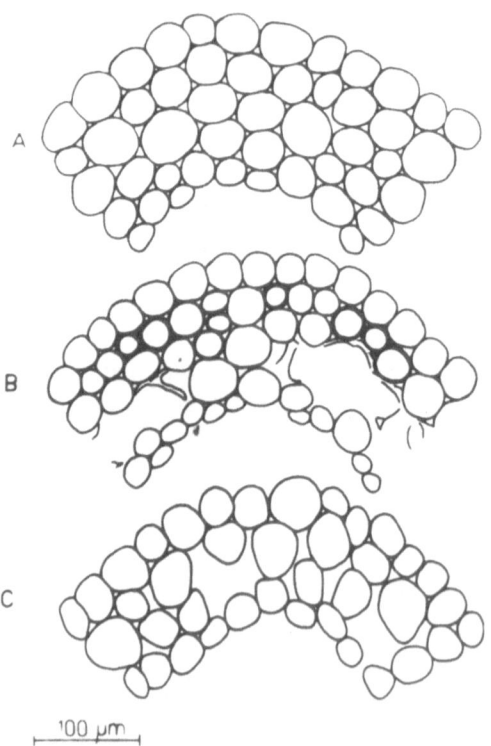

Fig. 3. Cross-section of wheat roots at the age of 10 days. **A** — proximal part from aerobically grown roots. **B, C** — hypoxically grown roots: **B** — partial lysis of cells and thickening of cell walls in proximal part of roots, **C** — radial enlargement of cells in medial and distal parts of roots.

O_2 transport from the upper part of the plant (Wiedenroth and Erdmann, 1985). The thickening is probably caused by mechanical resistance of the sand substrate. Root hypoxia causes, of course, adaptive changes in the whole plant that affects almost all the metabolic processes and thus also involves the formation of the emergency energy reserves for the maintenance of the plasma structures. The measurement of the capacitance and conductance of the root tissue may naturally give us only an integral criterion for the tissue density and the state of the membrane system, as well as for ionic equilibria in the cells. In interpretation of the results it may be borne in mind that the method only records combinations of all the conductances and capacitances and cannot account individually for the degree of development of aerenchyma, for the changes in the energy state of the cytoplasm, *etc*. On the other hand, it is advantageous that the sum of the changes caused by the lack of O_2 within the roots can be determined globally and instantly recorded.

From this work it can be concluded that the roots survive this stress as a result of the formation of aerenchyma with gas channels but the sorption of ions is considerably decreased. Because of higher permeability of membranes with the same ion concentration under anaerobic conditions the conductivity of the plant tissue increases. On the other hand the accumulating ability during adaptation to hypoxia decreases, this causes the decrease of both capacitance and conductivity.

The roots can adapt to the hypoxia by forming aerenchyma but this cannot fully compensate for the loss of the ability to absorb ions.

References

Dvořák M, Černohorská J and Janáček K 1981 Characteristic of current passage through plant tissue. Biol. Plant 23(4), 306–310.

Erdmann B, Hoffmann P and Wiedenroth E M 1986 Changes in the root system of wheat seedlings following root anaerobiosis. I. Anatomy and respiration in *Triticum aestivum* L. Ann. Bot. 58, 597–605.

Wiedenroth E M and Erdmann B 1985 Morphological changes in wheat seedlings (*Tricitum aestivum* L.) following root anaerobiosis and partial pruning of the root system. Ann. Bot. 56, 307–316.

B. C. Loughman et al. (Eds.), Structural and functional aspects of transport in roots, 97–99.
© 1989 by Kluwer Academic Publishers.

Abscisic acid promotes passive fluxes of vacuolar solutes in excised sunflower roots

ZVI GLINKA and NOEMI ABIR
Department of Botany, The George S. Wise Faculty of Life Sciences, Tel Aviv University, Tel Aviv, 69978, Israel

Key words: abscisic acid, exudation, passive flux, permeability, sunflower, thiourea

Introduction

It is well established that abscisic acid (ABA) strongly promotes both volume flow and ion release to the xylem in excised root systems (Fournier *et al.*, 1987; Glinka, 1980; Karmoker and Van Steveninck, 1978).

It has been reported recently that under conditions which considerably reduced the exudation rate from excised sunflower roots (*e.g.* deficient roots or aged excised roots), the magnitude of the effect of ABA on ion flux to the xylem was kept constant or in some cases, even increased (Glinka and Abir, 1983).

Since, probably, under these conditions the metabolic activity of the cells decreased, this fact indicates that the effect of ABA on ion flux into the xylem is a permeability- rather than a metabolically-dependent phenomenon. It was of interest, therefore, to check the influence of ABA on passive fluxes of a neutral solute in the excised root system. By observing the fluxes of a neutral molecule one overcomes some difficulties associated with the interpretation of measured ion fluxes. Thiourea molecules seemed to be well suited for such an observation (Glinka, 1974).

Materials and methods

Four-week-old sunflower plants (*Helianthus annuus* L. cr. Habad), grown in a growth chamber were used throughout. The plants were grown in 1-l jars containing half-strength Hoagland nutrient solution as described earlier (Glinka, 1980). On the morning of the experiment the plants were detopped above the transition zone and pieces of tightly fitting rubber tubing were attached to the cut stem. At selected time intervals the exuding sap was collected by means of a syringe and subjected to the appropriate measurements. The fresh weight of root systems ranged from 8 to 12 g.

Thiourea labelled by ^{35}S was added to the nutrient solution, either after the plant was detopped or 24 h earlier for pre-loading, to a final concentration of 5 mM. Abscisic acid (mixed isomer Sigma) was added from freshly prepared stock solution to a final concentration of 4 μM.

In order to obtain a hydrostatic pressure gradient, the root systems were placed in a chamber, through the lid of which the cut stem protruded and in which the pressure surrounding the roots could be raised.

Results and discussion

In Figs. 1 and 2 preliminary experiments showing the course of thiourea movement into and out of excised sunflower roots are summarized. When an exuding root system was immersed in a nutrient solution containing labelled thiourea, a quick appearance of radioactivity was detected in the exudate and a high rate of increase in the concentration of thiourea in the exuding sap took place. Thiourea concentration build-up in the cells of the roots, on the other hand, was much slower (Fig. 1), indicating that the main pathway of thiourea from the medium to the xylem by-passed the vacuoles. These results support previous find-

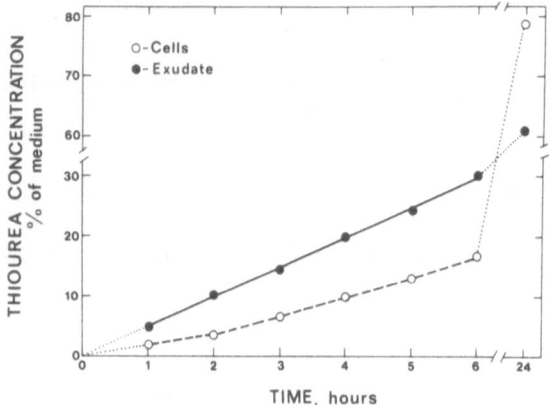

Fig. 1. The course of ^{35}S-thiourea build-up in the cells and in the exudate of excised sunflower roots.

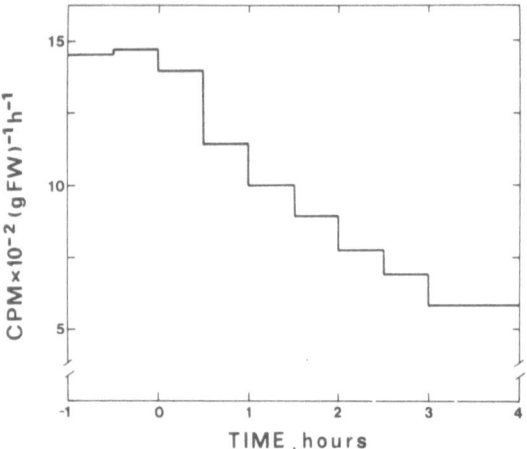

Fig. 2. The course of thiourea release from cells into the xylem of excised sunflower roots. The roots were previously immersed for 24 h in labelled thiourea solution. At time 0 they were transferred into thiourea-free solution.

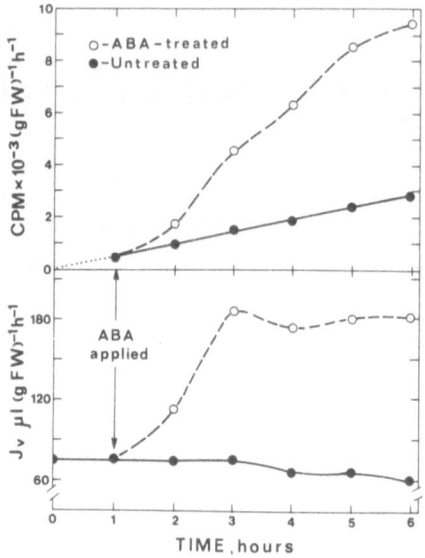

Fig. 3. The effect of ABA on volume exudation rate (J_v) and on thiourea transfer from the medium into the xylem of excised sunflower roots. Thiourea was added to the medium at time 0.

dependent on the rate of water flow through the root, the observed increase in thiourea flux could be an indirect effect resulting from the ABA-induced increase in exudation rate.

In order to distinguish between the effect of ABA and the effect of increased water flow on thiourea flux, the influence of ABA was compared to that of a similar increase in water flow which was obtained by application of a hydrostatic pressure gradient, in

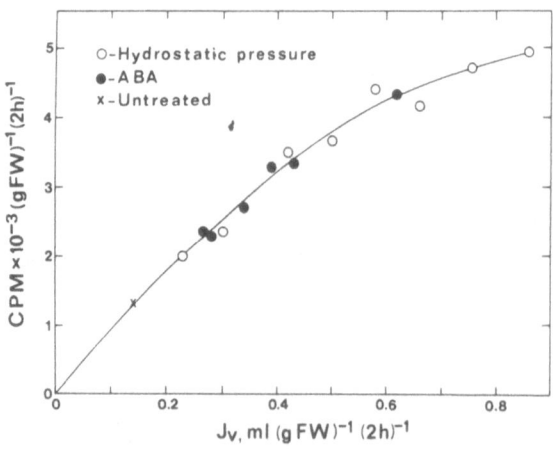

Fig. 4. The relation between volume exudation rate and the rate of thiourea transfer from the medium into the xylem. Increased exudation rates were obtained by application of either ABA or hydrostatic pressure. Cumulative values for the first 2 h after transferring the roots into thiourea solutions are given.

ings (Glinka, 1974) which indicated a relatively low permeability of the tonoplast, compared to that of the plasmalemma, towards thiourea.

Figure 2 describes the course of the decline in the rate of thiourea release into the xylem of preloaded roots which were transferred into thiourea-free nutrient solution. The thiourea which was present in the exudate came mainly from the vacuoles of the root cells since, reasonably, most of the thiourea in the preloaded roots was located there.

Application of ABA to the root medium strongly increased the rate of transfer of thiourea into the xylem (Fig. 3). However, since it is reasonable to assume that the movement of the thiourea may be

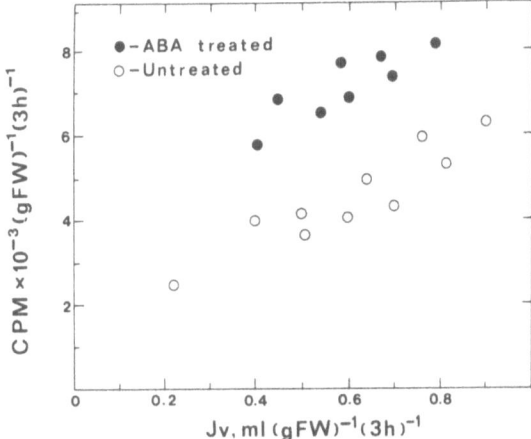

Fig. 5. The relation between volume exudation rate and the rate of thiourea release from the root cells into the xylem. Increased exudation rates were obtained by application of either ABA or hydrostatic pressure. Cumulative values for the first 3 h after transferring roots into thiourea-free solution are given.

the range between 25 and 75 KPa, across the root.

A clear dependence of the rate of thiourea transfer on that of water flow can be seen (Fig. 4). However, no difference between ABA treatment and hydrostatic pressure application could be detected whatsoever. The flux of thiourea from the medium into the xylem was identical in ABA-treated and untreated roots at comparable water flows.

In order to test the effect of ABA on thiourea transfer from the vacuoles of root cells into the xylem, ^{35}S-thiourea preloaded roots were used. The roots were transferred to thiourea-free solution and again either ABA or comparable hydrostatic pressure was applied to the roots. From Fig. 5 it can be seen that also in these conditions an increase in water flow across the root stimulated thiourea release into the xylem. However, the relation thiourea release *vs.* water flow was much higher in ABA-treated roots than in the untreated ones which were exposed to the hydrostatic pressure gradient. It appears therefore that ABA elevated thiourea flux from the vacuoles to the xylem independently of its effect on water flow.

The data presented in this paper indicate that ABA elevates the permeability of cell membranes

towards solutes promoting passive fluxes across them. These results are supported by the findings showing ABA-induced increase in permeability of liquid bilayers to non-electrolytes and cations (Stillwell and Hester, 1984; Wassal *et al.*, 1985) and of unilamellar liposomes to K$^+$ (Harkers *et al.*, 1986).

The fact that the effect of ABA was obtained when thiourea came from vacuoles but not when its route by-passed the vacuoles, indicates that mainly the tonoplast was affected. However, the apparent lack of influence of ABA on flux of thiourea from the medium to the xylem does not exclude the possibility that the permeability of the plasmalemma was also affected. An ABA-induced increase in the permeability of the membrane to water, which apparently exists (Glinka and Reinhold, 1971), may partly mask an increased flux of thiourea.

References

Fournier J M, Benlloch M and de la Guardia M D 1987 Effect of abscisic acid on exudation of sunflower roots as affected by nutrient status, glucose level and aeration. Physiol. Plant. 69, 675–679.

Glinka Z 1974 Fluxes of a nonelectrolyte and compartmentation in cells of carrot root tissue. Plant Physiol. 53, 307–311.

Glinka Z 1980 Abscisic acid promotes volume flow and ion release to the xylem in sunflower roots. Plant Physiol. 65, 537–540.

Glinka Z and Abir N 1983 Effect of abscisic acid on exudation from deficient and aged sunflower roots. Physiol. Plant. 59, 208–212.

Glinka Z and Reinhold L 1971 Abscisic acid raises the permeability of plant cells to water. Plant Physiol. 48, 103–105.

Harkers C, Hartung W and Gimmler H 1986 Abscisic acid-mediated K$^+$ efflux from large unilamellar liposomes. J. Plant Physiol. 122, 385–394.

Karmoker J L and Van Steveninck R F M 1978 Stimulation of volume flow and ion flux by abscisic acid in excised root system of *Phaseolus vulgaris* L. cv. Redland Pioneer. Planta 141, 37–43.

Stillwell W and Hester P 1984 Abscisic acid increases membrane permeability by interacting with phospatidylethanolamine. Phytochemistry 23, 2187–2192.

Wassall S R, Hester P and Stillwell W 1985 Abscisic acid increases lipid bilayer permeability to cations as studied by phosphorus-31 NMR. Biochim. Biophys. Acta 815, 519–522.

B. C. Loughman et al. (Eds.), Structural and functional aspects of transport in roots, 101–105.
© 1989 by Kluwer Academic Publishers.

The role of maize root tissues in sulphate absorption and radial transport

MARGITA HOLOBRADÁ and STEFAN KUBICA
Institute of Experimental Biology and Ecology, CBES, Slovak Academy of Sciences, CS-81434 Bratislava, Czechoslovakia

Key words: absorption, maize roots, microautoradiography, radial transport structure, sulphate

Introduction

The function of roots as nutrient transporting organs is closely related to the structure and function of root cells and tissues. Nutrient transport through roots is a polarized flow initiated by the uptake by root cells followed by centripetal radial transport *via* apoplast and/or symplast from epidermis to xylem.

Although the existence of both pathways of radial ion transport is generally accepted the relative rates of apoplasmic and symplasmic transport are not readily determined. The problem is often studied by using elements thought to demonstrate apoplastic pathway because they do not enter the symplast (Peterson *et al.*, 1986) or using specific apoplastic and symplastic fluorescent dyes (Peterson and Perumalla, 1984). The studies of radial ion transport examined with labelled tracers and X-ray analysis are not very numerous (Canning and Kramer, 1958; Clarkson and Sanderson, 1969; Läuchli, 1973; Peterson *et al.*, 1986; Sasaki *et al.*, 1982). Very little is known about the radial transport of sulphate. There are little available data, Biddulph (1967), Weigl and Lüttge (1962) and as recently reported by Peterson (in press).

Labelling methods were used to provide physiological and microautoradiographical anatomical evidences for the role of different root tissues in ^{35}S-sulphate uptake and radial transport during root growth and differentiation.

Material and methods

Three-day-old seedlings of *Zea mays* (cv. VIR 17, CE-330) were used for our experiments. After washing with tap water the seeds were germinated in moist rolled filter paper in the dark at 25°C. Intact seedlings with equal root length were treated with ^{35}S-labelled 1/5 Knop solution (Schropp, 1951). The labelling was 74 MBq $Na_2^{35}SO_4 l^{-1}$.

After 30 min absorption the roots were washed 20 times with tap water and used for physiological and anatomical determination of ^{35}S-sulphate.

Uptake and accumulation of ^{35}S-sulphate along the root length were directly determined in 1-mm segments (Frič and Holobradá, 1983). The rate of ^{35}S-sulphate distributed in the differentiated tissues was separately investigated in peripheral and stelar tissues. The apical part up to 12 mm behind root apex, including the root cap, was only cut into 1-mm segments as separation of tissues was not possible. This part was only investigated for radial transport by microautoradiographic technique.

Microautoradiography was used for direct determination of ^{35}S-sulphate uptake and transport in the cells of different root tissues along the whole root length. Roots, washed with tap water were cut into 1-cm segments, immediately frozen in liquid nitrogen and dehydrated by lyophilization. Paraffin infiltration was carried out with dried segments in vacuo. Embedded segments were sectioned at 10 μm. Sovietic 'M' emulsion was used for autoradiographs preparation and after 100 days exposure autoradiographic plates were developed and reduced Ag-grains above cells were counted.

Results and discussion

Variations in ion uptake and transport along the root axis have been documented experimentally using radionuclides applied to the root medium or

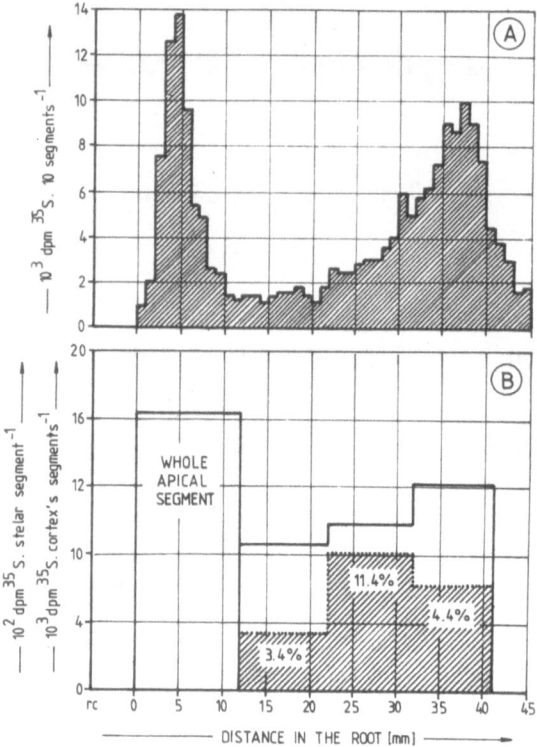

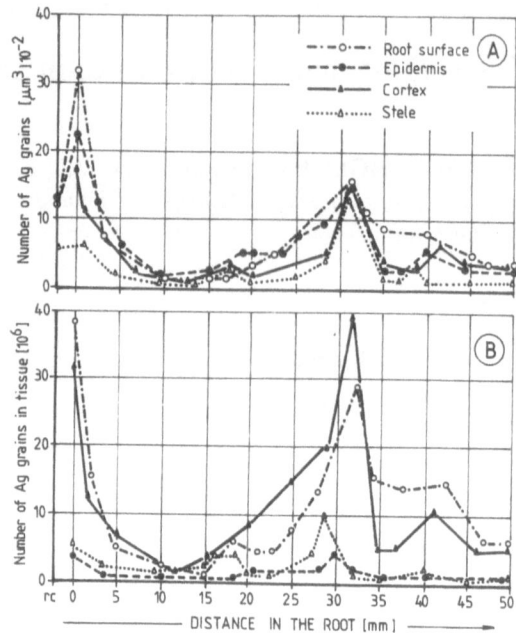

Fig. 1A. ³⁵S-sulphate uptake along the length of primary roots of maize (cv. CE 330).

Fig. 1B. Radial transport of ³⁵S-sulphate in differentiated tissues of maize primary roots (cv. CE 330). □ — peripheral tissues; ▨ — stelar tissues.

Fig. 2. ³⁵S-sulphate uptake by the different tissues of maize primary roots (cv. VIR 17). **A** Relative content of ³⁵S per volume unit. **B** Absolute contents of ³⁵S in different tissues.

to the different root zones (Biddulph, 1967; Clarkson and Sanderson, 1970; Eshel and Waisel, 1972; Holobradá, 1977; Holobradá *et al.*, 1980; Marschner and Richter, 1973; Rovira and Bowen, 1968; Wiebe and Kramer, 1954). However, little is known about the participation of root tissues in radial and longitudinal transport of sulphate.

Both processes, uptake and transport of ions, take place in a structural complex consisting of various tissues at different stages of growth and development along the length of the root. Among many factors the structural and functional variations along the length of the root are most responsible for the observed diversity of ion uptake and transport.

The zonality of absorbed ³⁵S-sulphate with two peaks of high uptake which was determined in whole 1-mm segments of intact primary maize roots of cv. CE 330 (Fig. 1A) correlates very well with microautoradiographic studies with VIR 17

presented in Fig. 2. The first peak corresponds with the apical growing region of the root and the second one with the region of root hair formation.

High ³⁵S-uptake in the apical part seems to be related to metabolic function of sulphur reported by Biddulph (1967) who observed ³⁵S-labelled nuclei already 2.5 to 3 mm behind the root tip of bean. The highest ³⁵S-protein synthesis has been found in the apical region when studying the relation between uptake and metabolism of ³⁵S-sulphate in primary maize roots (Holobradá, 1985). As ³⁵S-sulphur, incorporated into the protein fraction, appeared to be only a small portion of total ³⁵S-sulphate absorbed, we were interested in the pattern of ³⁵S-distribution across the roots. Of course by investigation of the whole segments it could not be determined whether all labelled sulphur was absorbed by root tissues or whether it was mainly concentrated at the root surface as has been reported for ⁴⁶Sc (Clarkson and Sanderson, 1969). Their autoradiographs showed that most of the ⁴⁶Sc in the root tip of *Allium cepa* was concentrated in a peripheral belt corresponding with the mucigel layer of the root cap. Our results obtained by microautoradiography confirm the above view as the highest reduction of Ag grains in response of

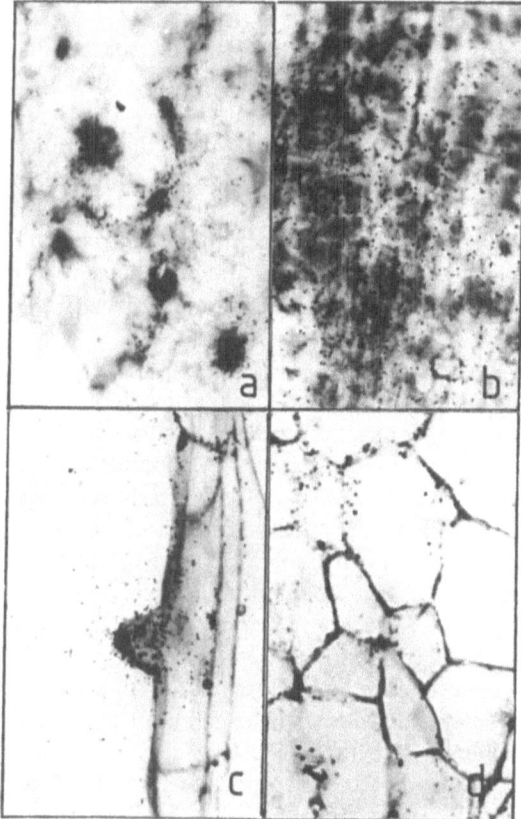

Fig. 3. **a.** ^{35}S-labelled nuclei in root cap cells, **b.** homogenous ^{35}S-labelling of root meristem, **c.** ^{35}S-labelling of hair cell, **d.** cortical cells where only the cytoplasm is labelled with [35]S. Magnification was approximately 210 × .

developing by radioactive ^{35}S atoms was found at the apical root surface (Fig. 2). In the meristematic region as much as 50% of the total ^{35}S-sulphate taken up was absorbed at the surface in mucigel. ^{35}S-sulphur taken up by meristematic cells was homogenously distributed through the whole tissue (Fig. 3b). We could not confirm the results of Lüttge and Weigl (1962) who showed that in the very tip region of the roots, ^{35}S-sulphate penetrated up to the border between periblem and plerom. It did not penetrate beyond, even though there was no visible barrier. Our autoradiograph does not suggest any barrier of sulphate transport in maize meristem. Some cells of the epidermis in the meristematic region and also the cells of the root cap accumulated ^{35}S in the nuclei (Fig. 3a).

Basipetally, behind the meristematic region the surface labelling decline up to 13 mm distance from the root tip never exceeding 10% of the apical

labelling. More basipetally an increase, resulting in the second maximum of peripheral tissue, very probably mucigelous, at the region of root hairs was observed (Fig. 2A). The ^{35}S-accumulation at the root surface may be due to sulphate affinity for mucigel root excretions. The mucigel layer obviously facilitates the uptake of ions as preferential adsorption of ions was demonstrated in the mucigel with subsequently more intensive radial transport than in the regions with lower surface ^{35}S-labelling (Fig. 2A). The very basal part of the root was low in surface ^{35}S, but, was the highest in the root hairs region; it represented up to 60% of total uptake. The sulphate absorbed by root tissues, followed the curve of those in the mucigel layer (Fig. 2).

The rate of centripetally transported sulphate, expressed per unit of tissue volume decreased in the direction — epidermis, cortex, stele (Fig. 2). Absolutely expressed amount of ^{35}S distributed in the tissues showed that approximately 40% of ^{35}S taken up was accumulated in the cortex and only 6% was transported into the stele. The lowest, only 4%, accumulation was observed in the epidermis. These anatomical findings are in accordance with investigations of peripheral and stelar tissues where also very low rates of transport of ^{35}S-sulphate into stele were found. The rate of accumulation of peripheral tissues and transport into the stele along the root axis is given in Fig. 1B. Of course, by this method, cortical, epidermal or even that absorbed at the root surface could not be determined. Only microautoradiographic studies have enabled us to widen our knowledge about the rate of radial transport in more detail. It was shown that ^{35}S-sulphate transport in epidermal cells up to the region of root hair formation was only symplasmic. Similar results have been obtained by using the apoplastic and symplastic dyes in maize roots (Peterson and Perumalla, 1984; Perumalla and Peterson, 1986; Peterson *et al.*, 1982).

It is suggested that the plasmalemma of all epidermal cells and those of cortex is able to transfer ions. But, the limiting factors can result from the structural differences of this outermost cell layer. The question has often arisen whether the ions are equally absorbed by all epidermal cells or whether the uptake is mainly concerned with the root hair region. Ultrastructural evidence of plasmodesmatal density support the view of the dominant role of root hair epidermal cells compared with epidermal cells without root hairs (Vachmistrov, 1981). Our results confirmed that the presence of root hairs increases the uptake capacity of epidermal cells (Fig. 3c) and in root hair cells the active uptake of sulphates was also observed. The autoradiographs suggest the transfer of sulphate from apoplast into symplast already at root hair epidermal cells. The greater part of ^{35}S taken up by root hair epidermal cells was concentrated in its apical, cytoplasm rich part (Fig. 3a). The active uptake with subsequent symplastic pathway of ^{35}S-transport also takes place in sub-epidermal cells of the cortex (Fig. 4a), as was observed in the above mentioned uptake and transport of dyes in the differentiating region 5 cm behind the root apex.

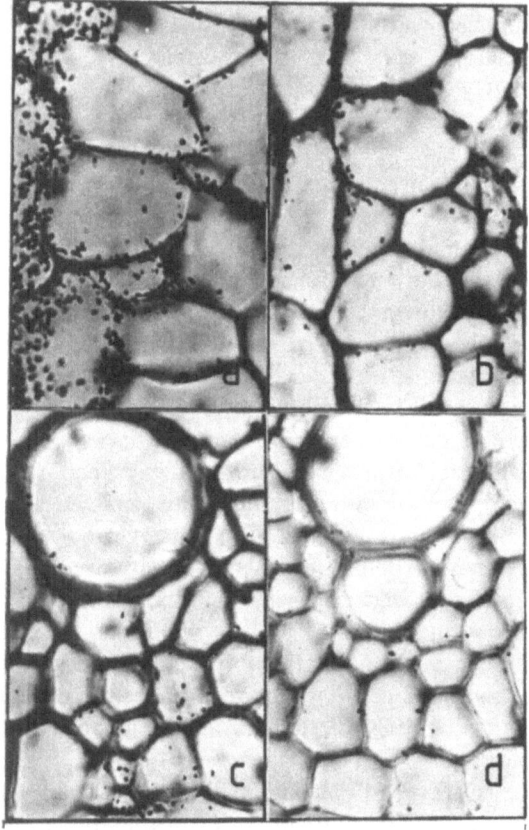

Fig. 4. **a.** Labelling of epidermal cells in the differentiated region; the root hair cell is labelled with ^{35}S only in the apoplast, **b.** the border between cortex and stele in the non-differentiated region, **c.** the border between cortex and stele in the proximal differentiated region where apoplast and symplast were labelled, **d.** the border between cortex and stele in the basal differentiated region where only the apoplast was labelled. Figures **b, c, d** indicate root parts associated with xylem. Magnification was approximately 420 ×.

The accumulation of ^{35}S in cortical cytoplasmic pool is well shown in Fig. 3d. In epidermal cells without root hairs ^{35}S-sulphate was found in cell walls (Fig. 4a).

It seems to be evident that the symplastic pathway of centripetal transport of sulphate must play an important role in the process of passing the sulphate anion from cortex into the central cylinder and the endodermal cells situated adjacent to the xylem poles are clearly of an importance (Fig. 4b, c, d).

Further microautoradiographic studies suggest that in stelar tissues both apoplastic and symplastic pathways of sulphate transport can take place up to the 4th cm behind the apex: ^{35}S-sulphate was distributed in stelar apoplast as well as in stelar symplast (Fig. 4b, c). In the most basal part of the root the conditions of stelar symplasmic transport deteriorated. As the ^{35}S-sulphur was localized only in the apoplast (Fig. 4d), no symplasmic transport could be expected.

The results show the close relation of anatomical characteristics with knowledge about the function of different structures to the uptake and transport processes. In radial transport of ions the importance of traversed root tissues is also clear. In addition it is obvious that at the same time the changes of tissue along the length of the root in response to differentiation and maturation must be accepted.

References

Biddulph S F 1967 A microautographic study of Ca^{45} and S^{35} distribution in the intact bean roots. Planta 74, 350–367.

Canning R E and Kramer P J 1958 Salt accumulation in various regions of roots. Am. J. Bot. 45, 378–382.

Clarkson D T and Sanderson J 1969 The uptake of a polyvalent cation and its distribution in the root apices of *Allium cepa*: Tracer and autoradiographic studies. Planta 89, 136–154.

Eshel A and Waisel Y 1972 Variations in sodium uptake along primary roots of corn seedlings. Plant Physiol. 49, 585–589.

Frič F and Holobradá M 1983 Liquid scintillation counting of ^{35}S in small segments of maize (*Zea mays* L.) roots. Radiochem. Radioanal. Letters, 56, 327–332.

Holobradá M 1977 Changes in sulphate acucmulation along the primary roots during tissue differentiation. Biol. Plant. 19, 331–337.

Holobradá M 1985 Uptake and assimilation of sulphur in plants. (In Slovak). Ed. Treatises on Biology, Veda, Bratislava.

Holobradá M, Mistrík I and Kolek J 1980 The relationship between root growth, temperature and anion uptake. Biológia (Bratislava), 35, 251–257.

Läuchli A 1973 Investigation of ion transport in plants by electron probe analysis: Principles and perspectives. *In* Ion Transport in Plants. Ed. W Anderson. pp 1–11. Acad. Press, London/New York.

Lüttge U and Weigl J 1962 Mikroautoradiographische Untersuchungen der Aufnahme und des Transportes von $^{35}SO_4$ und ^{45}Ca in Keimwurzeln von *Zea mays* L. Planta 58, 113–126.

Marschner H and Richter C 1973 Akkumulation und Translokation von K^+, Na^+, und Ca^{2+} bei Angebot zu einzelnen Wurzelzonen von Maiskeimpflanzen. Z. Pflanzenernaehr. Bodenkd. 135, 1–15.

Perumalla C J and Peterson C A 1986 Deposition of Casparian bands and suberin lamellae in the exodermis and endodermis of young corn and onion roots. Can. J. Bot. 64, 1873–1878.

Peterson C A, Emanuel M E and Wilson C 1982 Identification of a Casparian band in the hypodermis of onion and corn roots. Can. J. Bot. 60, 1529–1535.

Peterson C A and Perumalla C J 1984 Development of the hypodermal Casparian band in corn and onion roots. J. Exp. Bot. 35, 51–57.

Peterson T A, Swanson E S and Hull R J 1986 Use of lanthanum to trace apoplastic solute transport in intact plants. J. Exp. Bot. 37, 807–822.

Rovira A D and Bowen C D 1968 Anion uptake by apical region of seminal wheat roots. Nature, 218, 685–686.

Sasaki Y, Arima Y and Kumazawa K 1982 Studies on the radial transport and metabolism of phosphate in corn roots using the ^{32}P and ^{33}P double labelling method. Soil Sci. Plant Nutr. 28, 141–145.

Schropp W 1951 Methodenbuch, VIII. Der Vegetationsversuch. 1. Die Methodik der Wasserkultur höherer Pflanzen, Neuman Verl.

Vakhmistrov D B 1981 Specialization of root tissues in ion transport. *In* Structure and Function of Plant Roots. Eds. R Brouwer *et al.* pp. 203–208. Nijhoff, Dordrecht, The Netherlands.

Weigl J and Lüttge U 1962 Mikroautoradiographische Untersuchungen über die Aufnahme von $^{35}SO^{--}$ durch Wurzeln von *Zea mays* L. Die Funktion der primären Endodermis. Planta 58, 15–28.

Wiebe H H and Kramer P L 1954 Translocation of radioactive isotopes from various regions of roots of barley seedlings. Plant Physiol. 29, 342–348.

B. C. Loughman et al. (Eds.), Structural and functional aspects of transport in roots, 107–110.
© 1989 by Kluwer Academic Publishers.

Effect of light intensity on root growth, mycorrhizal infection and phosphate uptake in onion (*Allium cepa* L.)

C.L. SON[1], F.A. SMITH and S.E. SMITH[1]
Departments of Agricultural Biochemistry[1] and Botany[2], University of Adelaide, Adelaide, South Australia

Key words: *Allium*, light intensity, mycorrhiza, phosphorus, root growth

Introduction

It is well established that the growth responses of plants to mycorrhizal infection are influenced by the amount of phosphorus (P) supplied in the soil. In addition previous work has shown that at low light intensity (LI) plants have lower rates of root growth and smaller responses to mycorrhizal infection than plants grown at high LI (Bethlenfalvay and Pacovsky, 1983; Daft and El-Giahmi, 1978; Hayman, 1974; Tester *et al.*, 1985). The decrease in growth enhancement of roots at low LI's results from a higher investment by the plant in its leaves (Bethlenfalvay and Pacovsky, 1983). This adjustment in source-sink relationships follows decreases in photosynthesis with decreasing LI. However, in mycorrhizal plants there is a further confounding factor: the increased demand on the host plant for carbohydrates used by the fungus. Certainly, differences in carbon allocation between mycorrhizal and non-mycorrhizal plants have been noted (Pang and Paul, 1980; Snellgrove *et al.*, 1982), and may negate any beneficial effects on plant growth, resulting from P uptake via the fungal symbiont.

When P is readily available to the plant the positive growth response due to mycorrhizal infection is also reduced. Soil P has been shown to have a direct effect on source-sink relationships by lowering the dry matter in leaves of infected plants over non-infected controls (Bethlenfalvay and Pacovsky, 1983). Furthermore, it has been reported that high soil P levels and low LI's reduce percentage infection in mycorrhizal plants (Hayman, 1974; Mosse, 1973). It has been suggested that this may reflect a self-regulatory mechanism within the host plant preventing infection from developing to any appreciable extent under conditions where it would not be advantageous to the higher plant symbiont. However, while this may be true, it has also been found that unless infection is reduced to almost zero a mycorrhizal effect on P uptake may still be observed (Smith, 1982; Tester *et al.*, 1985).

Thus, in this work, we investigated the interaction of P supply, LI and mycorrhizal infection on growth of roots and shoots and on rates of P uptake by roots of onion plants.

Materials and methods

Allium cepa L. cv. Torrens White was grown in pots containing 600 g of a soil:sand mixture. Soil from the Caliph region of South Australia low in P (Smith and Smith, 1981) was autoclaved at 120°C for 40 min and mixed with steamed sand in the ratio of 1:9 soil:sand. Mycorrhizal plants were grown in a similar soil:sand mixture with the addition of root inoculum of *Glomus mosseae* (Nicol and Gerd) Gerdeman and Trappe, where pieces of infected root together with adhering soil were placed in the planting hole beneath each seed. Plants were thinned to five per pot after the first harvest. Application of a basic nutrient (Smith *et al.*, 1985) solution lacking P was also begun at 2 weeks. Plants were watered each day with deionized water to 12% w/v.

Two soil phosphorus (P) levels were chosen: no additional P (P_0) and $1.0 \, mmol \, P \, kg^{-1}$ soil (P_2). P was added to each pot as a solution of $0.1 \, M$ NaH_2PO_4 and mixed thoroughly through the soil:sand mixture prior to planting.

Plants were grown under glasshouse conditions during May–June (southern hemisphere autumn).

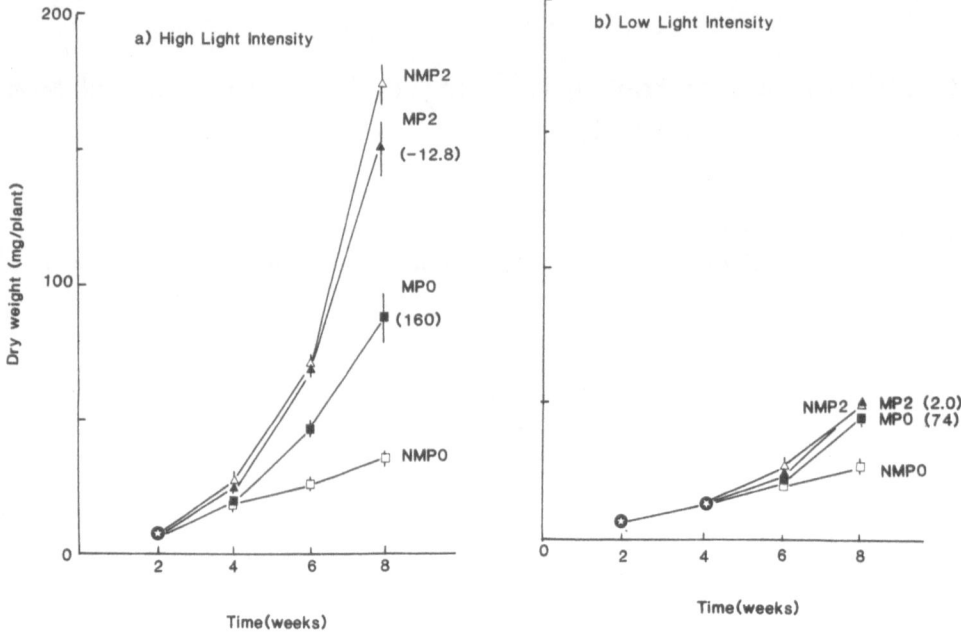

Fig. 1. Total dry weight of plants grown at (a) $600 \, \mu mol \, m^{-2} s^{-1}$ and (b) $250 \, \mu mol \, m^{-2} s^{-1}$. Figures in brackets are % growth response to mycorrhizal infection at 8 weeks.

LI's were adjusted using shadecloth mounted over wooden frames 48 cm above the soil level in the pots, to give mean irradiances of 600 ('high') and 250 ('low') $\mu mol \, m^{-2} s^{-1}$. Day length varied between 9.9 and 11 h. Temperatures were maintained between 15° and 25°C.

Plants were harvested at 2, 4, 6 and 8 weeks. Growth, P concentration, root length and vesi-

cular-arbuscular (V.A.) mycorrhizal infection were determined as described previously (Smith *et al.*, 1985). Mycorrhizal growth response was calculated on a dry weight basis as

$$\frac{DW \text{ M plants} - DW \text{ NM plants}}{DW \text{ NM plants}} \times 100$$

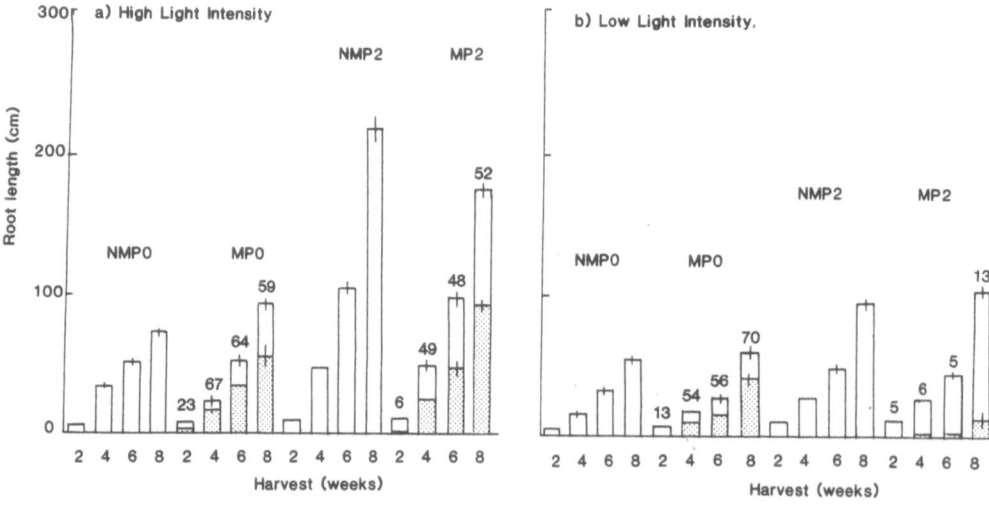

Fig. 2. Root length and mycorrhizal infection of plants grown over 8 weeks at (a) $600 \, \mu mol \, m^{-2} s^{-1}$ and (b) $250 \, \mu mol \, m^{-2} s^{-1}$. Stippled areas show length of root infected. Figures above columns indicate % infection.

Results and discussion

Under low LI dry weight (DW) of mycorrhizal (M) and non-mycorrhizal (NM) plants was reduced, compared with plants grown at high LI (Fig. 1). This confirms previous work on clover plants (Tester *et al.*, 1985). The mycorrhizal response to infection with low P (P_0) was less under low LI (compare Fig. 1a and 1b). When soil P was high (P_2) mycorrhizal plants had lower DW than equivalent non-mycorrhizal plants (Fig. 1a).

Compared with high LI, root growth was reduced at low LI in all treatments (see also Tester *et al.*, 1986) and the effect of P supply in increasing root growth was also reduced (Fig. 2). Mycorrhizal

infection had little effect on root length at either LI except at week 8 where a slight negative effect on MP_2 plants at high LI was observed (Fig. 2a). Percentage infection was not affected by decreasing LI alone but was markedly reduced by the interaction of low LI and the addition of P (Fig. 2b). This indicates that there were interactions between P nutrition, LI and mycorrhizal infection.

Mycorrhizal infection or additional P (or both) decreased the root:shoot ratio at high LI (Fig. 3a). At low LI all root:shoot DW ratios were reduced and there was little difference between treatments (Fig. 3b). This would support the suggestion (Bethlenfalvay and Pacovsky, 1983), that with decreased photosynthesis the plant allocates a greater share of its resources away from the roots and to the growing leaves. This, together with the lower mycorrhizal growth response, may reflect an increased significance of carbohydrate utilization by the fungus as previously considered (Tester *et al.*, 1985; Bethlenfalvay and Pacovsky, 1983).

P concentrations in the plant (not shown) were adequate to luxurious for plant growth in all but the NMP_0 treatments (range 1.47–4.98 μg P mg^{-1} DW). Thus neither P deficiency nor P toxicity could account for the observed depression in mycorrhizal response (Hayman, 1974; Mosse, 1973).

At high LI, P inflow (mol P absorbed cm^{-1} root length s^{-1}) was consistently increased both by mycorrhizal infection and by additional soil P (Fig. 4). Low LI had little effect on inflow in non-mycorrhizal roots, but severely reduced inflow in

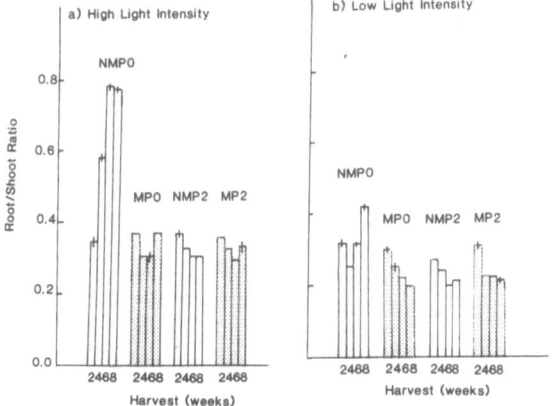

Fig. 3. Root:shoot dry weight ratios of plants grown over 8 weeks at (a) 600 μmol m^{-2} s^{-1} and (b) 250 μmol m^{-2} s^{-1}.

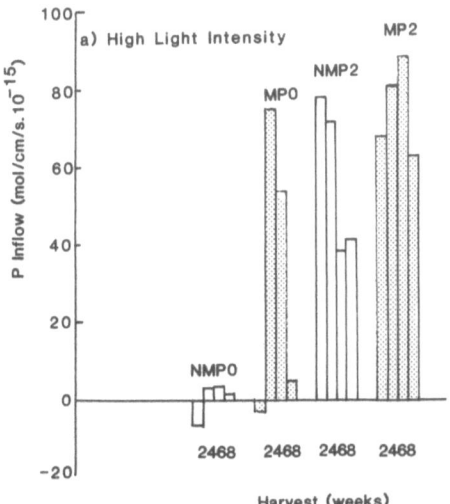
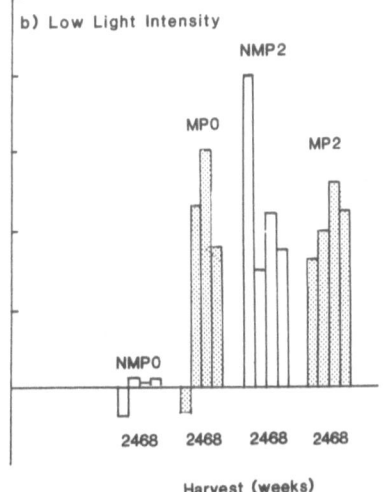

Fig. 4. Inflow of P into mycorrhizal and non-mycorrhizal *Allium cepa* at (a) 600 μmol m^{-2} s^{-1} and (b) 250 μmol m^{-2} s^{-1}.

mycorrhizal roots (Fig. 4b). The biggest decrease in P inflow at low LI was in MP$_2$ plants. In this experiment this was accompanied by a large decline in infection. However in a second experiment similarly decreased inflow occurred in the absence of any differences in infection (Son and Smith, 1988).

Therefore, we conclude that mycorrhizal infection influences P acquisition by roots over a wide range of soil P and light intensities. Nevertheless, the mycorrhizal contribution to uptake is more sensitive to low light and this, together with effects on root:shoot ratio supports the hypothesis that mycorrhizal roots require more photosynthate than non-mycorrhizal roots. This is a factor to be taken into account in relation to the current resurgence of interest in the physiology of roots in the field. In view of this, we may expect shade plants to be less likely to show mycorrhizal responses, unless they have evolved physiological adaptations to improve efficiency of photosynthesis and transfer of photosynthate to the fungus.

Acknowledgements

Financial support is provided by the Australian Research Grants Scheme.

References

Bethlenfalvay G S and Pacovsky R S 1983 Light effects in mycorrhizal soybeans. Pl. Physiol. 73, 969–972.

Daft M J and El-Giahmi A A 1978 Effect of arbuscular mycorrhiza plant growth. VIII. Effect of defoliation and light on selected hosts. New Phytol. 80, 365–372.

Hayman D S 1974 Plant growth responses to vesicular-arbuscular mycorrhiza. VI. Effect of light and temperature. New Phytol. 73, 71–80.

Mosse B 1973 Advances in the study of vesicular-arbuscular mycorrhiza. Ann. Rev. Phytopathol. 11, 170–196.

Pang P C and Paul E A 1980 Effects of vesicular-arbuscular mycorrhiza on ^{14}C and ^{15}N distribution in nodulated Fababeans. Can. J. Soil Sci. 60, 241–250.

Son C L and Smith S E 1988 Mycorrhizal growth responses: Interactions between photon irradiance and phosphorus nutrition. New Phytol. 108, 305–314.

Smith S E 1982 Inflow of phosphate into mycorrhizal and non-mycorrhizal *Trifolium subterraneum* at different levels of soil phosphate. New Phytol. 90, 293–303.

Smith F A and Smith S E 1981 Mycorrhizal infection and growth of *Trifolium subterraneum*: use of sterilized soil as a control treatment. New Phytol. 88, 299–309.

Smith S E, St John B J, Smith F A and Nicholas D J D 1985 Activity of glutamine synthetase and glutamate dehydrogenase in *Trifolium subterraneum* L. and *Allium cepa* L.: Effects of mycorrhizal infection and phosphate nutrition. New Phytol. 99, 211–227.

Snellgrove R C, Splittstoesser W E, Stribley D P and Tinker P B 1982 The distribution of carbon and the demand of the fungal symbiont in leek plants with vesicular-arbuscular mycorrhizas. New Phytol. 92, 75–87.

Tester M, Smith F A and Smith S E 1985 Phosphate inflow into *Trifolium subterraneum* L.: Effects of photon irradiance and mycorrhizal infection. Soil Biol. Biochem. 17, 807–80.

Tester M, Smith S E, Smith F A and Walker N A 1986 Effects of photon irradiance on the growth of shoots and roots, on the rate of initiation of mycorrhizal infection and on the growth of infection units in *Trifolium subterraneum* L. New Phytol. 103, 375–390.

B. C. Loughman et al. (Eds.), Structural and functional aspects of transport in roots, 111–113.
© 1989 by Kluwer Academic Publishers.

The participation of the primary maize root on the assimilation of NH_4^+ ions

MIROSLAVA LUXOVÁ

Institute of Experimental Biology and Ecology, CBES, Slovak Academy of Sciences, CS-814 34 Bratislava, Dúbravská 14, Czechoslovakia

Key words: glutamate dehydrogenase, glutamine synthetase, root, *Zea mays*

Introduction

It has been suggested recently that in plants ammonia obtained by reduction of nitrate or absorbed directly from the culture medium is incorporated primarily into organic compounds via glutamine synthetase (GS) and glutamate synthase (GOGAT) in the GS/GOGAT pathway (Arima and Kumazawa, 1977; Lewis *et al.*, 1983; McNally and Hirel, 1983; Miflin and Lea, 1976). This is in contradiction to the previous assumption about ammonia incorporation via reductive amination of 2-oxoglutarate catalyzed by glutamate dehydrogenase (GDH).

In this study we wanted to focus on the participation of anatomically different regions of the primary maize root in the processes of NH_4^+ ions incorporation into organic compounds. The activity of GS and GDH was studied in four different parts of the primary root for this reason. The present work is a continuation of our previous study of nitrate reductase and nitrite reductase in the same material (Luxová and Gašparíková, 1984).

Material and methods

The 6–7 cm long primary maize root (*Zea mays* L., cv. CE-330) of three-day-old seedlings were used as experimental material. The seeds were germinated between wet filter paper in the dark at 25°C. The roots were divided into the following anatomically defined regions: a) 2-mm long segments from the root apex (root cap removed); b) the consecutive 10-mm segments; c) the peripheral tissues (epidermis + cortex) from a distance of 12–65 mm and d) stele from the same distance as the previous sample.

Glutamate dehydrogenase activity was determined spectrophotometrically by measuring the oxidation rate of NADH at 340 nm using a spectrophotometer Specord UV-Vis (Gašparíková *et al.*, 1976). Enzyme activity was expressed as the amount of reduced substrate in $nkat \cdot mg^{-1}$ protein or in $nkat \cdot g^{-1}$ fr. wt.

The activity of glutamine synthetase was measured with a transferase assay by the modification of the method of Sahulka and Lisá (1979). The production of hydroxamate was determined spectrophotometrically at 540 nm. Enzyme activity was expressed in nkat hydroxamate $\cdot mg^{-1}$ protein or in $nkat \cdot g^{-1}$ fr. wt.

Proteins were determined directly in the supernatant by the method of Bradford (1976) with Coomasie Brilliant Blue G-250.

Results and discussion

The specific activity of GDH was 2–8 fold higher in the older tissues of the stele and cortex than in the growing part of the root—in the meristematic and elongation region (Fig. 1A). Similar results were obtained by Wallace (1973), who observed more than five times higher GDH activity in the last 50–60 mm segments of three-day-old maize roots in comparison with the root apex. Sahulka and Lisá (1978) studied the level of GDH in isolated pea roots. They found out that the increasing level of GDH reached a peak after 24 h in the apex

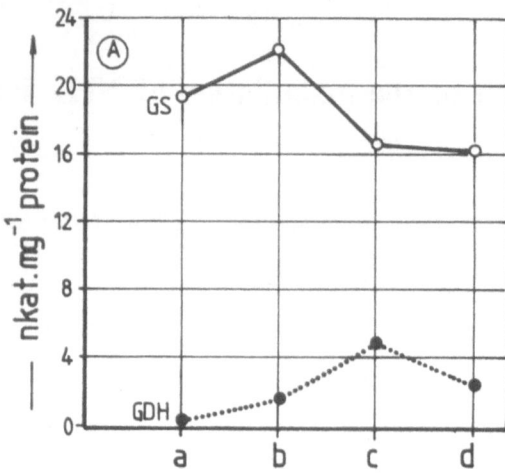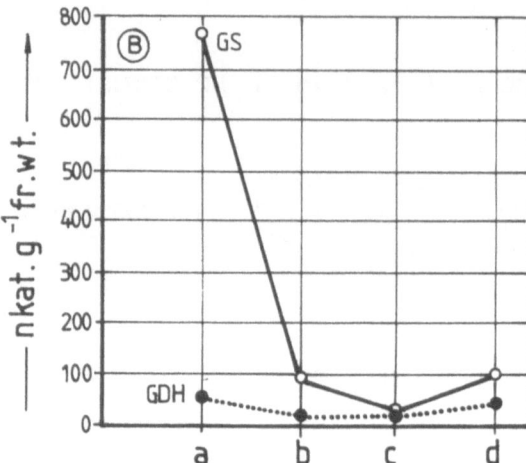

Fig. 1. Activities of glutamine synthetase (GS) and glutamate dehydrogenase (GDH) in different regions of primary maize root: $nkat \cdot mg^{-1}$ protein (**A**); $nkat \cdot g^{-1}$ fr. wt. (**B**). (**a**: 0–2 mm from apex; **b**: 2–12 mm; **c**: epidermis + cortex 12–65 mm; **d**: stele 12–65 mm).

but this increase continued for at least 72 h in the older parts of the root.

On the other hand the total activity of GDH was highest in the meristematic region (Fig. 1B). Different values of specific and total activity obtained in the meristematic region of the root are similar also for the enzyme GS and are apparently caused by a characteristic of meristematic cells. In general these cells are characterized by low vacuolation, low water content and a low fresh weight but by a high level of protein.

The pattern of the specific activity of GS was completely different. The highest activity was found in the first two segments — in the meristem and in the region of elongation. In the ontogenetically older tissues the specific activity of GS was about 25% lower with virtually no difference between the stele and the cortex. In the primary maize root, the results obtained indicate that the activity of GS is several times (3–32 fold) higher than the activity of GDH and it is effectively involved in the assimilation of NH_4^+ ions into the organic compounds.

Izmaylov *et al.* (1982) obtained similar results in three plant species — sugar beet, maize and pea — differing in the place of nitrate assimilation. The authors studied the activity of GS and GDH in the different organs of these plants (root, leaf, mesocotyl, hypocotyl, shoot, germinating seed). They found out that the activity of GS was several times higher than the activity of GDH in all the organs examined.

After the discovery of the utilization of the GS/GOGAT pathway in the assimilation of NH_4^+ ions a question arose about the role of GDH. Because of the high Km for ammonia in comparison with GS it is unlikely that GDH is a major factor in the assimilation of NH_4^+ ions under normal environmental conditions. Some authors (Givan, 1979; Miflin and Lea, 1976; Rhodes *et al.*, 1976) supposed that GDH may play an important role in the assimilation of NH_4^+ ions only in extremely high concentrations of this ion — when a rapid detoxification of ammonia is needed.

Lewis *et al.* (1983) reported that in the root of barley NH_4^+ ions absorbed from the soil were assimilated exclusively through the GS/GOGAT pathway and GDH played only a very limited role in this process. They supposed that GDH from higher plants probably plays an important role in a reverse reaction leading to oxidative deamination of glutamate and not in the amination of 2-oxoglutarate to glutamate. However Loyola-Vargas and Jimenez (1984) investigating the activity of the aminating and deaminating reactions of GDH in different tissues of maize (in root, leaf and calus) expressed the view that maize roots are a major site of the assimilation of NH_4^+ ions in normal conditions and this assimilation is achieved via GDH. Because the authors didn't investigate the activity of GS in these tissues they couldn't compare both assimilation routes.

Our results indicate a higher probability for the GS/GOGAT pathway in normal conditions in the primary maize roots. In agreement with Oaks and Hirel (1985) we suppose that a definite assessment of the primary route in the assimilation of NH_4^+ ions in higher plants can be obtained only if a GDH or GS deficient mutant becomes available.

The specificity of different root tissues in the assimilation of NH_4^+ ions was studied by comparing the activities of glutamine synthetase and glutamate dehydrogenase in anatomically defined regions of the primary maize root. Both enzymes were active in the whole primary root. Some differences in the activity were found between the growth region and the ontogenetically older root tissues. The specific activity of GS was higher in the growth region while the activity of GDH was considerably higher in the older parts of root. Comparing the specific activities of GDH to GS, several times (3–32 fold) higher GS activity than GDH was observed. These results imply that, NH_4^+ ions absorbed by roots from the culture medium or produced by the reduction of NO_3^- ions in maize roots under normal conditions are primarily assimilated into organic compounds by the GS/GOGAT cycle and not by GDH.

Acknowledgement

I wish to thank Dr O Gašparíková for helpful discussion and kind reading of the manuscript.

References

Arima Y and Kumazawa K 1977 Evidence of ammonium assimilation via the glutamine synthetase–glutamate synthase system in rice seedling roots. Plant and Cell Physiol. 18, 1121–1129.

Bradford M 1976 A rapid and sensitive method for the quantitation of microgram quantities of protein utilizing the principle of protein-dye binding. Anal. Biochem. 72, 248.

Gašparíková O, Pšenáková T and Nižňanská A 1976 Influence of various nitrogen sources on the activity of nitrate and nitrite reductase and glutamate dehydrogenase in *Zea mays* roots. Biológia (Bratislava) 31, 527–535.

Givan C V 1979 Metabolic detoxification of ammonia in tissues of higher plants. Phytochem. 18, 375–382.

Izmaylov S F, Baskakova S Yu, Aseyeva K B, Cyupa G P and Smirnov A M 1982 Distribution of glutamine synthetase and glutamate dehydrogenase activity in organs of different plants (*In Russian*). Izv. Akad. Nauk 3, 321–331.

Lewis O A M, Chadwick S and Withers J 1983 The assimilation of ammonium by barley roots. Planta 159, 483–486.

Loyola-Vargas V M and Jimenez E S 1984 Differential role of glutamate dehydrogenase in nitrogen metabolism of maize tissues. Plant Physiol. 76, 536–540.

Luxová Mir and Gašparíková O 1984 Characterization of nitrate reductase, nitrite reductase and glutamate dehydrogenase distribution in the maize root (*Zea mays* L.). Biológia (Bratislava) 39, 265–272.

McNally S F and Hirel B 1983 Glutamine synthetase isoforms in higher plants. Physiol. Vég. 21, 761–769.

Miflin B J and Lea P J 1976 The pathway of nitrogen assimilation in plants. Phytochem. 15, 873–885.

Oaks A and Hirel B 1985 Nitrogen metabolism in roots. Ann. Rev. Plant Physiol. 36, 345–365.

Rhodes D, Rendon G A and Steward G R 1976 The regulation of ammonia assimilating enzymes in *Lemna minor*. Planta 129, 203–210.

Sahulka J and Lisá L 1978 The influence of sugars on nitrate reductase induction by exogenous nitrate or nitrite in excised *Pisum sativum* roots. Biol. Plant. 20, 359–367.

Sahulka J and Lisá L 1979 Regulation of glutamine synthetase level in isolated pea roots. I. Differential effects of ammonium salts in sugar supplied roots. Biochem. Physiol. Pflanzen 174, 646–652.

Wallace W 1973 The distribution and characteristics of nitrate reductase and glutamate dehydrogenase in the maize seedlings. Plant Physiol. 52, 191–196.

B. C. Loughman et al. (Eds.), Structural and functional aspects of transport in roots, 115–121.
© 1989 by Kluwer Academic Publishers.

A simulation model of root and shoot growth at different levels of nitrogen availability

LOUISE SPEK and MARCEL VAN OIJEN[1]
Department of Plant Ecology, University of Utrecht, Lange Nieuwstraat 106, 3512 PN, Utrecht, The Netherlands. [1] Correspondence address: Foundation for Agricultural Plant Breeding (SVP), Droevendaalsesteeg 1, 6700 AC Wageningen, The Netherlands

Key words: C-metabolism, growth, *Zea mays*, N-metabolism, simulation model

Abbreviations: AA, amino acids; C, soluble carbohydrates; SC, structural carbon (cell wall carbohydrates); SN, structural nitrogen (proteins); L, light intensity

Introduction

A simulation model is presented for root and shoot growth, in the vegetative stage, based on current knowledge of the nitrogen and carbon metabolism. Although the model is of general applicability, values for the various parameters are based on experiments with maize plants (van Oijen et al., 1986). Our prime interest in building the model was the elucidation of mechanisms underlying growth, N (re)distribution between plant organs and the regulation of the root–shoot ratio. Application of the model in ecophysiology and plant breeding was an objective in constructing the model. The model calculates the dynamics of uptake and transport processes, biochemical conver-

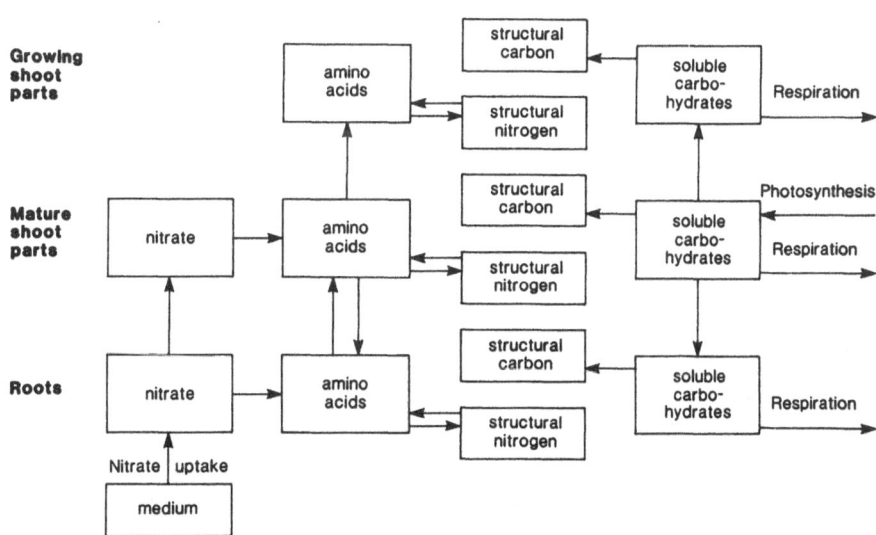

Fig. 1. Relational diagram of the model, showing all model quantities and processes, except those involved in extension growth (see Spek and van Oijen. *In press*). See text for further explanation.

sions, respiration and the concentrations of five groups of chemical compounds in the tissues, and the consequences for growth of roots and shoot at different levels of N-supply. As far as we are aware, no other models on this level of organization of biological systems have been presented. A more detailed description of the model is included in Spek and van Oijen (in press).

Model structure

Relational diagram

Figure 1 shows the basic structure of the model. The model plants are divided in three parts: growing shoot, mature shoot and roots. In the model program, the root system is divided into two equivalent parts, thus allowing the simulation of split-root experiments. Five different chemical compounds are distinguished (rectangles): nitrate (not in the growing shoot parts), amino acids (AA), structural nitrogen (SN = proteins), soluble carbohydrates (C) and structural carbon (SC = cell wall carbohydrates). Chemical conversions are indicated by horizontal arrows. The different plant parts are connected through the xylem and phloem (vertical arrows). The conversion and transport processes are integrated, giving the amounts of the compounds in the different organs. Input processes are nitrate uptake and photosynthesis. The photoperiod is $16\,h \cdot d^{-1}$, with a light intensity of $60\,W \cdot m^{-2}$. Temperature and relative humidity are kept constant for reasons of simplicity.

Basic model assumptions

A general equation, repeatedly used is:

$$d[y]/dt = Vmax \times minimum\ (f1, f2, \dots fn)$$

(1)

y = end product. Change in [y], due to the mentioned process only. Vmax = maximum rate of end product formation, chosen to be about twice as high as the actual rate, found under normal conditions. One or more functions (fi) in the second term, are described by a *Michaelis-Menten* equation, with a Km value, corresponding with concentrations in real plants. One of the functions (fi) may be a *feedback* by accumulated end products.

Quantification by means of Michaelis-Menten equations is according to generally accepted theories (Novoa, 1979; Novoa and Loomis, 1981).

a) Nitrate (net) uptake. Nitrate uptake rate has been made dependent on the nitrate concentration in the medium and the concentration of C in the root tissue.

$$d[NO_3^-]/dt = Vmax \times min.$$
$$(f1[C], f2[NO_3^-]medium)$$

see equation (1).
f1 is described by a Michaelis-Menten equation; f2: nitrate uptake is directly proportional to the nitrate concentration in the medium up to a certain low boundary level. Above this boundary level uptake rate is independent of the nitrate concentration.

b) Photosynthesis. The formulation of a light response curve of CO_2 assimilation after de Wit *et al.* (1978) was used. Photosynthesizing leaf area is calculated from mature shoot fresh weight by multiplication with a constant Specific Leaf Area.

c) Synthesis of amino acids. AA can be synthesized only in the mature shoot parts and in the roots. The synthesis has been made dependent on the NO_3^- the C and the AA concentration in the tissue, and in the mature shoot parts on the light intensity also.

$$d[AA]/dt = Vmax \times min.$$
$$(f1[NO_3^-], f2[C], f3[AA], f4[L])$$

see equation (1)
f1, f2 and f4 are described by Michaelis-Menten equations. f3 describes a linear relationship between AA synthesis and AA concentration. It represents a feedback mechanism.

d) Synthesis of structural nitrogen. Synthesis of SN takes place in all three plant parts. It has been made dependent on AA and C concentration in the tissues.

$$d[SN]/dt = Vmax \times min.(f1[AA], f2[C])$$

see equation (1)
f1 and f2 are described by Michaelis-Menten equations.

e) Synthesis of structural carbon. In all three plant parts, synthesis of SC can take place and it has been made dependent on the concentration of C and SC.

$$d[SC]/dt = Vmax \times f1[C] \times f2[SC]$$

f1 is described by a Michaelis-Menten equation. f2 is a linear function between the formation rate and SC concentration. It represents a feedback mechanism.

f) Degradation of structural nitrogen. $d[AA]/dt = k \times SN$, k = a constant specific protein degradation rate (gN/g/h).

g) Transport of nitrate and amino acids through the xylem. $d[y]/dt =$ transpiration rate $\times$ f[y], y = NO_3^- or amino acids, f is a linear function; transpiration = leaf area $\times$ f(L); f(L) is a Michaelis-Menten function of light.

h) Transport of soluble carbohydrates through the phloem. $d[C]/dt = f([C]source - [C]sink)$; source = mature shoot parts, sink = growing shoot parts or roots; f is a linear function.

i) Loading of amino acids into the phloem, for transport to the growing shoot parts and roots. $dAA/dt = AA(mature\ shoot) \times f1[C] \times f2[AA]phloem$; f1 is nonlinear, f2 is linear.

j) Unloading of amino acids from the phloem into the growing shoot parts and the roots. The AA transport has been coupled to the transport of C.

$$dAA/dt = f(amount\ of\ AA\ in\ the\ phloem,$$

transport rate of C in the phloem)

f is linear.

k) Respiration. This includes costs of C for:
(1) biochemical conversions
    $d[C]/dt$ = conversion rate $\times$ costs per g converted matter
(2) transport processes
    $d[C]/dt$ = transport rate $\times$ costs per g transported matter
(3) maintenance of gradients
    $d[C]/dt = f([C]tissue, dry\ matter)$, f is linear
Calculations for energy costs of the mentioned processes are based on calculations and data of Penning de Vries (1974) and Veen (1980).

l) Increase in dry weight. The sum of the five distinguished chemical components (NO_3^-, AA, SN, C and SC), is a measure for tissue dry weight.

Results and discussion

Three model experiments were performed in order to test the adequacy of the model. At the start of the simulation, two weeks old maize plants, well supplied with nutrients, were distributed over three different nitrogen treatments:

A. High nitrate supply ($+NO_3^-$).

B. Intermediate nitrate supply: a split-root system with only one half receiving nitrate ($+/-NO_3^-$).

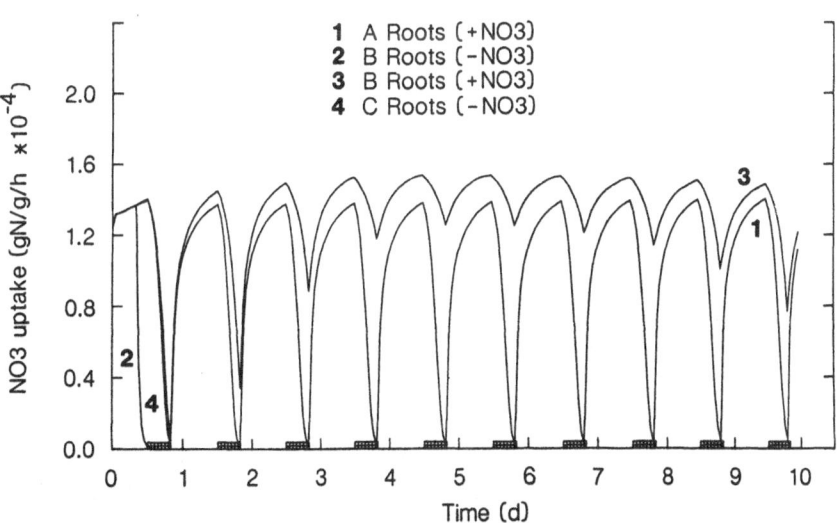

Fig. 2. Simulated nitrate uptake rate ($gN\,g^{-1}\,h^{-1}$), for three different levels of N availability.

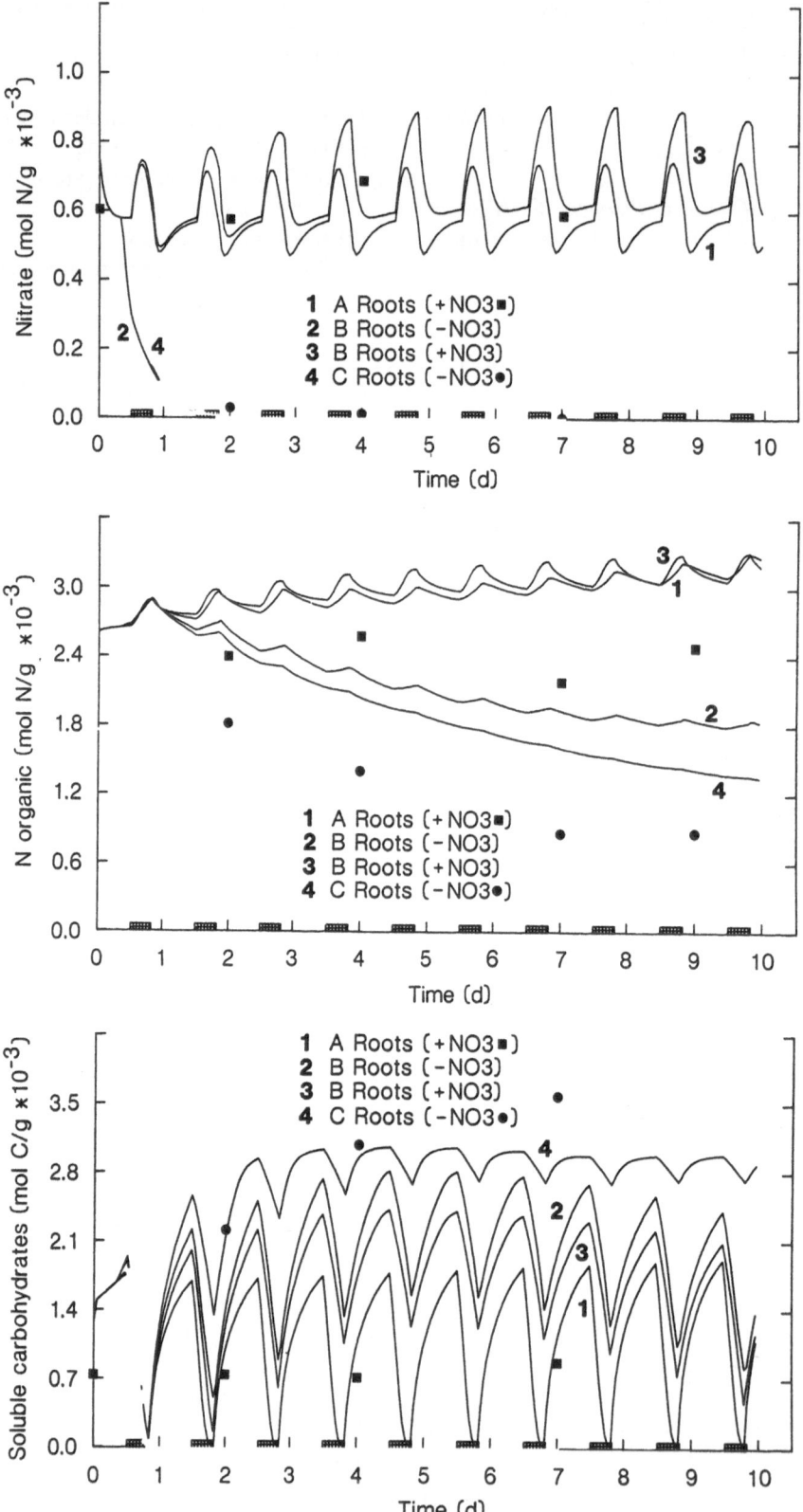

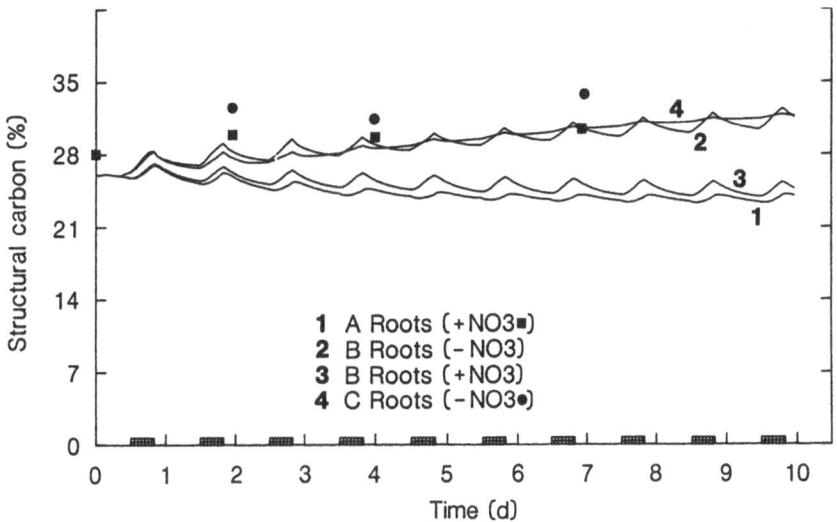

Fig. 3–6. Time courses for concentrations of NO_3^- (Fig. 3), organic N (Fig. 4), soluble carbohydrates (Fig. 5), and the percentage of structural carbon (dry matter, Fig. 6). Simulation: *continued lines*; experiment: ■ = $+ NO_3^-$; ● = $- NO_3^-$.

C. Nitrate omitted from the nutrient solution ($- NO_3^-$). In the C plant and in the $+ NO_3^-$ roots of the B plant, the NO_3^- accumulated was utilized within one day. The dynamics of all model processes and quantities were followed for ten days. Effects of separate parameter changes on total plant growth, are given.

Model experiments

A characteristic of the model results is the presence of pronounced daily fluctuations, the origin of which will be discussed.

Daily fluctuations in the NO_3^- uptake rate (Fig. 2, curve 1 and 2) are caused by daily fluctuations in C concentrations, as a consequence of variable transport of C to the roots. Without NO_3^- in the medium (Fig. 2, curve 2 and 4), uptake rate decreases to zero within two days. An increased NO_3^- uptake rate in the $+ NO_3^-$ roots (curve 3) of the B plant, is caused by an increased concentration of C in these roots.

Daily fluctuations in the transport rate of C are caused by photosynthesis. An increased transport rate of C to the $+ NO_3^-$ roots of the B plant, is due to an increased difference in C concentration

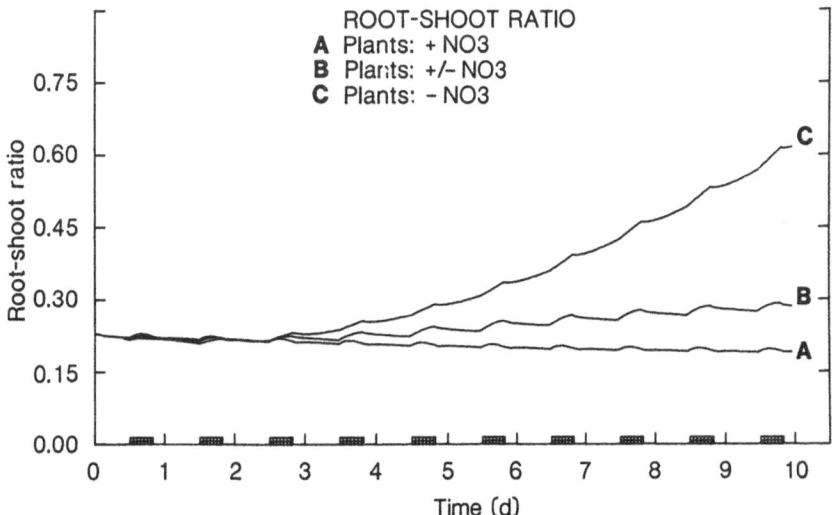

Fig. 7. Simulated root–shoot ratio, for three different levels of N availability.

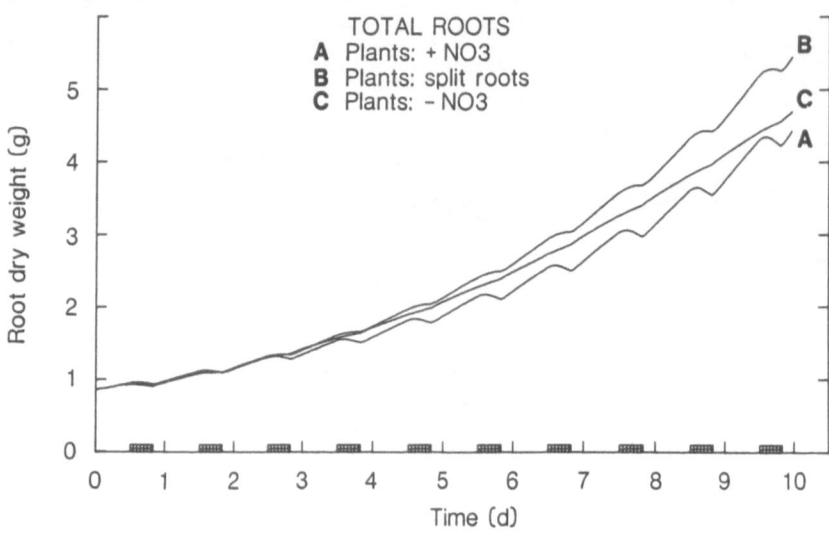

Fig. 8. Simulated root growth (dry weight), for three different levels of N availability.

between source and sink. A decreased difference in C concentration between source and sink caused a decrease in transport rate of C to the $- NO_3^-$ roots of the B plant and to the roots of the C plant. The NO_3^- taken up, can be transported to the mature shoot parts, assimilated into AA, or accumulated. One pool of NO_3^- and one pool of AA are distinguished. The transport rates of NO_3^- and AA in the xylem are generally highest during the light period,

due to the higher transpiration rate and adequate replacement of NO_3^- by uptake and of AA by nitrate reduction. Transport of NO_3^- from the $+ NO_3^-$ roots of the B plant is increased, due to an increased NO_3^- concentration in the roots. The $- NO_3^-$ roots of the B plant transport some AA in the xylem because of cycling and redistribution of AA in the plant, as contrasted with the C plant.

In the mature shoot, NO_3^- can accumulate or be

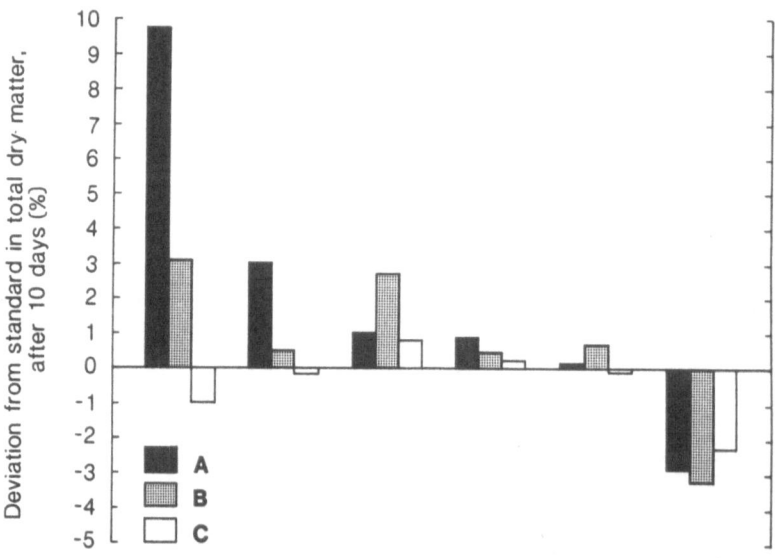

Fig. 9. Sensitivity analysis. Effects of six separate parameter changes on total dry weight. Increase with 10% of the maxima of: photosynthesis (**1**), NO_3^- uptake (**3**), amino acid formation (**5**) and structural nitrogen formation (**6**). Increase with 10% of the diffusion coefficient for transport of soluble carbohydrates through the phloem (**4**) and a decrease of 10% for respiration costs (**2**). The % deviation from the standard, ten days after the start of the simulation and parameter changes, is given. Standard values: A = 27.7 g, B = 24.5 g and C = 12.3 g.

assimilated into AA. The transported AA enter the AA pool directly. The AA pool, both in roots and mature shoot parts, is composed of AA derived from NO_3^- assimilation, imported AA and AA derived from SN degradation. In the growing shoot parts, the AA pool is composed of imported AA and AA derived from SN degradation, and there is no export of AA.

The dynamics in distribution and redistribution of N and C compounds, are mirrored by the dynamics in chemical conversions in the model plants. This leads to concentrations of NO_3^- organic nitrogen, C and percentages of SC in the roots, as given in Figs. 3–6. These figures show besides simulation results, data for real plants in a $+NO_3^-$ and $-NO_3^-$ situation.

Fresh weight growth of roots and shoot has been made dependent on the concentrations of SN and C, with a higher sensitivity for SN (at lower SN concentrations) in the shoot. This resulted in an increase in root–shoot ratio under $-NO_3^-$ conditions (Fig. 7).

The increase in dry weight in the root system under different levels of nitrogen supply, is shown in Fig. 8. An increased root growth in the $-NO_3^-$ plant, is also found in real experiments. Shoot growth in this model plant is strongly reduced, while in real plants, shoot growth was 50% reduced compared with the control. The increased root growth of the B plant, is due to an increased root growth of the $+NO_3^-$ roots, and a continued growth in the $-NO_3^-$ roots. These $-NO_3^-$ roots behave like a root system, completely deprived of NO_3^-. The shoot growth of the N plant is somewhat reduced.

Sensitivity analysis

Figure 9 shows the effect of six separately changed model parameters. The introduced parameter changes can be interpreted as changes in genetic properties of the plants. Most parameter changes cause an increase in total dry matter compared with the standard plants, but for the different N treatments, not in the same extent. Such model experiments can be used in ecophysiology to study growth reactions of different species at different levels of nutrient supply, and in plant breeding for selection of genotypes for certain environments.

Conclusions

The model calculations correspond in a qualitative way rather well with experimental data and the model appears to be quite stable. This means that a) the division into three plant parts (growing and mature shoot parts and roots) and b) the mechanisms simulated (chemical conversions and transport processes) can form a good basis for describing plant growth.

Acknowledgements

We gratefully remember Prof Dr R Brouwer, whose original ideas and enthusiastic support were of essential importance for the accomplishment of this simulation model. We thank Prof Dr H Lambers for critically reading the manuscript.

References

Novoa R 1979 A Preliminary Dynamic Model of Nitrogen Metabolism in Higher Plants. Thesis.

Novoa R and Loomis R S 1981 Nitrogen and plant production. Plant and Soil 58, 177–204.

Oijen M van, Spek L Y and Brouwer R 1986 A simulation model of growth and C and N metabolism in young maize plants. *In* Fundamental Ecological and Agricultural Aspects of Nitrogen Metabolism in Higher Plants. Eds. H. Lambers *et al.* pp. 323–327. Martinus Nijhoff Publishers, Dordrecht.

Penning de Vries F W T, Burnsting A H M and Laar H H 1974 Products, requirements and efficiency of biosynthesis: A quantitative approach. J. Theor. Biol. 45, 339–377.

Spek L Y and Oijen M van 0000 A simulation model of plant growth at different levels of nitrogen availability. *In press.*

Veen B 1980 Energy cost of ion transport. *In* Genetic engineering of osmoregulation. Impact on plant productivity for food, chemicals and energy. Eds. D W Rains *et al.* pp. 187–195. Plenum Press, New York, London.

Wit C T de *et al.* 1978 Simulation of assimilation, respiration and transpiration of crops. Simulation Monographs. PUDOC Wageningen.

B. C. Loughman et al. (Eds.), Structural and functional aspects of transport in roots, 123–125.
© 1989 by Kluwer Academic Publishers.

The relationship between phosphate absorption and root length in nine wheat cultivars

W. RÖMER, J. AUGUSTIN and G. SCHILLING
Martin-Luther-Universität Halle-Wittenberg, Sektion Pflanzenproduktion, Wissenschaftsbereich Agrochemie, DDR-4020 Halle/S., Adam-Kuckhoff-Str. 17 b, GDR

Key words: intensity, P uptake, pot experiment, root length, wheat cultivars

Introduction

Because the diffusion coefficients of potassium and especially of phosphate ions in the soil are very small (Barber, 1974) the uptake of these nutrients is extensively dependent on the concentration gradient and the diffusion conditions (*e.g.* water content) of the soils (Römer and Schilling, 1987). For this reason the nutrients mentioned can only be exploited from a small soil cylinder around the roots (Jungk, 1984). Therefore plant species with a large root system like grasses are superior to others *e.g.* leguminous plants with regard to phosphate uptake (Steffen, 1984).

Hence the question arises whether the phosphate uptake of different wheat idiotypes correlates with their root surface or whether the uptake intensity per surface unit — *e.g.* by a smaller Michaelis-constant of the carrier system — can additionally compensate for the disadvantages of small root systems. As the P-supply influences root growth (Böhm, 1974), these relations should be investigated under conditions of low as well as high P-supply.

Materials and methods

From each of the 9 cultivars 4 groups of 10 plants were cultivated in polyethylene vessels (50 cm deep, diameter 5.6 cm) on quartz sand in a phytotron. The seed weight was 45 ± 5 mg. Nutrients (6 mg P·pot^{-1} and 44 mg P·pot^{-1} respectively, 0.2 g N, 0.3 g K, 0.07 g Mg, micro-nutrients) were mixed with quartz sand (1560 g). Phosphorus was applied as $CaHPO_4 \cdot 2H_2O$.

After 3 weeks the following parameters were determined: dry matter of shoots and roots, their P-content and the total root length, both main and of lateral roots (Augustin, 1984). The P taken up by plants within 3 weeks is the difference between the P-content of the plants and that of the seeds. Hence, the P-uptake intensity per cm root length in 3 weeks could be computed (μg P·cm^{-1}.[3 weeks]$^{-1}$).

Results

Table 1 shows that a higher P-supply causes a higher P-uptake in all cultivars. The concentration of P in shoots on the low P-level ranged from 0.32 to 0.41% P in dry matter and on the high P-level

Table 1. P uptake (μg P plant^{-1}) of 9 wheat cultivars within 3 weeks

Cultivar	Low P level (6 mg P·pot^{-1})		High P level (44 mg P·pot^{-1})	
	μg P	relative	μg P	relative
Fanal	365	81	838	100
Remus	369	82	696	83
Iljitschjowka	351	78	672	80
Compal	341	76	669	80
Durum wheat	450	100	616	74
14/44	324	72	554	66
Almus	304	68	540	64
Fakta	362	80	514	61
15080	396	88	504	60
x̄	362	(100)	622	(172)
LSD 5% Tukey	43	10	77	9

Table 2. The dependence of total root length (cm/plant) and P-uptake intensity (μg P·cm^{-1} root. [3 weeks]$^{-1}$ in 9 wheat cultivars (high P-level = 44 mg P·pot^{-1}, low P-level = 6 mg P·pot^{-1})

Cultivar	Total root length				P-uptake intensity			
	High P level		Low P level		High P level		Low P level	
	cm·plant^{-1}	rel.	cm·plant^{-1}	rel.	μg·cm^{-1}	rel.	μg·cm^{-1}	rel.
Fanal	229	75	359	69	3.7	100	1.04	100
Remus	299	97	415	79	2.3	63	0.89	80
Iljitsch.	302	98	444	85	2.3	61	0.81	78
Compal	257	83	360	69	2.6	71	0.96	92
Durum w.	309	100	522	100	2.0	55	0.87	84
14/44	280	91	426	82	2.0	53	0.78	75
Almus	230	75	345	66	2.4	64	0.90	87
Fakta	226	73	363	69	2.3	62	1.04	100
15080	241	75	464	89	2.1	57	0.86	83
$\bar{x}$	263	(100)	410	(156)	2.4	(100)	0.91	(38)
LSD 5% Tukey	45	15	73	14	0.32	9	0.2	18

from 0.62 to 0.96% respectively. Therefore the experiment was principally suitable for testing the reaction of the cultivars with respect to their root growth and phosphate efficiency.

Table 2 shows that the P-supply significantly influenced the total root length and the uptake of P by the 9 cultivars. The root length of all cultivars increased at the lower P-supply at an average of 147 cm (from 263 to 410 cm), or 56%. Contrary to this the P-uptake intensity (μg P·cm^{-1}. [3 weeks]$^{-1}$ decreased by more than 60%. The influence of these relations on the total P uptake of plants was tested by a linear correlation analysis (Table 3). At the high P-level the relation between total P-uptake and uptake intensity was close (r = 0.88) while the root length was relatively less important (r = 0.39). At the low P-level the significance of the root length clearly increased, r increased from 0.39 to 0.5 while the relation between total P-uptake and uptake intensity per root unit became unimportant (r = 0.23).

These general tendencies were modified by the behaviour of the cultivars. It was interesting that some cultivars adapted themselves well to the low P-supply while others were incapable of doing so. Although the absolute P-uptake was generally reduced in case of the low P-supply, the cultivars Durum wheat and 15080 were able to increase their total P-uptake in comparison with other cultivars like Fanal or Compal (Table 1). Under the same conditions Fanal reduced its P-uptake from 100% to 81%. The other cultivars responded in a less distinct manner.

Fakta, Durum wheat and 15080 increased their absolute root length by 61 to 93% under conditions of low P-supply. Besides, all 3 cultivars increased their P uptake compared to Fanal (Table 1). Fanal increased its absolute root length only by 57%. Consequently it lost its leading position regarding the total P-uptake at the low P level in favour of Durum wheat.

Discussion and conclusions

From these results we must conclude that the root length per plant is not important for phosphate uptake at a high P-supply. Obviously high P-uptake intensities per root and time unit enable sufficient P-uptake even when the root system is small. But these results are only relevant for 3 week old plants. Nevertheless, for high yields a high P-uptake up to the shooting phase is decisive

Table 3. Relations between P uptake per plant and the 2 parameters root length and P-uptake intensity per cm root length (linear correlation coefficient r)

y	x	High P level	Low P level
Absorbed P amount per plant	Total root length (cm)	0.39	0.50[a]
	Intensity of P uptake (μg P cm^{-1} root length · 3 weeks)	0.88[a]	0.23

[a] Statistically significant at 1% level.

(Römer and Schilling, 1986). Under conditions of decreasing P-availability the P-absorbing root surface (root length) becomes more and more important. The cultivar spectrum showed that there are some varieties which can adapt themselves to this situation by extending their root system and/or developing a more effective P-uptake per root unit. If the heritability of such characteristics were high, types could be selected with a lower requirement for P-supply. Of course the adaptation capability should not be combined with a low yield level.

References

Augustin J 1984 Untersuchungen zur Beeinflussung des Wurzelwachstums von Weizenkeimpflanzen durch exogene und endogene Faktoren. Diss. A der Landw. Fakultät Halle/S.

Barber S A 1974 Influence of the plant root on ion movement in soil. *In* The Plant Root and its Environment, pp. 525–564. University Press of Virginia.

Böhm W 1974 Phosphatdüngung und Wurzelwachstum. Die Phosphorsäure 30, 141–157.

Jungk A 1984 Phosphatdynamik in der Rhizosphäre und Phosphatverfügbarkeit für Pflanzen. Die Bodenkultur 35, 99–107.

Römer W and Schilling G 1986 Phosphorus requirement of the wheat plant in various stages of its life cycle. Plant and Soil 91, 221–229.

Römer W and Schilling G 1987 Das diffusible Phosphat des Bodens und die Abhängigkeit des Phosphatdiffusionskoeffizienten von exogenen Faktoren. Arch. Acker. Pflanzenbau Bodenk. 32, 115–122.

Steffen D 1984 Wurzelstudien und Phosphat-Aufnahme von Weidelgras und Rotklee unter Feldbedingungen. Z. Pflanzenernaehr. Bodenk. 147, 85–97.

B. C. Loughman et al. (Eds.), Structural and functional aspects of transport in roots, 127–129.
© 1989 by Kluwer Academic Publishers.

Characteristics of NO₃⁻ uptake in Lemna and Pisum

P. OSCARSON, B. INGEMARSSON, M. AF UGGLAS[1] and C.-M. LARSSON

Department of Botany, University of Stockholm, S-106 91 Stockholm, Sweden. [1] Research Institute of Physics, Frescativ. 24, S-104 05 Stockholm, Sweden

Key words: efflux, influx, kinetics, net uptake, nitrate

Abbreviations: RGR, relative growth rate; R_N, relative nitrogen addition rate

Introduction

The fluxes of NO₃⁻ over the plasmalemma influence both NO₃⁻ reduction and storage, as well as the overall capacity of the plants to utilize external NO₃⁻. In earlier papers on Pisum and Lemna (Ingemarsson et al., 1984; Ingemarsson, 1987; Oscarson and Larsson, 1986) we have shown that actual nitrate reduction correlates closely with changes in net uptake. Morgan et al. (1985), using $^{15}NO_3^-$, showed that recently reduced NO₃⁻ originated mainly from NO₃⁻ recently taken up, whereas previously stored NO₃⁻ was predominantly translocated or effluxed. Thus, the flux of NO₃⁻ over the plasmalemma seems to be of great importance for further processes involved in NO₃⁻ assimilation.

It is often reported that when deprived of external NO₃⁻, plants respond with an initially increased capacity for uptake, which again declines with time (Clarkson, 1986). Variations in uptake have been attributed to regulation of carrier activity of intracellular NO₃⁻ (Jackson et al., 1976), changes in number of active carriers (Clarkson, 1986; Lee and Drew, 1986), or an altered relation between the influx and efflux components of net NO₃⁻ uptake (Breteler and Nissen, 1982; Deane-Drummond and Glass, 1983; Deane-Drummond, 1984).

Concentration kinetics of NO₃⁻ uptake rates reveal the capacity of the plant to utilize external NO₃⁻. However, the interpretation is often equivocal due to an unknown efflux component during uptake. This paper describes the nitrate uptake properties in Lemna and Pisum, following long term acclimation of the plant to different levels of N limitation. Attention has been paid both to net uptake and influx characteristics.

Materials and methods

Plant culture

Seeds of *Pisum sativum* L. cv. Marma were germinated for six days, the cotyledons were removed and the plants were then placed in a liquid growth medium containing all necessary ions nutrients except for NO₃⁻ for ten days. From day 16 and onwards the plants were given daily additions of NO₃⁻, calculated to exponentially increase the total N-content of the plant by a preset relative nitrogen addition rate (R_N). The daily addition was calculated from the formula $N_t = N_o \cdot e^{R_N \cdot t}$, where N_t is the N-content after one day, N_o the initial N-content, and R_N the chosen addition rate. Thus, the daily addition is given by subtracting N_o from N_t. About ten days after the start of the NO₃⁻ additions the relative growth rate (RGR) is stable (Oscarson and Larsson, 1986).

Cultures of *Lemna gibba* L. were grown according to the same principle as used for Pisum. In this manner, Lemna cultures can be maintained at stable growth rates for several months (Ingemarsson et al., 1984). For further details concerning culturing see Ingemarsson *et al.* (1984; Lemna), and Oscarson and Larsson (1986; Pisum).

NO_3^- uptake measurements

Net uptake of NO_3^- was measured spectrophotometrically as the difference in absorbance between 202 and 250 nm. Influx was measured in short-term experiments using radioactive $^{13}NO_3^-$ as a tracer. Both influx and net uptake could be recorded continuously and simultaneously in depletion experiments (see Ingemarsson *et al.*, 1987; Oscarson *et al.*, 1987). Net uptake kinetics were calculated employing Lineweaver-Burke plots of net uptake rates at six different external concentrations of NO_3^- (10, 15, 25, 50, 75 and 100 μM).

Results and discussion

Growth

At suboptimal relative nitrogen addition rates, RGR was well correlated with R_N, as can be seen in Fig. 1. In the figure can also be seen that growth of Lemna deviates from the expected rate at the highest R_N levels used, where growth rather indicates the maximal RGR under prevailing conditions, *i.e.* when some other factor other than N limits growth. It is also apparent that with de-

Table 1. Root to shoot (frond) dry weight ratios in Pisum and Lemna at different R_N. Also shown are the net uptake rates in percent of influx, measured at 40–60 μM NO_3^- with $^{13}NO_3^-$ as an influx tracer (flux experiments from Oscarson *et al.*, 1987 concerning Pisum; Ingemarsson *et al.*, 1987 concerning Lemna)

Pisum			Lemna		
R_N	Root: shoot	Net uptake, % of influx	R_N	Root: frond	Net uptake, % of influx
0.03	1.19		0.05	0.83	
0.06	1.08	96	0.10	0.57	
0.09	0.98		0.15	0.52	
0.12	0.81	83	0.20	0.41	98
0.15	0.80		0.30	0.31	
0.18	0.80		0.40	0.20	

creased R_N, the root/shoot ratio (or root/frond in the case of Lemna) increases, which is a well known response to N limitation (Table 1).

Uptake of nitrate

The capacity for net NO_3^- uptake, *i.e.* V_{max}, increases in both species with increasing R_N; however, at the highest R_N used for Lemna, a decline in V_{max} was noted (Fig. 2). There was no marked tendency for the K_s values to be dependent on R_N in either species. The data on net uptake are compared with previous estimates of the unidirectional flux components in Table 1. In Lemna growing at R_N 0.20 day^{-1}, there is virtually no differences between influx and net uptake of NO_3^- when assayed at external NO_3^- concentrations of about 50 μM, indicating a negligible efflux component. Similar results were obtained with high NO_3^--grown Lemna, and with NO_3^--preloaded or non-loaded Lemna with WO_4^{2-}-inactivated nitrate reductase (Ingemarsson *et al.*, 1987). It is, thus, concluded that plasmalemma transport of NO_3^- in Lemna at external concentrations approaching saturation is essentially unidirectional, irrespective of plant N-status. In Pisum, influx clearly exceeds net uptake when measured at about 50 μM external NO_3^-, demonstrating an efflux component which increases with R_N. However, the relation between V_{max} and R_N in Pisum cannot be explained by differences in efflux, but must involve a major response of influx to R_N.

Thus, long-term acclimation of Pisum and Lemna to constant relative rates of N-supply

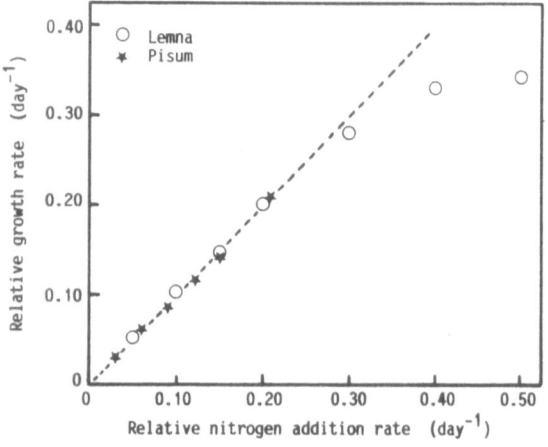

Fig. 1. Relative growth rate (RGR) of Lemna and Pisum plotted against the relative nitrogen addition rate (R_N). RGR of Pisum is calculated in the interval 8–12 days after start of NO_3^- additions. *The dotted line* represents the ideal response.

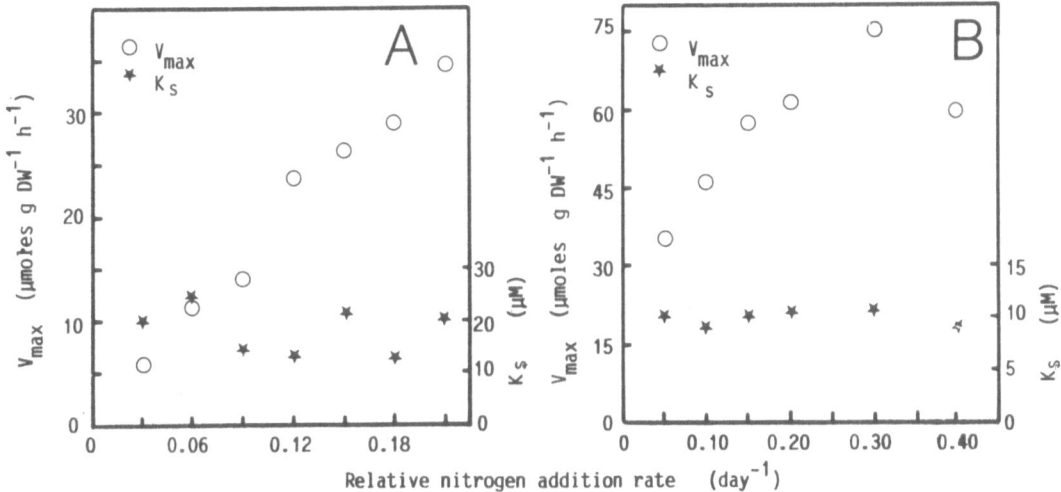

Fig. 2. V$_{max}$ and K$_s$ values for net NO$_3^-$ uptake in Pisum (**A**) and Lemna (**B**) at different relative nitrogen addition rates. Note the different scales.

affects the uptake system in a manner where the uptake capacity becomes proportional to R$_N$ and RGR, mainly as a consequence of changes in influx. In both species, the decreasing V$_{max}$ at decreasing R$_N$ may reflect a decreased number of active carriers in the plasmalemma. As to the drop in V$_{max}$ in Lemna when R$_N$ > RGR, we have no clear explanation yet. The affinity of the nitrate carrier does not seem to change with changed R$_N$.

Acknowledgement

This work was supported by the Swedish Natural Science Research Council.

References

Breteler H and Nissen P 1982 Effects of exogenous and endogenous nitrate concentration on nitrate utilization by dwarf bean. Plant Physiol. 79, 754–759.

Clarkson D T 1986 Regulation of the absorption and release of nitrate by plant cells. A review of current ideas and methodology. *In* Developments in Plant Soil Sciences Vol. 19. Eds. H Lambers *et al.* pp. 3–27. Martinus Nijhoff Publishers, Dordrecht.

Deane-Drummond C E 1984 Mechanism of nitrate uptake into *Chara corallina* cells: Lack of evidence for obligatory coupling to proton pump and a new NO$_3^-$/NO$_3^-$ exchange model. Plant Cell Environ. 7, 317–323.

Deane-Drummond C E 1986 Nitrate uptake into *Pisum sativum* L. cv. Feltham First seedlings: Commonality with nitrate uptake into *Chara corallina* and *Hordeum vulgare* through a substrate cycling model. Plant Cell Environ. 9, 41–48.

Deane-Drummond C E and Glass A D M 1983 Short-term studies of nitrate uptake into barley plants using ion specific electrodes and ^{36}ClO$_3^-$. I. Control of net uptake by NO$_3^-$ efflux. Plant Physiol. 73, 105–110.

Ingemarsson B 1987 Nitrogen utilization in Lemna. I. Relations between net nitrate flux, nitrate reduction, and in vitro activity and stability of nitrate reductase. Plant Physiol. 85, 856–859.

Ingemarsson B, Johansson L and Larsson C-M 1984 Photosynthesis and nitrogen utilization in exponentially growing nitrogen-limited cultures of *Lemna gibba*. Physiol. Plant. 62, 363–369.

Ingemarsson B, Oscarson P, af Ugglas M and Larsson C-M 1987 Nitrogen utilization in Lemna. II. Studies of nitrate uptake using ^{13}NO$_3^-$. Plant Physiol. 85, 860–864.

Jackson W A, Kwik K D, Volk R J and Butz R G 1976 Nitrate influx and efflux by intact wheat seedlings. Effects of prior nutrition. Planta 132, 149–156.

Lee R B and Drew M C 1986 Nitrogen-13 studies of nitrate fluxes in barley roots: Effect of plant N-status on kinetic parameters of nitrate influx. J. Exp. Bot. 37, 1768–1779.

Morgan M A, Jackson W A and Volk R J 1985 Concentration dependence of the nitrate assimilation pathway in maize roots. Plant Sci. 38, 185–191.

Oscarson P and Larsson C-M 1986 Relations between uptake and utilization of NO$_3^-$ in Pisum growing exponentially under nitrogen limitation. Physiol. Plant. 67, 109–117.

Oscarson P, Ingemarsson B, af Ugglas M and Larsson C-M 1987 Short-term studies of NO$_3^-$ uptake in Pisum using ^{13}NO$_3^-$. Planta 170, 550–555.

B. C. Loughman et al. (Eds.), Structural and functional aspects of transport in roots, 131–135.
© 1989 by Kluwer Academic Publishers.

Variation in the rate of root respiration of two *Carex* species: A comparison of four related methods to determine the energy requirements for growth, maintenance and ion uptake

HANS LAMBERS and ADRIE VAN DER WERF
Department of Plant Ecology, University of Utrecht, Lange Nieuwstraat 106, 3512 PN Utrecht, The Netherlands

Introduction

A significant portion (12–29%) of the carbohydrate produced daily in photosynthesis is respired in the roots. This value varies with age of the plants, growth conditions and the species under investigation (Lambers, 1987). Not only the proportion of photosynthates used in root respiration, but also the rate of root respiration, measured as oxygen uptake, varies two- to threefold between species, even when plants are grown under similar conditions (Lambers *et al.*, 1983). This variation does not correlate with that of the alternative path in root respiration, though there are major variations of this parameter as well (Lambers *et al.*, 1983). What then could explain these differences in the rate of respiration?

In this paper we will analyse the data on root respiration, root growth and anion uptake for two *Carex* species, known to differ in their rate of root respiration. In an attempt to explain such differences in respiration rate we have measured the rate of accumulation of fresh and dry weight and of total nitrogen in both roots and shoots and correlated the parameters derived from this analysis with the rate of root respiration. This allows the determination of the energy requirement for growth, maintenance and ion uptake.

One of the methods used in the present paper is that developed by Lambers and Steingröver (1978) and Szaniawski (1981), who attempted to separate two components of respiration only: the growth and the maintenance component. A slight modification of this method, taking account of variation in the contribution of the alternative, non-phosphorylating path in root respiration, has been described by Blacquière (1987). A more elegant

method, separating the three components of respiration, namely growth, maintenance and ion uptake, has been introduced by Veen (1980). A slight modification of this method has subsequently been described in detail by Van der Werf *et al.* (1988). It takes account of any ontogenetic drift in the contribution of the alternative, non-phosphorylating path in root respiration.

The aim of the present paper is to compare the two-component-approach as used in Lambers and Steingröver (1978) and in Szaniawski (1981), with the three-component-approach introduced by Veen (1980). In addition we will compare results obtained using the methods used by Blacquière (1987) and by Van der Werf *et al.* (1988) with those based on a somewhat simpler approach, ignoring the participation of the alternative path.

Theory

The rate of ATP production in the roots, to be estimted from the rate of respiration and the contribution of the alternative path, depends on three major energy-requiring processes, *i.e.* maintenance of root biomass, root growth and ion uptake. The overall equation can be described as:

$$r_{ATP} = m_{ATP} + 1/Y_{ATP}^{GR} \cdot RGR + 1/U_{ATP}^{I} \cdot NIR$$

where r_{ATP} is the rate of ATP production in root respiration; m_{ATP} is the rate of ATP consumption for the maintenance of root biomass; $1/Y_{ATP}^{GR}$ is the ATP requirement for the synthesis of cell material; RGR is the relative growth rate of the roots; U_{ATP}^{I} is the ATP requirement for ion uptake; NIR is the net rate of uptake of anions. (The terminology is based upon that given in De Visser and Lambers

Table 1. Definitions of terms used in calculations of the rate of oxygen consumption and ATP production for maintenance, synthesis of root cell material and anion uptake. The terms used here have been described by De Visser and Lamberg (1983) or Blacquière (1988), or derived therefrom

Term	Description	Units
r_{ATP}	Rate of ATP-production in respiration	$mmol\,ATP\,(g\,FW)^{-1}\,d^{-1}$
m	Rate of O_2-consumption to produce ATP for maintenance of biomass	$mmol\,O_2\,(g\,FW)^{-1}\,d^{-1}$
m_{ATP}	Rate of ATP-consumption in respiration for the maintenance of biomass	$mmol\,ATP\,(g\,FW)^{-1}\,d^{-1}$
$1/Y_{EG}$	The amount of oxygen consumed to produce ATP for synthesis of cell material	$mol\,O_2\,(g\,FW)^{-1}$
$1/Y_{ATP}^{GR}$	ATP-requirement for the synthesis cell material	$mol\,ATP\,(g\,FW)^{-1}$
RGR	Relative growth rate	$mg\,(FW)\,(g\,FW)^{-1}\,d^{-1}$
$1/U_I$	O_2-requirement for ion uptake	$mmol\,O_2\,(mmol\,ions)^{-1}$
$1/U_{ATP}^I$	ATP-requirement for ion uptake	$mmol\,ATP\,(mmol\,ions)^{-1}$
NIR	Net rate of anion uptake	$mmol\,(g\,FW)^{-1}\,d^{-1}$

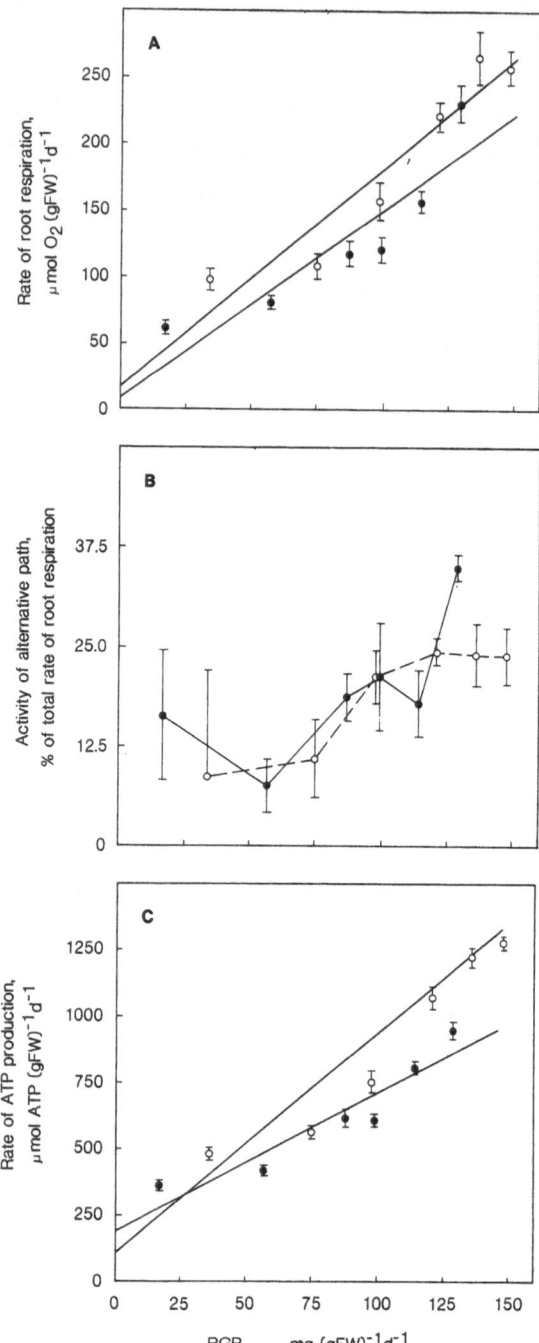

Fig. 1. The rate of root respiration (**a**), the relative contribution of the alternative, non-phosphorylating respiratory pathway (**b**), and the rate of ATP-production (**c**), calculated from the data in a and b, of *Carex acutiformis* (●) and *C. diandra* (○). *The slope of the lines* in a and c yield $1/Y_{EG}$ and $1/Y_{ATP}^{GR}$, respectively; the intercepts with the y-axis yield m and m_{ATP}, respectively. For further explanation, see Theory. Based on original data, presented in Van der Werf *et al.* (1988).

(1983); units of the various terms in this equation are included in Table 1.)

The rate of ATP production for maintenance of root biomass is assumed to be linearly related to the fresh weight of the root biomass to be maintained. Similarly, it is assumed that the rate of ATP production for anion uptake is proportional to the amount of anions taken up, whereas that for root growth is assumed to be proportional to the relative growth rate of the roots, provided the chemical composition of the root biomass is constant during the experimental period.

Based on these assumptions, the rate of ATP production in roots can be related to the root weight, the relative growth rate of the roots and the rate of ion uptake by the roots, using a multiple regression analysis (Van der Werf *et al.* 1988).

Table 2. Mean values for m (μmol O$_2$ (g FW)$^{-1}$ d^{-1}) and m$_{ATP}$ (μmol ATP (g FW)$^{-1}$ d^{-1}), for $1/Y_{EG}$ (mmol O$_2$ (g FW)$^{-1}$) and $1/Y_{ATP}^{GR}$ (mmol ATP (g FW)$^{-1}$), and for $1/U_I$ (mmol O$_2$ (mmol ions)$^{-1}$) and $1/U_{ATP}^I$ (mmol ATP (mmol ions)$^{-1}$) (SE in brackets). *Top left part*: two components, without correction for the alternative path; *top right part*: two components, with correction for the alternative path; *bottom left*: three components, without correction for the alternative path; *bottom right*: three components, with correction for the alternative path. For further explanation, see text

Item	Species		Item	Species	
	C. acutiformis	*C. diandra*		*C. acutiformis*	*C. diandra*
m	10.0 (32.7)	19.0 (37.2)	m$_{ATP}$	193 (108)	110 (150)
$1/Y_{EG}$	1.42 (0.16)	1.64 (0.16)	$1/Y_{ATP}^{GR}$	5.36 (0.54)	7.74 (0.68)
$1/U_I$	–	–	$1/U_{ATP}^I$	–	–
r^2	0.71	0.75	r^2	0.77	0.79
m	17.9 (31.4)	62.8 (33.1)	m$_{ATP}$	221 (103)	322 (134)
$1/Y_{EG}$	0.91 (0.31)	0.75 (0.32)	$1/Y_{ATP}^{GR}$	3.7 (1.0)	3.4 (1.3)
$1/U_I$	1.39 (0.73)	1.29 (0.41)	$1/U_{ATP}^I$	3.4 (1.8)	4.7 (1.2)
r^2	0.75	0.81	r^2	0.80	0.85

Materials and methods

Carex acutiformis Ehrh. and *Carex diandra* Schrank were grown from cuttings in nutrient solutions as fully described by Van der Werf *et al.* (1988). Respiration was measured polarographically and the contribution of the alternative path in root respiration was derived from the inhibition by the specific inhibitor salicylhydroxamic acid (SHAM). SHAM was applied in a concentration which was carefully chosen on the basis of preliminary experiments using a range of concentrations between 0–25 m*M*. This method, and details of chemical analyses are included in Van der Werf *et al.* (1988).

Results and discussion

Figure 1a gives an analysis of root respiration as dependent on both RGR of the roots and root weight for the two *Carex* species in its simplest form, as carried out before by Szaniawski (1981). The slope of such a graph yields $1/Y_{EG}$, whilst the intercept with the y-axis gives m. (For an explanation of the terms, see Table 1). The values obtained with this method, including further statistical information, is included in Table 2. This method does not yield separate values for $1/U_I$; costs for ion uptake are comprised in those for growth.

Since the contribution of the alternative path decreases with increasing age of the plant, and thus with decreasing RGR of the roots (Fig. 1b), the rate of ATP production in the roots is not proportional to that of root respiration. At a high RGR less ATP is produced per mol of oxygen consumed. Thus, the slopes of the lines in Fig. 1a are actually too steep, in comparison with those expected if the decrease in the contribution of the alternative path with age is taken into account.

Figure 1c presents an analysis similar to the one in Fig. 1a, but now the ontogenetic drift of the relative contribution of the alternative path in root respiration is taken into account. Such an analysis has been carried out before with roots of *Plantago lanceolata* and *P. major* (Blacquière, 1987). The slopes of the lines in Fig. 1c yield $1/Y_{ATP}^{GR}$ and the intercepts give m$_{ATP}$. This correction increases r^2 of the regression equation (Table 2; top part). No values can be obtained for $1/U_{ATP}^I$; values for ion uptake are included in $1/Y_{ATP}^{GR}$.

If six moles of ATP are produced per mol of oxygen consumed, the ratio of the values for $1/Y_{ATP}^{GR}$ and $1/Y_{EG}$ might be expected to be 6. In reality, lower ratio's are found, illustrating the point made above, that the slope of the lines in Fig. 1a is too steep.

When three components of respiration are used in the regression analysis, as explained in Theory, information may be obtained about the respiratory requirement for anion uptake (Table 2). In comparison with the method used to obtain the values in the left top part of the table, r^2 has improved, but is still fairly low. This method is essentially the same as used by Veen (1980). The values for m, obtained with this method, are higher than those obtained with the method used in Fig. 1a (Table 2, left part). This is due to the fact that anion uptake and growth were poorly correlated, especially in *C. diandra*. When the RGR was low, the net rate of anion uptake was relatively high; it decreased with increasing RGR, reaching a minimum around RGR = 75 mg g^{-1}d^{-1}, but increased again at higher RGR. This yields the somewhat curvilinear relationship, as found in both Fig. 1a and Fig. 1c. A linear regression analysis may thus result in underestimated values for the maintenance requirement. The ontogenetic change in the ratio of RGR and ion uptake was not as great in *C. acutiformis*, so that m, obtained with the two methods, was more similar than in *C. diandra*. Whenever the ratio of RGR of the roots and the rate of anion uptake by the roots is approximately constant throughout the experimental period, m should be the same no matter which method of the two is used. Moreover, if such a very good correlation exists, it is not possible to separate the components for growth and for ion uptake with this method.

The best analysis is given in the right bottom part of Table 2, where three components are taken into account and any ontogenetic drift of the alternative path has been acknowledged. This method is not only better because it has a sounder physiological basis, avoiding any biasses discussed above, but also since the statistics have improved considerably compared with any of the simpler approaches. The analysis is graphically presented in a three-dimensional graph in Fig. 2, for the data of *Carex diandra*. The plane in Fig. 2 gives the rate of ATP production in root respiration (y-axis) as a function of both the relative growth rate of the roots (z-axis) and the net rate of anion uptake by the roots (x-axis). Since the ATP requirement for maintenance, expressed per g root fresh weight, is assumed to be constant (see Theory), it appears as the intercept of the plane with the y-axis in Fig. 2. The plane intersects both the y-z plane and the y-x plane. The slope of the plane with the z-axis and the x-axis gives $1/Y_{ATP}^{GR}$, and $1/U_{ATP}^{I}$, respectively.

Figure 3 gives the rate of ATP production for growth and for ion uptake as dependent on RGR. Due to one of the assumptions in this analysis, ATP production for root growth increases linearly with increasing RGR of the roots. Since the slow growing, older plants absorbed anions at a much slower rate, the rate of ATP production for anion uptake was considerably less at low than at high RGR, especially in *C. diandra*. The ratio of ATP consumption for growth and that for anion uptake showed a maximum for both *Carex* species. This is not due to any change in the specific *costs* for either of these two processes (these were assumed to be constant in the present analysis), but rather to

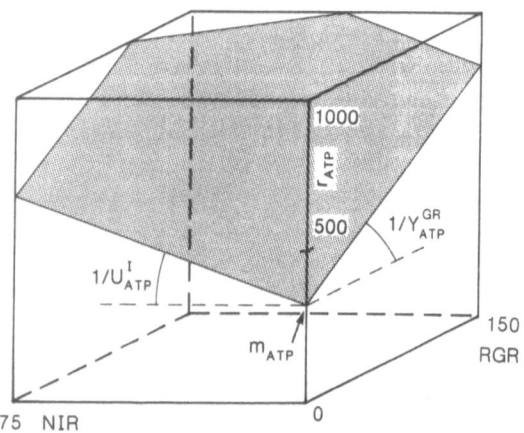

Fig. 2. The total rate of ATP production (μmol (g FW)$^{-1}$)d^{-1}) in the roots of *Carex diandra* as related to both the relative growth rate (RGR) of the roots (mg g^{-1}d^{-1}) and the net rate of anion uptake (NIR) by the roots (μmol g^{-1}d^{-1}). The intercept of the shaded plane with the y-axis gives m$_{ATP}$. *The slope of the shaded plane with the y-z plane gives $1/Y_{ATP}^{GR}$*; when projected on the x-y plane, the slope gives $1/U_{ATP}^{I}$. For further explanation, see text.

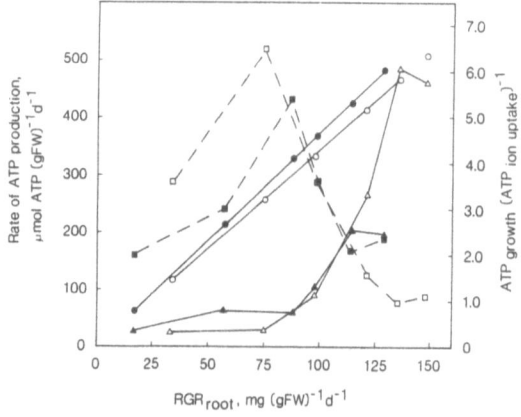

Fig. 3. The rate of ATP production for root growth (O and ●) and for ion uptake by the roots (△ and ▲) in *Carex acutiformis* (*closed symbols*) and *C. diandra* (*open symbols*). *The broken line* gives the ratio of the rate of ATP production for growth and that for ion uptake. Values were obtained with the method of Van der Werf *et al.* (1988), referred to in the right bottom part of Table 2.

changes in the specific *rates* of growth and anion uptake.

Concluding remarks

The method of Van der Werf *et al.* (1988), more so than the other three, gives further information on the relative costs of energy requiring processes in roots of higher plants. The method might be further elaborated and include more major energy requiring components, *e.g.* N_2-fixation, provided the rate of this process and that of any of the three components used so far, do not strongly correlate throughout the experimental period. Clearly, the basic data on respiration, growth and anion uptake need to be very reliable for any reliable coefficients to be derived.

It should be kept in mind that the present method, as any other used so far in this area of research, is based on a number of assumptions. Reasonable as these assumptions may appear, it should be stressed that no experimental data are available to support that the maintenance requirement and the specific costs for ion uptake do not vary with age. If they do, they will bias the results in a manner which is not quite predictable. Further investigation of the basic assumptions, as well as a comparison of the data obtained with the present approach with those obtained in a completely different way, is therefore a necessity.

Acknowledgements

We wish to thank our colleagues Tadaki Hirose, Henk Konings and Hendrik Poorter for valuable comments on this manuscript.

References

Blacquière T 1987 Ammonium nutrition in *Plantago lanceolata* L. and *P. major* L. ssp. *major*. II. Efficiency of root respiration and growth. Comparison of measured and theoretical values of growth respiration. Plant Physiol. Biochem. 25, 775–785.

De Visser R and Lambers H 1983 Growth and the efficiency of root respiration of *Pisum sativum* as dependent on the source of nitrogen. Physiol. Plant. 58, 533–543.

Lambers H 1987 Growth, respiration, exudation and symbiotic associations: The fate of carbon translocated to the roots. *In* Root Development and Function — Effects of the Physical Environment. Eds. J Gregory *et al.* pp 125–145. Cambridge University Press, Cambridge.

Lambers H, Day D A and Azcón-Bieto J 1983 Cyanide-resistant respiration in roots and leaves: Measurements with intact tissues and isolated mitochondria. Physiol. Plant. 58, 148–154.

Szaniawski R K 1981 Growth and maintenance respiration of shoots and roots in Scots pine seedlings. Z. Pflanzenphysiol. 101, 391–398.

Van der Werf A, Kooijman A, Welschen R and Lambers H 1988 Respiratory costs for the maintenance of biomass, for growth and for ion uptake in roots of *Carex diandra* and *Carex acutiformis*. Physiol. Plant. 72, 483–491.

Veen B W 1980 Energy costs of ion transport. *In* Genetic Engineering of osmoregulation: Impact on Plant Productivity for Food, Chemicals and Energy. Eds. D W Rains *et al.* pp 187–195. Plenum, New York.

Session 3

Absorption and transport of water

B. C. Loughman et al. (Eds.), Structural and functional aspects of transport in roots, 139–145.
© 1989 by Kluwer Academic Publishers.

Water transport in roots

ERNST STEUDLE

Lehrstuhl für Pflanzenökologie, Universität Bayreuth, Universitätsstraße 30, D-8580 Bayreuth, FRG

Key words: hydraulic conductivity, permeability coefficients, pressure probe, reflection coefficients, root pressure, water relations

Introduction

The hydraulic conductance of the root ($Lp_r \cdot A_r$; Lp_r = hydraulic conductivity per m^2 of root surface, A_r) is a rather complex and variable parameter which depends on the pattern by which different parts of the root or root zones contribute to the overall conductance in relation to their developmental state. The different tissues in the root (rhizodermis, cortex, endodermis, and stele) as well as the xylem system are arranged in series and should add to the overall resistance. Furthermore, in each of the tissues water may move in three parallel pathways, namely, in the apoplasmic, the symplasmic, and the transcellular path, whereby the latter two components should be summarized as the cell-to-cell path, because they cannot be measured separately to date. Experimental evidence indicates that water flow in the root is also related to active or passive solute (nutrient) flow (see reviews: Anderson, 1976; Pitman, 1982; Weatherley, 1982). The interactions between flows could, in principle, be due to either changes in the osmotic gradients created by solute flow at the different barriers within the root or to direct interactions between flows (*e.g.*, by solvent drag). Thus, effects of inhibitors, hormones, pollutants *etc.* on water transport may result either from direct effects on Lp_r (*e.g.*, by affecting the Lp of cell membranes) or from indirect effects *via* a coupling between water and solute flow.

The formalism of irreversible thermodynamics may be used for a quantitative description of the phenomena. Series and parallel arrangements of membrane-like barriers such as in the root may, in principle, be also described by this theory (*cf.* Kedem and Katchalsky, 1963a; b). However, this application involves adequate techniques for measuring force/flow relations in roots. Recently, the root pressure probe has been introduced which permits the determination of root pressure in excised roots and the concomitant measurement of water flows between root xylem and the root surroundings (Steudle and Jeschke, 1983; Steudle *et al.*, 1987). Permeability (P_{sr}) and reflection (σ_{sr}) coefficients of roots can also be determined besides water movements and, in principle, also active solute uptake into roots can be followed. The following article summarizes some of the results obtained so far using the new technique.

Materials and methods

In the root pressure probe technique an entire root system or part of a root is tightly connected with a pressure probe (Fig. 1) which is similar to the probe used for measuring water relations of individual cells (*cf.* Steudle and Zimmermann, 1984; Zimmermann and Steudle, 1978). The whole system is carefully filled with silicone oil and water (or $0.5\,mM$ $CaSO_4$ solution) and should be as rigid as possible, *i.e.*, the elastic modulus of the apparatus (ε_s) should be as high as possible. Under these conditions, the root pressure builds up in the system and can be measured for 1 to 2 days or longer. The silicone seal around the excised root is prepared from liquid silicone material (Xantopren plus from Bayer, Leverkusen, FRG). It should be pressure-tight but should not constrain the flow in the xylem. These requirements can be tested experimentally. In principle, two types of experiments

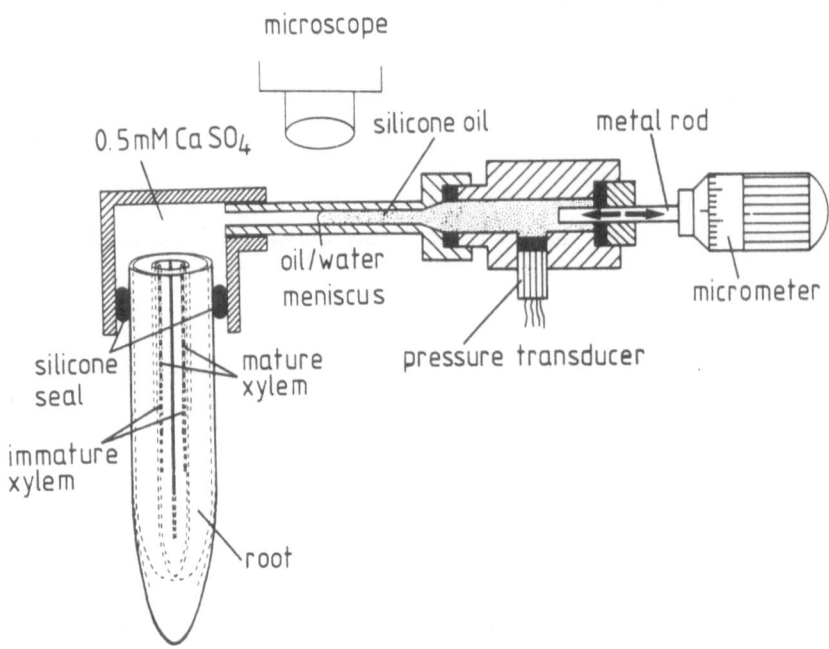

Fig. 1. Root pressure probe for measuring root pressure (P_r) and water (J_{vr}) and solute flow in roots (*schematical*). The excised root is tightly connected to the probe by a silicone seal. The stationary root pressure built up in the system can be changed either by moving the metal rod ('*hydrostatic experiments*') or by changing the osmotic pressure of the medium ('*osmotic experiments*') to induce hydrostatic or osmotic water flows (Figs. 2 and 3). In order to determine the elasticity of the measuring system instantaneous changes in volume (ΔV) are produced with the aid of the metal rod and the resulting changes in root pressure are measured. ΔV is calculated from the diameter of the capillary and the movement of the oil/water meniscus.

can be performed using the root pressure probe analogous to the experiments with isolated plant cells using the conventional cell pressure probe, namely, hydrostatic and osmotic experiments. The direction of water flow can be also changed to be out of ('*exosmotic experiments*') or into ('*endosmotic experiments*') the roots. Usually, relaxations of root pressure are measured (as for cells), and from the half-time Lp_r is evaluated. To do this, the elastic coefficient of the measuring system (apparatus plus xylem) has to be also determined (see legend to Fig. 1 and Steudle *et al.*, 1987).

Although the procedure for evaluating Lp_r from root experiments are analogous to those used for cells, there are differences, since with the root probe, ε_s rather than the elastic coefficient of the xylem (ε_x) determines the rate of water exchange during the measurements (besides Lp_r and the root geometry). Furthermore, xylem solution may be moved by convection across the cut surface into the pressure apparatus. The theory assumes that the xylem is a well-stirred compartment which is separated from the medium by a single homo-

genous membrane. This is, of course, a simplification of the real situation and some of the observations (*e.g.*, the polarity of water movement) do indicate that unstirred layers as well as the complex composite structure of the barrier play a role (see below).

The root pressure probe offers a possibility of measuring the hydraulic resistance of roots and to determine solute flows at the same time. When permeable solutes are supplied to a root, the decrease in water potential first causes a reduction in root pressure due to an exosmotic water flow ('water phase'). After a minimum in root pressure has been reached (P_{rmin}), the original root pressure is, however, re-attained, because the permeable solute enters the xylem and decreases the water potential, thus, causing an endosmotic water flow (Fig. 3). The theory for the biphasic relaxations obtained with roots in the presence of permeable solutes has been given elsewhere (Steudle *et al.*, 1987). The equivalent single membrane model has been again used to evaluate water and solute flows. Deviations from the theory may be expected,

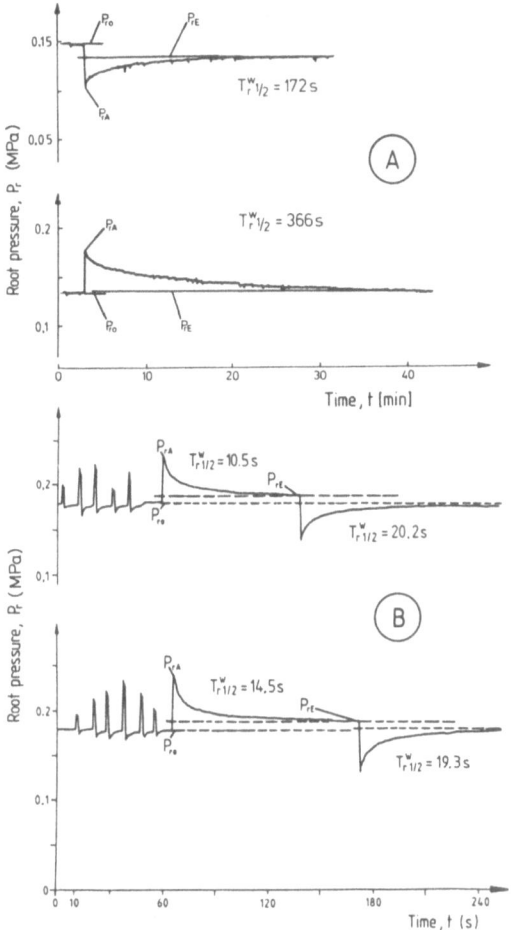

Fig. 2. Hydrostatic root pressure relaxations of end segments of seminal barley (**A**) and corn (**B**) roots as obtained with the root pressure probe (Fig. 1). For barley, the half-times of the pressure relaxations ($T_{r1/2}^W$) were rather long indicating a fairly low hydraulic conductivity (Lp_r), whereas for maize $T_{r1/2}^W$ was short and Lp_r high.

because of concentration polarization phenomena and because of osmoregulatory effects. However, the experimental results indicate that the simple model basically describes the situation at least for rather permeable solutes for which the half-time of solute equilibration may be short compared with the time necessary to produce significant changes in root pressure due to active processes. Preliminary results indicate that active solute uptake may be also followed with the technique. This could offer a possibility to measure changes in nutrient uptake after changes in the root environment (*e.g.*, after applying pollutants, inhibitors, anoxia, a cold temperature *etc.*) which may affect active pumping.

It should be emphasized that in the root pressure probe technique small parts of roots as well as entire root systems may be analyzed. Experiments can be performed on roots grown in hydroculture as well as in soil, and the technique can be also used outdoors. As with the cell pressure probe measurements on tissue cells, the determination of $Lp_r \cdot A_r$ from root pressure relaxations is not affected by hydraulic resistances in the soil which are in series with the root. Thus, soil and root hydraulic resistances can be separated and, if the total resistance of the system (soil plus root) can be measured by an independent experiment, also the soil resistances may be evaluated.

Results and discussion

Detailed results have been obtained so far for end segments of young seminal roots of barley and maize (Steudle and Jeschke, 1983; Steudle *et al.*, 1987). The roots of both species show similar transport properties but also differences in their hydraulic properties. In both species, the hydraulic resistance to radial water movement across the root cylinder was found to be rate limiting. This can be demonstrated by root pressure experiments in which a stationary root pressure is first established and the hydraulic conductance is measured as outlined above. When the root is successively cut back from the apex, the root pressure suddenly drops close to zero when developed xylem is hit which occurred about 20 mm behind the apex. Relaxations in the presence of open xylem were much faster, *i.e.*, the hydraulic conductance was much larger than in the closed system. This indicates that tissue transport (medium distance transport) of water rather than vascular transport (long distance transport) limits water uptake at least in the upper parts of the seminal roots of barley and maize.

The development of the endodermis as well as the suberization of older parts of the roots may influence the absolute value of Lp_r. The measured values of the root segments (length of the end segments for barley: 80 to 135 mm and for maize: 45 to 130 mm; age of the roots in days: barley: 6–13; maize: 5–13) are, therefore, overall values which should be examined for a variation along the developing root in more detail. This is possible with the technique, for example, if impermeable solutes

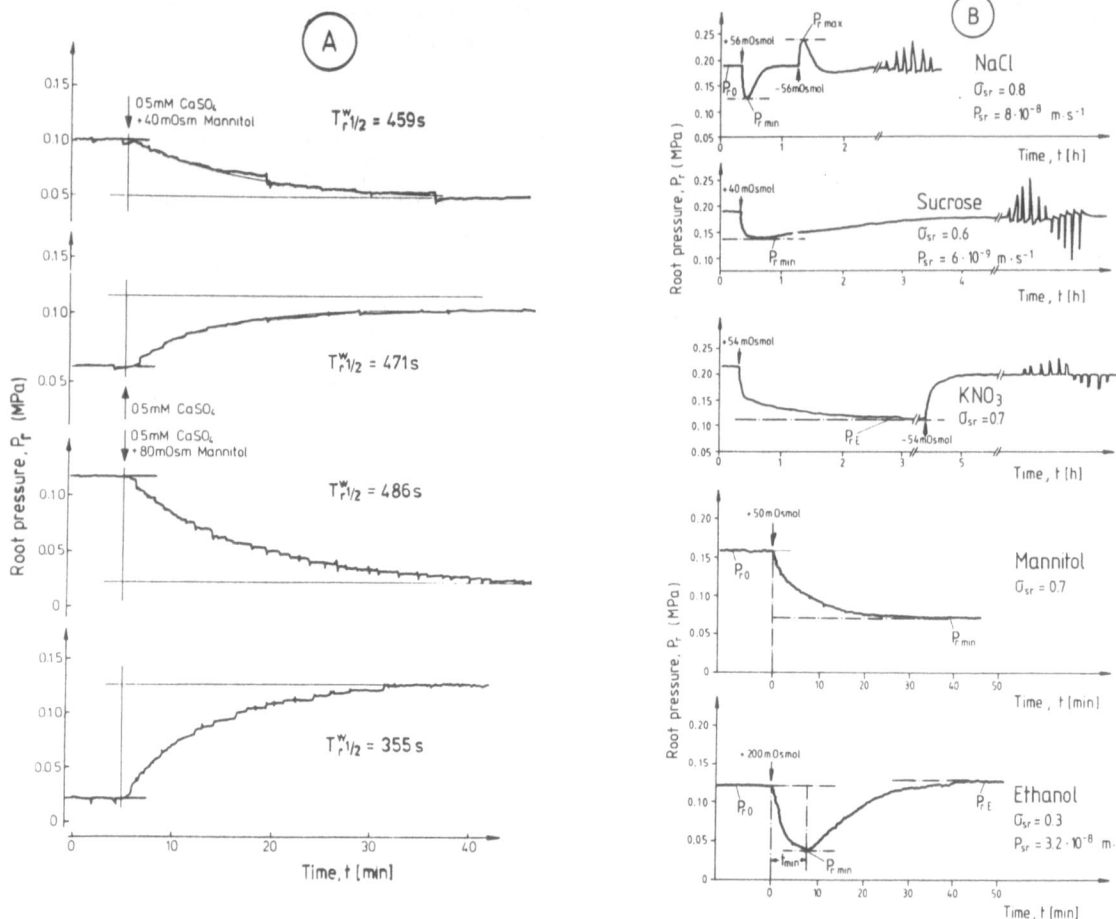

Fig. 3. Osmotic pressure relaxations for end segments of seminal barley (**A**) and maize (**B**) roots. Half-times of the water phases of the relaxations were of the same order of magnitude for both species and similar to the hydrostatic $T^W_{r1/2}$ for barley. For maize, biphasic relaxations are given for some permeable solutes (NaCl, sucrose, and ethanol). From the second phase ('*solute phase*'), permeability coefficients of the roots (P_{sr}) were calculated. For NaCl it is shown that the osmotic responses were reversible, *i.e.*, when the medium was replaced by the original solution, biphasic curves with a pressure maximum were obtained.

($\sigma_{sr} = 1$) are applied to narrow root zones and the resulting water flows are estimated. The technique should be also useful to follow the degree of suberization of the hypodermis. From studies with dyes confined to the apoplast it has been concluded that for maize the suberization of the hypodermis results in the formation of a Casparian band (Peterson and Perumalla, 1984; see also Peterson, these proceedings). The 'exodermis' formed may also reduce the permeability of the roots to water and other polar substances. In fact, the polarity observed for the water flow in young maize roots may result from the formation of such a structure (Steudle *et al.*, 1987) and may be caused by the series arrangement of endodermis and exodermis, if

it is assumed that both structures have different reflection coefficients. However, the large water flow within the apoplast of corn roots (or across secondary root initials) which is observed in the presence of hydrostatic gradients is obviously not significantly restricted by Casparian bands (see below).

Typical root pressure relaxations of maize and barley roots are shown in Figs. 2 and 3 for hydrostatic and osmotic experiments, respectively. Fig. 2 shows that in the presence of a hydrostatic gradient, the half-times for barley were by 1 to 2 orders of magnitude larger than for maize which resulted in a much smaller Lp_r for barley (Table 1). The absolute value of hydrostatic Lp_r for maize was

Table 1. Some literature data of the hydraulic conductivity of roots (Lp$_r$) measured by different techniques using either hydrostatic or osmotic gradients to induce water flows. Experiments performed with root pressure probe are denoted. Except for tomato and barley, the osmotic Lp$_r$ seems to be smaller than the hydrostatic

Species	Root hydraulic conductivity, Lp$_r \cdot 10^8$ (m·s^{-1}·MPa^{-1})		Cell Lp × 10^8 (m·s^{-1}·MPa^{-1})
	Osmotic flow	Hydrostatic flow	
Zea mays	5.7 (a)		–
	1–12*) (b, c)		–
	2.2 (d)	10–21 (h, i)	–
	4.0 (e)		–
	1.2 (f; root) pressure probe)	10 (f; root) pressure probe)	24 (f; cell) pressure probe)
Phaseolus vulgaris	0.56 (d)	1–5**) (j)	–
		8–61 (k, l)	–
Helianthus annuus	0.71 (d)	2–12**) (j)	–
Glycine max	1–14**) (g)	27 (m)	–
Lycopersicon esculentum	6.1 (d)	2–7**) (j)	–
Hordeum distichon	0.3–4.33 (n; root pressure probe)	0.3–4.0 (n; root pressure probe)	12 (n; cell pressure probe)

*) Lp$_r$ varies along the root
**) Lp$_r$ increases with increasing J$_{vr}$

(a) House and Findlay 1966; (b) Anderson *et al.*, 1970; (c) Pitman *et al.*, 1981; (d) Newman 1973; (e) Collins and Kerrigan 1974; (f) Steudle *et al.*, 1987; (g) Michel 1977; (h, i) Miller 1985a, b; (j) Salim and Pitman 1984; (k) Fiscus 1981; (l) Fiscus and Markhart 1979; (m) Fiscus 1977; (n) Steudle and Jeschke 1983.

Table 2. Some reflection (σ_{sr}) and permeability (P$_{sr}$) coefficients of roots (ref. g, h: root pressure probe; ref. a-f: other techniques). For maize, data are given for intact excised roots as well as for cortical root sleeves

Species	Osmoticum	Root reflection coefficient, σ_{sr}	Root permeability coefficient P$_{sr}$ × 10^8 (m·s^{-1})
Glycine max (a)	nutrients	0.90	–
Lycopersicon esculentum (b)	nutrients	0.76	–
Zea mays, cortical sleeves (c, d)	sucrose	0.90–1.00	–
	urea	0.85	–
	NaCl	1	–
	KCl (e)	0.05	–
Zea mays, intact excised roots (g)	nutrients (f)	0.85	–
	ethanol	0.27	4.7
	mannitol	0.74	–
	sucrose	0.54	1.2
	PEG 1000	0.82	–
	NaCl	0.64	5.7
	KNO$_3$	0.67	–
Hordeum distichon (h)	mannitol	$\simeq 0.5$	–

(a) Fiscus 1977; (b) Mees and Weatherley 1957; (c, d) Ginsburg and Ginzburg 1970a, b; (e) Collins 1974; (f) Miller 1985a; (g) Steudle *et al.*, 1987; (h) Steudle and Jeschke 1983.

similar to that of the cell Lp of cortical cells and, thus, it can be concluded that there must have been a predominant apoplasmic flow in the maize root under these conditions. For barley, on the other hand, the comparison between cell and root hydraulic conductivity demonstrated that there was at least a substantial contribution of the cell-to-cell path, because Lp$_r$ was by an order of magnitude smaller than the cell Lp (Table 1). These differences in the hydraulic properties of the roots point to differences in the tightness of the Casparian strips which could perhaps be variable and depend on the growing conditions (*cf.* Peterson, these proceeedings). For barley, the experiments suggest that the Casparian band in the endodermis interrupts the hydraulic flow in the apoplast, whereas for maize

this structure seems to be rather permeable. Another possibility to explain the large apoplasmic flow in maize could be the existence of secondary root initials which have been shown to be rather permeable to apoplasmic dyes in corn and broad bean (Peterson *et al.*, 1981).

These explanations do not contradict the findings of the osmotic experiments where similar half-times and Lp$_r$ values have been observed for both species (Fig. 3). Under 'osmotic conditions' the apoplasmic path should be rather ineffective, because the reflection coefficient of the wall material should be very low for low molecular weight solutes. This explains why water flow could depend on the nature of the driving force. For the maize root, Lp$_r$ was by nearly an order of magnitude smaller in osmotic than in hydrostatic experiments, and the polarity of water movement was reversed. This finding also suggests differences in the transport mechanisms. Using pressure probe techniques, similar differences have been obtained for other tissues (for references see Steudle *et al.*, 1987).

Table 1 shows that the absolute values of Lp_r measured with the aid of the probe are within the range of values obtained by other techniques. The table also indicates that at least some of the scatter in the literature data for Lp_r may be due to the fact that different driving forces have been used by applying common techniques such as root exudation, pressure chambers, and pressure jump techniques. Compared with these techniques the root pressure probe allows to measure Lp_r by using different forces and also to compare the results from the tissue (organ) level with those from the cell level.

The finding that the reflection coefficients of roots are significantly lower than unity (Fig. 3; Table 2) even for solutes for which cell membranes have a $\sigma \simeq 1$ is surprising if the traditional view of the root as a nearly ideal osmometer with the endodermis as the osmotic barrier is considered. However, if the Casparian band has a reasonable permeability for solutes, at least during some stages of root development, low σ_{sr} values would be understandable. In fact, there are also some literature data which indicate a $\sigma_{sr} < 1$ for other species and techniques (Table 2). The low σ_{sr} values do allow a proper function of the root, since the P_{sr} values (which are a measure of passive leakages across the root cylinder) seem to be sufficiently small. Some caution is needed before accepting the low σ_{sr} values because of effects of unstirred layers (*e.g.*, in the root cortex and stele, if the endodermis is the main osmotic barrier) which would tend to reduce the true σ_{sr}. However, these effects should be relatively small during the measurements with the probe, because the water flows induced are fairly small and the σ_{sr} values are corrected for solute flow which also incorporates unstirred layers, at least to some extent (Steudle *et al.*, 1987). If the result of low σ_{sr} values in roots holds in general it should be of some importance for the nutrient uptake into roots. If σ_{sr} is low, the solvent-drag component of solute (nutrient) uptake could become important at high transpiration rates depending on the absolute concentration of solutes within the root.

Acknowledgement

This work was supported by a grant from the Deutsche Forschungsgemeinschaft, Sonderforschungsbereich 137.

References

Anderson W P 1976 Transport through roots. *In* Encyclopedia of Plant Physiology (New Series), Vol. 2, Part B, Transport in Plants II. Eds. U Lüttge and M G Pitman. pp 129–156. Springer-Verlag, Heidelberg.

Anderson W P, Aikman D P and Meiri D P 1970 Excised root exudation: A standing gradient osmotic flow. Proc. Roy. Soc. London (B) 174, 445–458.

Collins J C 1974 Hormonal control of ion and water transport in excised maize roots. *In* Membrane transport in plants. Eds. U Zimmermann and J Dainty. pp 441–443. Springer-Verlag, Heidelberg.

Collins J C and Kerrigan A P 1974 The effect of kinetin and abscisic acid on water and ion transport in isolated maize roots. New Phytol. 73, 309–314.

Fiscus E L 1977 Determination of hydraulic and osmotic properties of soybean root systems. Plant Physiol. 59, 1013–1020.

Fiscus E L 1981 Effect of abscisic acid on the hydraulic conductance and the total ion transport through Phaseolus root systems. Plant Physiol. 68, 169–174.

Fiscus E L and Markhart A H 1979 Relationships between root system water transport properties and plant size in Phaseolus. Plant Physiol. 64, 770–773.

Ginsburg H and Ginzburg B Z 1970a Radial water and solute flow in roots of *Zea mays*. I. Water flow. J. Exp. Bot. 21, 580–592.

Ginsburg H and Ginzburg B Z 1970b Radial water and solute flow in roots of *Zea mays*. II. Ion fluxes across the root cortex. J. Exp. Bot. 21, 593–604.

House C R and Findlay N 1966 Water transport in isolated maize roots. J. Exp. Bot. 17, 344–354.

Kedem O and Katchalsky A 1963a Permeability of composite membranes. Part 2: Parallel elements. Trans. Far. Soc. 59, 1931–1940.

Kedem O and Katchalsky A 1963b Permeability of composite membranes. Part 3: Series array of elements. Trans. Far. Soc. 59, 1941–1953.

Mees G C and Weatherley P E 1957 The mechanism of water absorption by roots. II. The role of hydrostatic pressure gradients across the cortex. Proc. Roy. Soc. London (B) 147, 381–391.

Michel H E 1977 A model relating root permeability to flux and potentials: Application to existing data from soybean and other plants. Plant Physiol. 60, 259–264.

Miller D M 1985a Studies of root function of *Zea mays*. III. Xylem sap composition at maximum root pressure provides evidence of active transport into the xylem and a measurement of reflection coefficient of the root. Plant Physiol. 77, 162–167.

Miller D M 1985b Studies of root function of *Zea mays*. IV. Effects of applied pressure on hydraulic conductivity and volume flow through the excised root. Plant Physiol. 77, 168–174.

Newman E I 1973 Permeability to water of the roots of five herbaceous species. New Phytol. 72, 547–555.

Peterson C A, Emanuel M E and Humphreys G B 1981 Pathway of movement of apoplastic fluorescent dye tracers through the endodermis at the site of secondary root formation in corn

(*Zea mays*) and broad bean (*Vicia faba*). Can. J. Bot. 59, 618–625.

Peterson C A and Perumalla C J 1984 Development of the hypodermal Casparian band in corn and onion roots. J. Exp. Bot. 35, 51–57.

Pitman M G 1982 Transport across plant roots. Quart. Rev. Biophys. 15, 481–55.

Pitman M G, Wellfare D and Carter C 1981 Reduction of hydraulic conductivity during inhibition of exudation from excised maize and barley roots. Plant Physiol. 67, 802–808.

Salim M and Pitman M G 1984 Pressure-induced water and solute flow through plant roots. J. Exp. Bot. 35, 869–881.

Steudle E and Jeschke W D 1983 Water transport in barley roots. Planta 158, 237–248.

Steudle E, Oren R and Schulze E D 1987 Water transport in maize roots. Measurement of hydraulic conductivity, solute permeability, and of reflection coefficients of excised roots using the root pressure probe. Plant Physiol. 84, 1220–1232.

Steudle E and Zimmermann U 1984 Water relations of plant cells: Further development of the pressure probe and of techniques for measuring pressure-dependent transport. *In* Membrane transport in plants. Eds. W J Cram *et al.* pp 73–82. Academia, Prague.

Weatherley P E 1982 Water uptake and flow in roots. *In* Encyclopedia of Plant Physiology (New Series), Vol. 12B. Physiological Plant Ecology II. Eds. P S Lange *et al.* pp 79–109. Springer-Verlag, Heidelberg.

Zimmermann U and Steudle E 1978 Physical aspects of water relations of plant cells. Adv. Bot. Res. 6, 45–117.

B. C. Loughman et al. (Eds.), Structural and functional aspects of transport in roots, 147–150.
© 1989 by Kluwer Academic Publishers.

A biophysical model for water movement in roots: Root exudation and root pressure

KIYOSHI KATOU, TAKEHIDE TAURA and MUNEYOSHI FURUMOTO[1]
Biological Institute and [1] Department of Earth Sciences, Faculty of Science, Nagoya University, Nagoya 464, Japan

Key words: canal model, cell wall apoplast, root exudation, root pressure

Abbreviations: C^{ex}: the osmotic concentration of the xylem exudate (osmol m^{-3}), $C^{ex}(P^x)$: C^{ex} as a function of root pressure P^x (osmol m^{-3}). C^i: the osmotic concentration in the symplast (osmol m^{-3}), $C_s(x)$: the osmotic concentration of absorbable solute in the canal at x (osmol m^{-3}), $C_s(x, P^x)$: $C_s(x)$ as a function of root pressure (osmol m^{-3}), C_s^{av}: the average of $C_s(x)$ in the canal (osmol m^{-3}), $C_s^{av}(P^x)$: C_s^{av} as a function of root pressure (osmol m^{-3}), D: the solute diffusion coefficient in the canal (m^2 s^{-1}), I_s: the rate of net solute transport across the symplast cell membrane into the canal (osmol m^{-2} s^{-1}), $I_s(x, P^x)$: I_s as a function of x and root pressure (osmol m^{-2} s^{-1}), I_s^{ix}: the rate of passive solute leakage into the canal (osmol m^{-2} s^{-1}), I_s^{xi}: the rate of back diffusion of solute from the canal into the symplast (osmol m^{-2} s^{-1}), I_a: the rate of active solute excretion into the canal (osmol m^{-2} s^{-1}), $I_v(x)$: the rate of water secretion into the canal at x (m s^{-1}), $J_s(x)$: the rate of solute flow along x-axis across unit cross-sectional area of the canal at x (osmol m^{-2} s^{-1}), J_s^{ex}: the solute exudation per root surface (osmol m^{-2} s^{-1})$\cdot(= C^{ex} J_v^{ex})$, J_v^{ex}: the fluid exudation per root surface (m s^{-1}), L_p: the hydraulic conductivity of the symplast membrane (m s^{-1} Pa^{-1}), P^i: the intracellular pressure with reference to the bathing medium (Pa), P_o^x: P^x immediately after the stoppage of the volume exudation (Pa), P_s: the permeability coefficient of the symplast cell membrane for the solute (m s^{-1}), q: the width of a canal with a rectangular cross-section (m), R: the gas constant (J mol^{-1} K^{-1}), T: absolute temperature (K), $v(x)$: the linear velocity of fluid in the canal at x (m s^{-1}).

Introduction

The nature of the radial transport across roots has been the subject of controversy. Especially it was under debate whether the parenchyma cells in the stele are leaky or not. The accumulated findings, however, were in favour of the idea of metabolically active stele. The situation, however, remained poorly understood because of the inaccessibility of the symplast/xylem boundary in the stele.

The fluid exudation of an excised root depends on aerobic metabolism and doesn't seem fully osmotic (House and Findlay, 1966). Ginsburg (1971) interpreted this phenomenon introducing a double membrane system with different σs (Curran and MacIntosh, 1962). However, this explanation

seems unlikely because the σ of cell membranes for physiological solutes is now considered to be 1 (Hastings and Gutknecht, 1978).

The xylem perfusion experiments have directly shown that the xylem parenchyma cells are quite active in solute transport (De Boer and Prins 1985; Clarkson and Hanson, 1986). Therefore, it is inappropriate to assume stelar cells with membranes of low σ as an explanation for the transport in the root stele.

Results and discussion

A canal model in the root stele

In the parenchyma of the root stele, cells and cell walls are juxtaposed and the parenchyma cells form

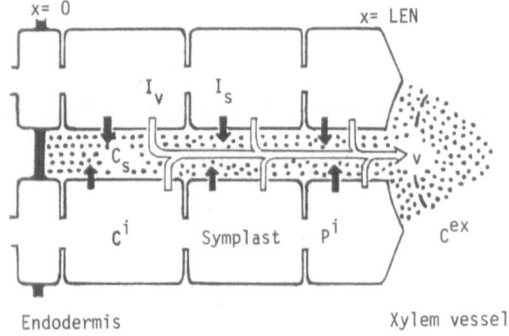

Fig. 1. Schematic representation of the apoplast canal system in the root stele which enables metabolism dependent water transport into vessels. Net solute excretion into the canal (I_s) increases the solute concentration within it (C_s), thus drawing water from the symplast into the canal. *Symbols* used in the figure: see abbreviations.

a symplast. Most of them do not make contact directly with the xylem vessels but with the apoplast of thin cell walls the solute concentration in which would easily vary depending on the cellular transport activity. We consider a single wall canal extending from the endodermis to a vessel (Fig. 1) as a simplest case. The endodermal end is plugged by a Casparian strip and the other end is open to the vessel. A net solute excretion into the canal would create a solute enrichment in the canal, thus drawing water from the symplast into the canal. Then the solutes and the water would simultaneously move into the vessel. This process can mathematically be described in the same way as reported previously (Diamond and Bossert, 1967; Katou and Furumoto, 1986).

Let the x-axis be the distance along the canal from the endodermal end at x = 0 to the open end at x = LEN. For solute flow in the canal along the x-axis

$$J_s = -D \frac{dC_s}{dx} + C_s v \qquad (1)$$

Water flow across the symplast cell membrane into the canal is governed by

$$I_v = L_p \{(C_s - C^i)RT - (P^x - P^i)\} \qquad (2)$$

where $\sigma = 1$ is assumed.

Within an infinitesimal rectangular solid segment of the canal extending from x to x + dx (Katou and Furumoto, 1986), flows of solute and water should be continuous. Then $qdJ_s = 2I_s dx$ and $qdv = 2I_v dx$. Therefore the equations (1) and (2)

become

$$\frac{dv}{dx} = \frac{1}{M}(C_s - K) \qquad (3)$$

$$\frac{d^2v}{dx^2} = \frac{1}{MD}(C_s v - J_s) \qquad (4)$$

$$\frac{d^3v}{dx^3} = \frac{v}{D}\frac{d^2v}{dx^2} + \frac{K}{MD}\frac{dv}{dx} + \frac{1}{D}\left(\frac{dv}{dx}\right)^2 - \frac{2I_s}{qDM} \qquad (5)$$

where $M = q/(2L_pRT)$ and $K = C^i - (P^i - P^x)/RT$.

Solutions for exudation in the stele

For simplicity, no longitudinal difference in exudation activity was assumed, and I_s was assumed to be constant throughout the canal. Although quasi-equilibrium between symplast cells and exudate was assumed, a small difference in water potential of ca. 63 kPa was estimated from Miller's experiment (1980). Therefore the value of $(C^i - C^{ex})RT + 63$ kPa was approximated as the estimated P^i. The boundary conditions are as follows, $J_s(0) = 0$, $v(0) = 0$ and $C_s(LEN) = J_s(LEN)/v(LEN) = C^{ex}$.

C^i, C^{ex} and J_s^{ex} in a model maize root (Table 1) were estimated from studies of exudation in *Zea mays* roots (Anderson and Reilly 1968; Dunlop and Bowling, 1971). Major osmotica in the model root were assumed to be K^+ and Cl^-. The effective canal length (LEN) was chosen as 120 μm based on microscopic observations. P^x was set at zero because an excised root was considered. The values of D and L_p were 3×10^{-11} m^{2}s^{-1} and 1×10^{-13} m s^{-1} Pa^{-1} respectively (Katou and Furumoto, 1986). Calculations were performed by the Runge-Kutta-Gill procedure programmed with TURBO PAS-CAL-87 Ver. 3.02A (Borland International, Inc., Scotts Valley, CA, USA) using a personal computer NEC PC9801VM2 (NEC Corp., Tokyo, Japan).

Exudation in 1 mol m^{-3} KCl medium was chosen as an arbitrary standard state (see Table 1). The I_s which fulfilled the boundary conditions of C_s (LEN) = J_s (LEN)/v(LEN) = C^{ex} = 39 osmol m^{-3} was calculated to be 246.9 nosmol m^{-2}s^{-1}. It was found that v increased linearly from the closed endodermal end to the opening end, and that the calculated profile of C_s was flat throughout

Table 1. Root exudation of a model maize root and results of calculations

KCl mol m^{-3}	^{a}C^i osmol m^{-3}	^{a}C^{ex} osmol m^{-3}	^{b}J$_v^{ex}$ nm/s	P^i MPa	I$_s$ nosmol m^{-2} s^{-1}	C$_s$(120) osmol m^{-3}	v(120) nm/s
0.1	190	34	3.4 (0.62)	0.450	133.1	27.8	574.5 (0.76)
1.0	220	39	5.5 (1.04)	0.512	246.9	39.0	759.6 (1.01)
10.0	240	58	5.7	0.514	380.5	58.8	776.9

^{a} Estimated from Dunlop and Bowling (1971).
^{b} Estimated from Anderson and Reilly (1968). All the media contain 0.1 mol m^{-3} CaCl$_2$. The osmotica in roots were assumed to be K$^+$ and Cl$^-$. Value in each parentheses is the ratio of the value of each parameter to that of 1.0 mol m^{-3} KCl condition.

the canal, *i.e.* C$_s$ = C^{ex} anywhere in the canal. The linear increase of v along the canal means that the solution secreted flows along the canal and eventually into a vessel.

I$_s$ should be proportional to J$_s^{ex}$ under the present condition. Therefore, from the ratio of J$_s^{ex}$ the theoretical value of I$_s$ for conditions other than the standard one can be calculated. For concentration of 0.1 and 10.0 mol m^{-3} KCl, I$_s$ values were 133.1 and 380.5 nosmol m^{-2} s^{-1} respectively. For each I$_s$ value, the canal equations were numerically solved for the boundary conditions of J$_s$(0) = 0, v(0) = 0 and C$_s$(LEN) = J$_s$(LEN)/v(LEN). The C$_s$(LEN) calculated is the C^{ex} predicted by the present canal model. For 0.1 and 10.0 osmol m^{-3} KCl, C$_s$(LEN)s were 27.8 and 58.8 osmol m^{-3} respectively. In spite of a rather simplified model, these values agree well with those observed (34 and 58 osmol m^{-3} respectively) (Table 1). It goes without saying that the ratio of the rate of canal exudation is also shown to agree well with that of the root exudation, for v(LEN) is equal to J$_s$(LEN)/C^{ex}.

The uptake process of solute and water in the root cortex can be described by the similar canal model.

Solutions for the development of root pressure

When volume exudation is prevented (J$_v^{ex}$ = 0), the symplast and the canal should be in equilibrium. Therefore the root pressure is expressed by

$$P^x = \{C_s^{av}(P^x) - C^i\}RT + P^i \quad (6)$$

It is assumed that P^i, C^i and I$_a$ are constant during the development of pressure.

Net solute transport into the canal would raise the osmotic concentration in the canal involving

increase in root pressure. However, the rate of net solute transport is given by I$_s$ = I^{ix} − I^{xi} + I$_a$. It is probable that the increased C$_s$ causes an increase in the passive back flow (I$_s^{xi}$) which is given by $\Delta I_s^{xi}(x, P^x) = P_s\{C_s(x, P^x) - C_s(P_o^x)\}$. Therefore, I$_s$ might be no longer constant but a function of x and P^x,

$$I_s(x, P^x) = I_s(P_o^x) - P_s\{C_s(x, P^x) - C_s(P_o^x)\} \quad (7)$$

I$_s$ in equation (5) should be substituted by I$_s$(x, P^x). The boundary conditions for root pressure development are J$_s$(0) = 0, v(0) = 0, v(LEN) = 0, and C$_s$(LEN) > C^{ex}(P$_o^x$).

The back diffusion fluxes of K$^+$ and Cl$^-$ into symplast can be calculated by the well known Goldman-Hodgkin-Katz equation for partial flux (Katz, 1966). Root pressure becomes a maximum when net solute excretion into the canal ceases. After that the root pressure would be in a relatively stable state. The maximum root pressure under the present condition is 0.329 MPa. Under 0.1 and 10.0 mol m^{-3} KCl conditions, the maximum root pressures are estimated to be 0.191 and 0.475 MPa respectively.

References

Anderson W P and Reilly E J 1968 The effect of temperature on the exudation process of excised primary roots of *Zea mays*. J. Exp. Bot. 19, 648–657.

Clarkson D T and Hanson J B 1986 Proton fluxes and the activity of a stelar proton pump in onion roots. J. Exp. Bot. 37, 1136–1150.

Curran P F and MacIntosh J R 1962 A model system for biological water transport. Nature 193, 347–348.

De Boer A H and Prins H B A 1985 Xylem perfusion of tap root segments of *Plantago maritima*: the physiological significance of electrogenic xylem pumps. Plant Cell Environ. 8, 587–594.

Diamond J M and Bossert W H 1967 Standing-gradient osmotic flow: A mechanism for coupling of water and solute transport in epithelia. J. Gen. Physiol. 50, 2061–2083.

Dunlop J and Bowling D J F 1971 The movement of ions to the xylem exudate of maize roots. II. A comparison of the electrical potential and electrochemical potentials of ions in the exudate and in the root cells. J. Exp. Bot. 22, 445–452.

Ginsburg H 1971 Model for iso-osmotic water flow in plant roots. J. Theoret. Biol. 32, 147–158.

Hastings D F and Gutknecht J 1978 Potassium and turgor pressure in plants. J. Theoret. Biol. 73, 363–366.

House C R and Findlay N 1966 Water transport in isolated maize roots. J. Exp. Bot. 17, 344–354.

Katou K and Furumoto M 1986 A mechanism of respiration-dependent water uptake enhanced by auxin. Protoplasma 133, 174–185.

Katz B 1966 Nerve, Muscle and Synapse. McGraw-Hill, New York.

Miller D M 1980 Studies of root function in *Zea mays*. I. Apparatus and methods. Can. J. Bot. 58, 351–360.

B. C. Loughman et al. (Eds.), Structural and functional aspects of transport in roots, 151–155.
© 1989 by Kluwer Academic Publishers.

Plant root water extraction studies using stable isotopes

F. N. DALTON

US Department of Agriculture, Agricultural Research Service, US Salinity Laboratory, Riverside, CA 92501, USA

Key words: isotopic abundance, isotopic ratio $^1H_1/^2H_2$, root, soil water extraction, tomato plant

Introduction

Soil water extraction by plant roots occurs in response to an environmental evaporative demand at the leaf surface. A reduction in water content of the leaf mesophyll cells is accompanied by a reduced hydraulic pressure in the plant's conducting tissue and is assumed to be transmitted to the xylem tissue in the roots. Water moves from one energy state in the soil to a relatively lower energy state in the plant. Due to the adsorption properties of the soil matrix, the energy state of soil water decreases as the water content diminishes and therefore the soil competes with the plant for available water.

When this qualitative picture of the transpiration process is quantified, then by necessity, one introduces concepts of plant resistance, water potential gradients and hydraulic pathways. Van den Honert (1948), using the electrical equivalence of Ohms law, defined the resistance to water flow in a plant as the ratio of the water potential difference between leaf and soil to the transpiration rate. Gardner (1960) was the first to quantitatively show how the water transport properties of soil could affect the potential for water extraction by plant roots. The rate limiting step in the transpiration process was thought to be controlled by either a 'plant resistance' or a 'soil resistance' and these important concepts formed the basis of a very active research effort (Boyer, 1974; Rawlins, 1963). A fundamentally different approach, (Dalton *et al.*, 1974), assumed that the roots exert the primary control of water and ion flow into the plant by the presence of an effective semi-permeable membrane located in the root (probably the endodermis) and

developed a model describing the dependent flows of water and ions based on physical and chemical transport properties of a root system. The predictive capabilities of these two models are quite different, but both models share the important assumption that root water extraction is a hydraulic phenomenon and all of the water enters the root in the liquid phase.

Because of root shrinkage and soil water depletion due to transpiration and drainage, vapor spaces are created in the root zone and portions of the root surface lose contact with liquid phase water (Herkelrath *et al.*, 1977; Huck *et al.*, 1970). Even in this situation, it is tacitly assumed that water is extracted in the liquid phase. Important consequences of the transpirational demand being met by liquid phase root water extraction, such as the effect of active ion transport on water uptake (Dalton and Gardner, 1978) and the effects of diffusion and mass flow on plant nutrient uptake, (Barber, 1969) have been studied in depth. From time to time however, it has been argued from circumstantial evidence that there is at least the possibility of an appreciable vapor flux to the plant root (Bonner, 1959; Cowan and Milthorpe, 1968; Philip, 1957). Others have argued sensibly against this notion (Bernstein *et al.*, 1959). In any case, either hypothesis has heretofore been untestable. The consequences of a significant vapor component to the water extraction process leads to important practical implications for water management in a saline environment and a thorough understanding of the physics of root water uptake phenomenona warrants further investigation. The purpose of this study therefore was the direct verification of the physical mechanism of water uptake by plant roots. The objectives were to design a root water extraction experiment which could distinguish between liquid or vapor phase water absorption by plant root tissue. This was accomplished by making use of the physical properties of the stable isotopes of hydrogen.

Hypothesis

Water will be symbolized by H_2O when both hydrogen atoms are the isotope $^1H_1$, and by HDO (D for deuterium) when one hydrogen atom is the isotope $^2H_1$. Measurements of isotopic ratios of hydrogen in natural waters are made with a mass spectrometer. The results are expressed in reference to the isotopic abundance of $^2H_1$, γ‰, as

$$\delta‰ = \left[\frac{R - R_0}{R_0} \right] \times 1000 \qquad (1)$$

where,

$$R = \frac{(HDO)}{(H_2O)} \qquad (2)$$

and $R_0 = 155.76 \times 10^{-6}$ is an arbitrary standard obtained from isotope ratio measurements of sea water. The isotopic abundance, γ‰, can be either positive ($R > R_0$) or negative ($R < R_0$) and increases as the concentration of HDO increases relative to H_2O.

The vapor pressure of the lighter H_2O molecule is slightly greater than that of the HDO molecule (Wahl and Urey, 1935). Therefore, vapor phase water in equilibrium with liquid phase water will have a slightly lower ratio $(HDO)/(H_2O)$ than that of the liquid phase. A vapor component to the water extraction process will manifest itself by depleting a larger fraction of the lighter H_2O molecule from the residual soil water, than the heavier HDO molecule. As a consequence, the isotopic ratio, R, of the residual soil water will increase as will the measure of isotopic abundance, δ‰. If there is no vapor component to root water extraction, the proportion of HDO in the residual soil water remains constant as will the measure of isotopic abundance. It is assumed that there is no fractionization of the soil water if it enters the root in the liquid phase. The experiment is therefore designed to remove water exclusively by the transpiration process and to measure any changes in the isotopic abundance of the residual soil water.

Materials and methods

Six tomato plants were grown from seedling to maturity in small columns of very fine sand. Each column was continuously leached with ten pore volumes of nutrient solution prior to the start of the root water extraction study. The columns were sealed to prevent any soil water evaporation, thus assuring that water loss from each column could only occur by root water extraction in response to the transpiration demand. Figure 1 shows the experimental chronology. Each one of the six tomato plants was allowed to transpire for a consecutively longer time period thus allowing various fractions of soil water to be extracted from the columns. After each specified time interval, the column was weighed, the plant de-topped and the column re-weighed to obtain total water loss and fresh top weight. For example, the first column was sampled on the second day and nineteen hours from the beginning of the experiment. The columns were immediately brought into the laboratory and emptied into the corner of a large polyvinyl bag which functioned as a humidity chamber and prevented any evaporation. A 10-cm section of the soil matrix with a majority of the roots, shown in crosshatch in Fig. 1, was sliced off. This section was passed through a #4 and #10 sieve to remove all roots from the soil. A complete separation of roots from soil could be obtained in less than five minutes. The soil sample was then placed in a stoppered one-liter vacuum flask and all residual soil water was collected by vacuum distillation. Figure 2 shows the schematic of the soil water extraction equipment which consists of a vacuum pump, safety trap, a cold finger immersed in a mixture of alcohol and dry ice maintained at a temperture of $-64°C$, and a vacuum flask for the soil. The soil flask is maintained at 104°C for over 12 hours. By this distillation process all of the vaporized soil water is transferred to the cold finger where it condenses. The water is eventually col-

EXPERIMENTAL CHRONOLOGY

VARIOUS FRACTIONS OF SOIL WATER ARE EXTRACTED FROM EACH COLUMN BY TRANSPIRATION - THEN SAMPLED

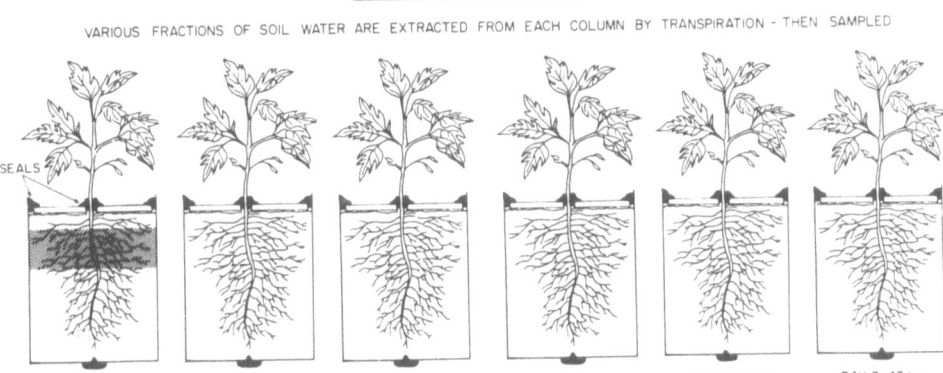

SAMPLING TIMES FROM START OF EXPERIMENT

Fig. 1. Schematic of experimental design and chronology of sampling for residual soil water.

lected and analyzed for the stable isotope abundance.

Results and discussion

Figure 3 shows the amount of water transpired by each plant just before the column was sampled for the residual soil water. The experiment was started at 1600 hrs on Day 1. On Day 2 at about 0900 hrs, the first plant was sampled and at later times on the same day, two or more plants were sampled. It is noted that the amount of water transpired on Day 2 is linearly related with the sampling time and that a constant rate of soil water extraction was obtained. By the end of Day 2, approximately 80% of the available water was extracted and the remaining plants were severely wilted. Consequently, there was no linear relationship between transpiration rate and time of sampling for Day 3 plants.

The isotope abundance measurements for the residual soil water in each column are shown in Fig. 4 as a function of transpired water. The negatively decreasing values on the vertical axis (*i.e.* increasing values of ∂‰) indicate a relative enrichment of the heavier isotope in the residual water. The enrichment is attributed to water vapor absorption by the plant roots. It is noted that during the time period when root water extraction proceeded at a constant rate, the heavy isotope enhancement is linearly related to the amount of water transpired, indicating an isotopic separation proportional to the fraction of water transpired. The results of the isotopic abundance measurements obtained the following day, Day 3, indicate an HDO depletion of the residual soil water. The isotopic depletion observed in these samples occurred following a night period for which there was no transpiration demand. During this period there was the possibility of reverse flow and a mixing of root water and soil water. The apparent enrichment of the residual

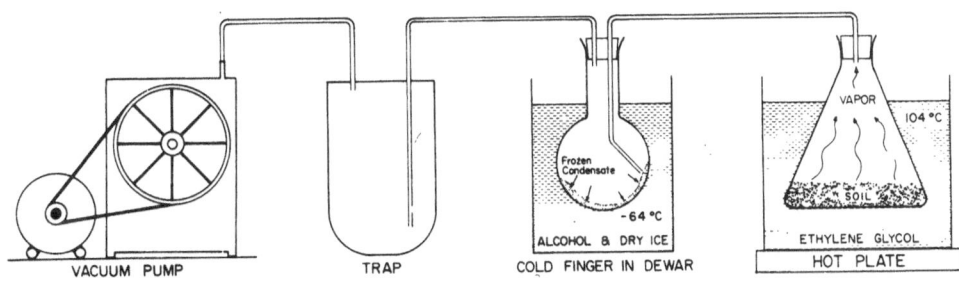

Fig. 2. Schematic of equipment used to recover residual soil water.

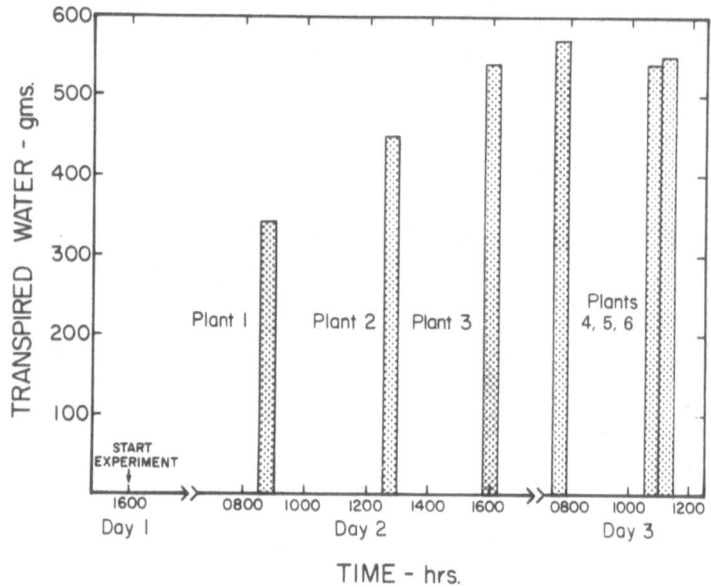

Fig. 3. Amount of water transpired for each plant just prior to sampling for residual soil water.

water following a night period is not readily explained and requires further study.

The difference between the isotopic abundance of the residual soil water and the water obtained from the corresponding root tissue in sample #2, is consistent with water vapor absorption by the root. That is, the root water has a higher fraction of H_2O than the soil water. Finally, it is noted that to within the accuracy of the measurements and with the exception of the anomalous data from Day 3, the residual soil water becomes increasingly enriched with the heavy isotope HDO, relative to that of the initial leachate. This is also consistent with water vapor absorption by plant roots.

In summary the experimental technique and isotopic ratio measurements are shown to be suf-

ficiently sensitive to test the hypothesis concerning liquid or vapor phase water absorption by plant roots. Experimental results indicate the possibility of a vapor component to the water extraction process and that this component increases in proportion to the amount of water transpired. If the component of vapor absorption is assumed to be described by a Rayleigh distillation process then the observed rate of increase in isotopic abundance of the residual soil water indicates a very small component of vapor absorption. While exact quantitative analysis requires further work, this novel experimental procedure offers the first direct method for studying previously unverifiable assumptions concerning the mechanism for water absorption by plant roots.

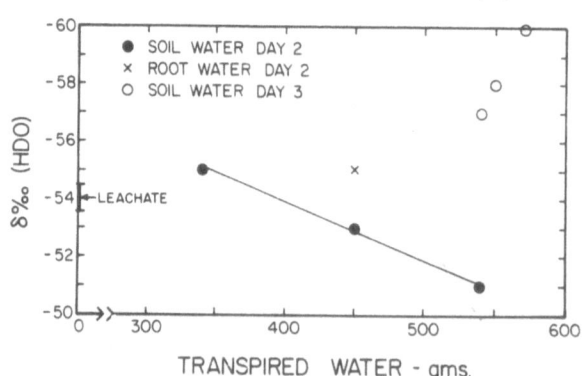

Fig. 4. Isotopic abundance δ‰ as a function of transpired water.

Acknowledgements

Appreciation is expressed to Dr Don Suarez for introducing the author to stable isotope techniques and to Dr Richard Nieman for many fruitful discussions.

References

Barber S A 1962 A diffusion and mass flow concept of soil nutrient availability. Soil Sci. 93, 39–49.

Bernstein L, Gardner W R and Richards L A 1959 Letters; Is there a vapor gap around plant roots? Science 129, 1750–1753.

Bonner J 1959 Water transport. Science 129, 447–450.

Boyer J S 1974 Water transport in plants: Mechanisms of apparent changes in resistance during absorption. Planta 117, 187–207.

Cowan I R and Milthorpe F L 1968 Plant factors influencing the water status of plant tissues. *In* Water Deficits and Plant Growth. Ed. T T Kozlowski. pp 137–193. Academic Press, New York.

Dalton F N and Gardner W R 1978 Temperature dependence of water uptake by plant roots. Agron. J. 70, 404–406.

Dalton F N, Raats P A C and Gardner W R 1974 Simultaneous uptake of water and solutes by plant roots. Agron. J. 67, 334–339.

Gardner W R 1960 Dynamic aspects of water availability to plants. Soil Sci. 89, 63–73.

Herkelrath W N, Miller E E and Gardner W R 1977 Water uptake by plants. II. The root contact model. Soil Sci. Soc. Am. J. 41, 1039–1043.

Huck M G, Klepper B and Taylor H M 1970 Diurnal variations in root diameter. Plant Physiol. 45, 529–530.

Philip J R 1957 The physical principles of soil water movement during the irrigation cycle. Third Congr., Int. Comm. Irrig. Drain. (San Francisco. CA0, Quest. 8, p. 8.125–8.154.

Rawlins S L 1963 Resistance to water flow in the transpiration stream. *In* Stomata and Water Relations in Plants. Ed. Israel Zelitch. pp 69–85. Conn. Agric. Ex. Sta. Bull. 664.

van den Honert T H 1948 Water transport in plants as caternary process. Discuss. Faraday Soc. 3, 146–153.

Wahl M H and Urey H C 1935 The vapor pressures of isotopic forms of water. J. Chem. Phys. vol 3, 411–414.

B. C. Loughman et al. (Eds.), Structural and functional aspects of transport in roots, 157–163.
© 1989 by Kluwer Academic Publishers.

Rectification of radial water flow in the hypodermis of nodal roots of *Zea mays*

M.G.T. SHONE[1] and D.T. CLARKSON[2]
[1] Department of Plant Sciences, University of Oxford, OX1 3PF, UK and [2] University of Bristol, Long Ashton Research Station, Bristol, BS18 9AF, UK

Key words: conductivity, hydraulic, hypodermis, polarity, roots, water, *Zea mays*

Introduction

The outer cells of the cortex in roots of some species may develop lamellar structure, usually assumed to contain suberin, to form a hypodermis immediately beneath the epidermal cells (Peterson, 1988). The hypodermis together with the epidermis and a few ranks of cortical collenchyma cells may be isolated from the central tissues of the root by enzymic digestion. Such hypodermal sleeves provide a convenient experimental preparation for studying the relationhip between the extent of deposition of lamellae and the radial permeability of the hypodermis to the diffusion of labelled ions and water (Clarkson *et al.*, 1978; Clarkson *et al.*, 1987).

The present experiments were originally designed to measure the hydraulic conductivity to mass flow of water in maize hypodermal sleeves since diffusional permeability is not necessarily directly related to hydraulic conductivity but depends on the porosity of the tissue and on the contribution of viscous flow through pores. In the course of these experiments we found that the hydraulic conductivity varied with flow rate and depended on the direction of flow in both sleeve preparations and in intact segments of living roots. Furthermore, flows induced by differences in osmotic potential were several orders of magnitude smaller than those arising from equivalent hydrostatic pressure differences. These findings are discussed and compared with observations on the endodermis and other plant and animal tissues.

Materials and methods

Hypodermal sleeves were prepared from the basal 50-mm of adventitious roots emerging from the second node of *Zea mays* plants (cv. LG 11) as previously described (Clarkson *et al.*, 1987). Most of the 50-mm root zone had been exposed to moist air above the level of the culture solution. These 'air-grown' sleeves were mechanically stronger than those from roots immersed in the solution ('solution-grown'). Sleeves 8 to 10 mm long were mounted on a perforated polythene tube and the ends sealed onto the tube with a 50:50 paraffin wax/resin mixture. The tube was then inserted and sealed through holes in the circumference of a cylindrical metal pressure chamber equipped with a gas-tight cover (Fig. 1). One end of the tube was closed with a clip, the other was sealed to a $100\,\mu$l 'Microcap' capillary. Movement of liquid (usually distilled water) in the capillary was measured with a travelling microscope; the minimum detectable volume change was $0.02\,mm^3$. Pressure or suction was applied by an air compressor, or a vacuum pump fitted with an Edwards USK 1B vacuum switch, in series with a $0.1\,m^3$ steel cylinder to buffer fluctuations and a mercury manometer. In the Results and discussion sections, axial pressure and suction refer respectively to positive and negative pressures applied to the capillary attached to the sleeve and radial pressure and suction those applied to the tube sealed into the cover of the pressure chamber (Fig. 1). The first reading in each series of measurements was neglected since the movement here was largely due to stretching or shrinkage of the sleeve on the supporting polythene tube. In experiments to examine the permeability to molecules of different size under solvent drag accompanying applied pressure, THO ($100\,\mathrm{Bq\,mm^{-3}}$), ^{3}H-glucose and ^{14}C-polyethylene glycol 4000 (PEG) ($10\,\mathrm{Bq\,mm^{-3}}$) or ^{14}C-urea ($40\,\mathrm{Bq\,mm^{-3}}$) were incorporated in the medium together with 0.1 per cent of the appropriate unlabelled organic compounds as carriers. To examine the effect of osmotic as opposed to hydrostatic pressure on radial water movement sucrose solutions ($-10\,\mathrm{bar}$, 12.3% w/v and $-40\,\mathrm{bar}$, 37.5% w/v) were introduced inside or outside the sleeve.

Intact root segments 20 to 25 mm long were excised at 50 to 70 mm from the tips of nodal roots. These segments, unlike most of those used for preparing sleeves, had been immersed during growth below the level of the culture solution. The segments were sealed into polythene tubes and mounted in the pressure chamber (Fig. 1) and surrounded by culture solution of the same com-

position as that used for growing the plants. Preliminary experiments, confirmed by sealing the cortex and stele separately with wax (Fig. 1), showed that when positive axial pressure was applied to these short segments, water moved axially largely through the cortex, presumably by infiltration of the intercellular air spaces, rather than through the xylem vessels.

Results

Hypodermal sleeves: rectification of flow

After closing one of the capillaries attached to the sleeve (Fig. 1) we measured radial flow in response to axial and radial pressure or suction on six sleeves that behaved qualitatively in the same manner, sleeve 5 (Fig. 2) being typical. A series of successive measurements of flow rate was made under a pressure difference of 0.1 bar at 2 minute intervals, starting with flow under radial pressure. This was followed by application of radial suction, axial pressure and axial suction. Only in this last treatment, the situation normally experienced in the transpiring plant, was there any substantial flow (Fig. 2a). When however these treatments were immediately repeated in the same order, flow rates were comparably large in all treatments (Fig. 2b). The same sleeve was then allowed to remain under atmospheric pressure for 48 hours at 5°C and the four treatments repeated, when once again only axial suction gave substantial flows (Fig. 2c). Blockage of radial flow caused by appression of the sleeve against the holes in the support could not account for these effects since under axial pressure and radial suction the sleeve was stretched away from the support. On other sleeves we further noted that, whereas with axial suction, flow rate under a given pressure difference remained constant over extended periods, with the other treatments, flows decreased with time, particularly at larger pressure differences (0.2 to 0.4 bar). The sleeves then remained relatively non-conducting until the hydraulic conductivity was initiated by axial suction. The rectifier-like properties of the sleeves thus depended not only on their previous treatment but also on the magnitude of the applied pressure difference.

Fig. 1. Apparatus for measuring radial water movement in hypodermal sleeves or intact root segments in response to pressure or suction. Hypodermal sleeves (*lower diagram*) were mounted on a perforated polythene tube and root segments (*upper diagram*) sealed into plastic tubes; sleeves or segments were then sealed into a metal pressure chamber (*upper diagram*). Axial pressure or suction could be applied to the tubes attached to the sleeves or segments and radial pressure or suction to the tube mounted in the cover of the pressure vessel.

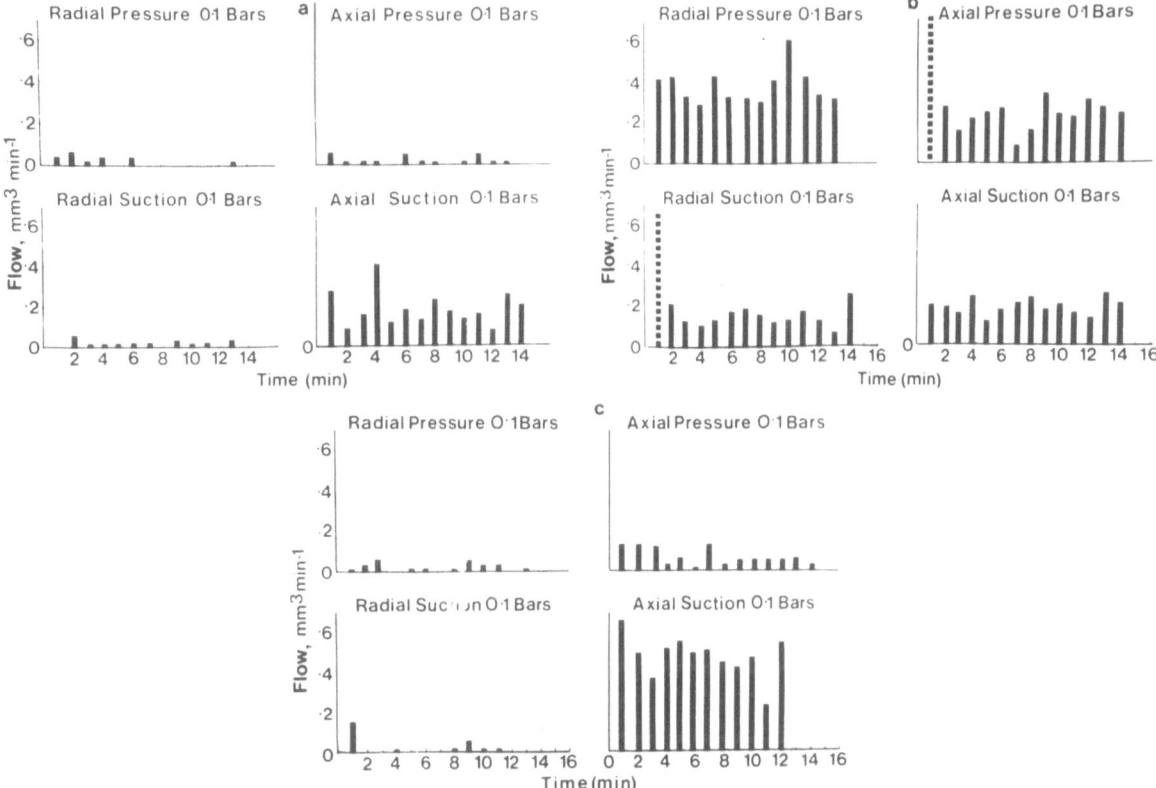

Fig. 2a, b, c. Radial flow-rates in hypodermal sleeves measured at 2 minute intervals under a pressure difference of 0.1 bar. Measurements were made in the order; Radial pressure, Radial suction, Axial pressure, Axial suction. Measurements in Fig. 2b immediately followed those in Fig. 2a. A period of 48 hours elapsed between measurements in Figs. 2b and 2c, during which the sleeve was kept under atmospheric pressure at 5 °C.

Hypodermal sleeves, hydraulic conductivity

Under rising axial suction, flow increased but in a non-linear manner, so that the volume hydraulic conductivity or conductance (ratio of flow rate to applied pressure difference) increased with rising suction or flow rate. On decreasing the suction, the conductance did not decline. Incorporation of THO and ^{14}C-labelled urea in the water surrounding the sleeve showed that the increase in volume hydraulic conductance was associated with an increase in the ratio of urea to THO moving radially through the sleeve under solvent drag (Fig. 3). This suggested that the enhancement of the conductivity was associated with the opening up of more pores, or the enlargement of existing pores in the sleeves, allowing a greater proportion of the larger urea molecules to pass through the walls of the sleeve (Fig. 3). By contrast, these 'air-grown' (see

Methods) sleeves were largely impermeable under solvent drag to ^{3}H-labelled glucose and ^{14}C-labelled PEG incorporated in the ambient medium (Table 1), from which we inferred that the reflection coefficient for these substances was approximately unity.

Over a range of axial suction (0.05 to 0.5 bar) the hydraulic conductivity per unit surface area (L_p) of four 'air-grown' sleeves varied from 0.4×10^{-4} to 4.3×10^{-4} mm sec^{-1} bar^{-1}. Most of the 'solution-grown' sleeves ruptured under applied suction but one sleeve over this range gave L_p of 2×10^{-2} to 4×10^{-2} mm sec^{-1} bar^{-1} or about 100 times greater than the 'air-grown' sleeves. For 'air-grown' sleeves that had not previously been subjected to axial suction, radial pressure gave L_p of around 10^{-5} mm sec^{-1} bar^{-1}, but as pressures greater than 0.25 bar were prolonged, L_p tended to decline and in some cases became immeasurably small.

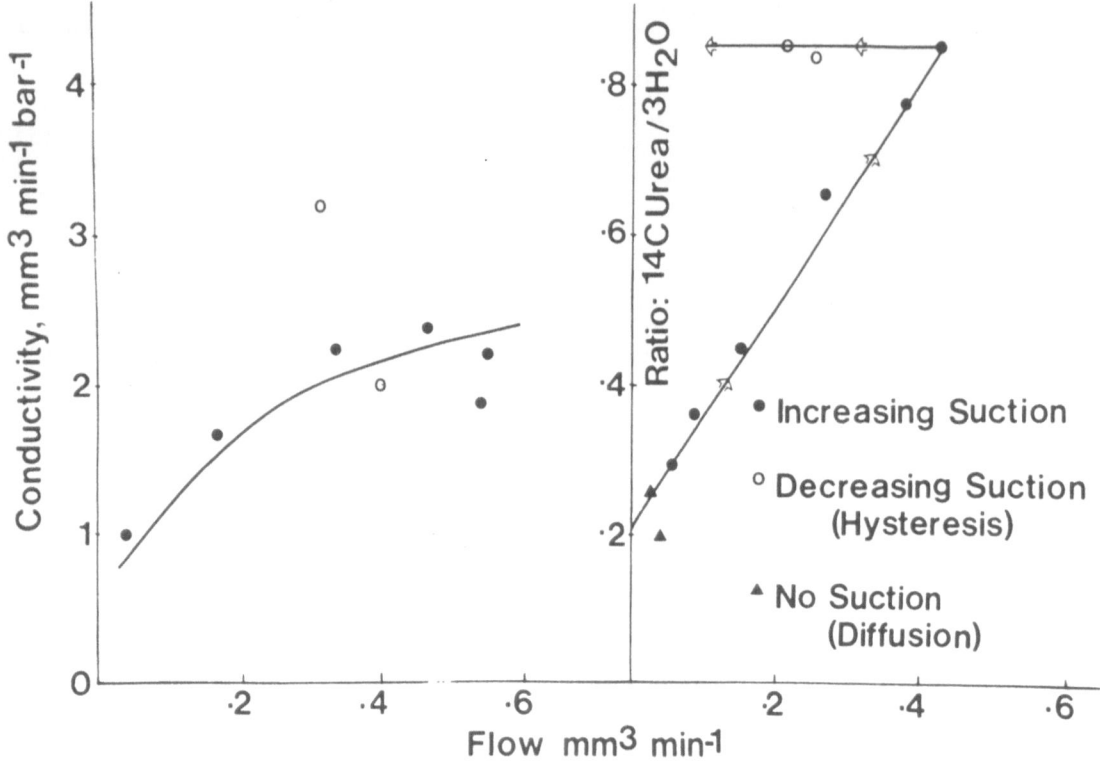

Fig. 3. Relationships between flow rate across a sleeve under axial suction, volume hydraulic conductivity (conductance) and the ratio of labelled urea to THO, relative to that in the ambient medium, transported across the sleeve by diffusion or under solvent drag.

Table 1. Ratio of concentration of labelled glucose and PEG-4000 inside lumen of sleeves to that in the ambient solution (R In/Out) after application of 0.2 bar axial suction

	Sleeve 2	Sleeve 3
Total water moved, mm^3	11.2	34.8
Flow rate, mm^3 sec^{-1} × 10^3	1.8	3.3
R In/Out: Glucose	0.018	0.017
PEG	0.001	0

Hypodermal sleeves, hydraulic conductivity under osmotic potential differences

The results shown in Table 1 suggested that the reflection coefficient for sucrose, intermediate in molecular size between glucose and PEG, would be approximately unity. However sucrose solutions (−10 and −40 bar) placed inside or outside 'air-grown' sleeves over periods of up to 90 minutes induced only very small flows that were difficult to measure accurately, but the L_p's for endosmosis and exosmosis respectively were not greater than 2×10^{-8} and 1×10^{-7} mm sec^{-1} bar^{-1}, or smaller

than the minimum hydraulic conductivity under hydrostatic pressure differences by a factor of at least 10^3.

Intact root segments

Sealing the cortex and stele separately with wax (see Fig. 1 and Methods) had established that most of the axial flow through the living root segments occurred through the intercellular spaces of the cortex. This tissue would therefore have contributed some resistance to the flow through the cortex and hypodermis, resulting in a pressure drop along the segment axis. Values of the apparent L_p quoted in the present section are based on the applied pressure difference and may therefore underestimate the L_p of the hypodermis. In most of these experiments we did not seal the cortex and hypodermis but one of the tubes attached to the segment (Fig. 1) was closed with a clip.

Measurements on 10 segments of flow in response to pressure or suction showed that they behaved in a manner qualitatively similar to the sleeves, although apparent hydraulic conductivities were generally larger in these 'solution-grown' roots than in the 'air-grown' sleeves. Fig. 4 shows a typical example; initially applied radial pressure over the range 0.1 to 0.5 bar gave only small flows (histograms 1 to 3). Subsequent axial suction (histograms 4 to 6) gave larger flows. When radial pressure was again applied (histograms 7 to 9), flows rose to up to four times the initial values. Subsequent axial pressure (histograms 10 to 12) and radial suction (histograms 13 to 15) gave even larger flows. Finally, axial suction was again applied (histograms 16 to 18) giving a small flow rate. This last result seemed to exclude the possibility that the increase in flow rate observed over treatments 1 to 15 has been due to progressive infiltration of the cortical intercellular spaces and hence a decrease in the axial resistance. In these experiments the apparent L_p ranged from 1.7×10^{-4} (histogram 1) to 3×10^{-3} mm sec^{-1} bar^{-1} (histogram 15). Fig. 5 shows experiments on another segment in which the flow rate again increased curvilinearly with applied axial suction but exhibited hysteresis on decreasing the suction. The apparent hydraulic conductivity thus rose with in-

creasing flow rate but remained constant when suction declined. This may be compared with Fig. 3 for a hypodermal sleeve although L_p for the sleeve was smaller by about an order of magnitude. In some experiments both of the tubes attached to the segment (Fig. 1) were left open and flow in both tubes was measured from the movement of an air bubble in each tube. When axial pressure was initially applied to one tube, between 95 and 100 per cent of the water emerged in the other tube, so that the axial resistance of the segment was very much smaller than the radial resistance of the hypodermis to outward flow. Incorporation of THO in the perfusing solution enabled radial permeability coefficients (P_d) to be determined for the segments assuming that radial movement of THO was by diffusion rather than by mass flow. For three segments taken at 50 to 70 mm from the root tip, P_d ranged from 4×10^{-4} to 1×10^{-3} mm sec^{-1} and for two basal segments 150 mm from the tip, 5×10^{-5} and 1×10^{-4} mm sec^{-1}. These P_d values are sufficiently low to be largely unaffected by unstirred-layer effects and agree reasonably well with those of Clarkson *et al.* (1987, Fig. 12) for 'solution-grown' hypodermal sleeves, and with measurements on living segments using a different

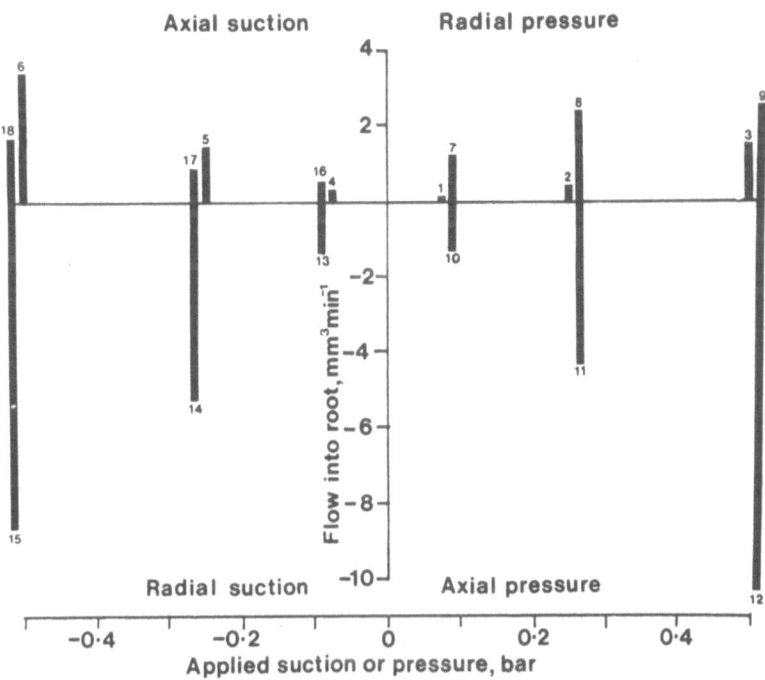

Fig. 4. Radial flow across the hypodermis of a maize root segment in response to pressure differences of varying magnitudes and directions. Successive measurements were made in the order denoted by the numbers above the histograms.

162   Shone and Clarkson

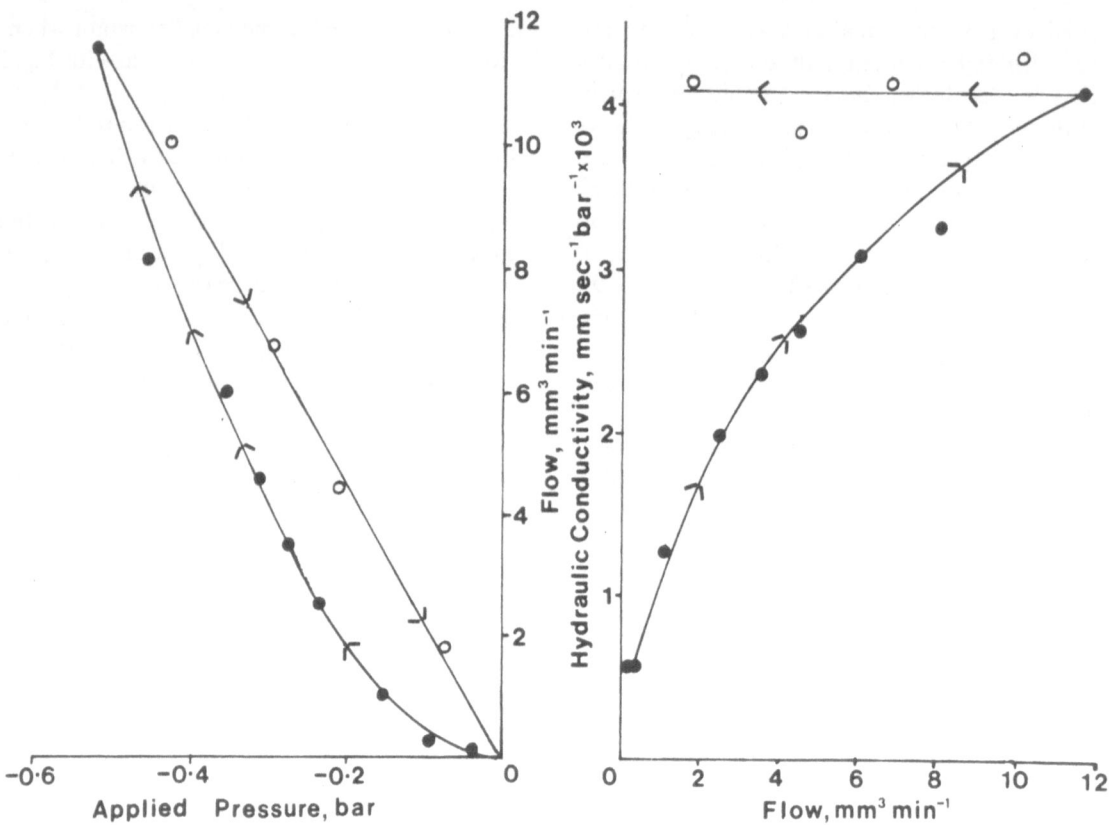

Fig. 5. Relationships between applied pressure and flow, and between flow and apparent hydraulic conductivity, for a maize root segment under axial suction. *Closed circles*, increasing suction; *open circles*, decreasing suction.

procedure (Shone *et al.*, 1984). We therefore conclude that isolation of the sleeves by enzymic digestion did not greatly alter their permeabilities.

By contrast with axial pressure, when axial suction was applied to one end of the segment, over 90 per cent of the water entered the root by radial inflow rather than through the distal end of the segment. Flow through the xylem vessels was thus very small as confirmed by separately sealing the cortex and stele (see Methods). Axial pressure applied immediately after axial suction, however, caused substantial radial flow through the hypodermis as found with sleeves (Fig. 2) and intact root segments (Fig. 4). No attempts were made to compare hydrostatic and osmotic hydraulic conductivities on root segments. Under axial suction, substantial quantities of labelled glucose and PEG were radially transported with water suggesting that the reflection coefficient under these circumstances was much smaller than unity.

We also made a few estimates of the stelar L_p, based on the external surface area of freshly excised root segments, by applying pressure or suction to the stele after sealing off the cortical axial pathway with wax (Fig. 1). Axial pressure (0.66 bar) gave $L_p = 2 \times 10^{-5} \, mm \, sec^{-1} bar^{-1}$, a value about twice that found by Steudle (1988) in maize using the pressure probe technique. The mean value of the stelar L_p from six sets of determinations ranging from 0.2 to 0.7 bar radial pressure on one root was $(1.03 \pm 0.17) \times 10^{-5} \, mm \, sec^{-1} bar^{-1}$. Axial suction on the stele however consistently gave larger values of L_p (1 to $2 \times 10^{-4} \, mm \, sec^{-1} bar^{-1}$). If movement under hydrostatic pressure is largely apoplastic (Steudle, 1988) and the cortical resistance is relatively small, our results suggest that in this zone of the root the endodermis and hypodermis have comparable L_p's when expressed per unit surface area of each tissue and that the L_p of the endodermis may also vary with the direction of flow and magnitude of the pressure difference.

Discussion

We wish to draw attention to three aspects of this work namely: (a) the observations that in both sleeves and living root segments the hydraulic conductivity depends on the direction of the applied pressure difference. This questions the interpretation of experiments in which the water relations of whole plants or detached roots are investigated by subjecting the root systems to pressures greater than atmospheric. (b) The evidence that in both sleeves and segments the hydraulic conductivity varies with the applied pressure difference in a manner that has frequently been observed in intact plants (Weatherley, 1982). (c) The evidence from hypodermal sleeves that hydrostatic pressure differences are far more effective than equivalent differences in osmotic potential as a driving force for radial flow.

These effects have been for some time familiar in some animal and plant tissues (Hakim and Lifson, 1969; Spyropoulos, 1983; Steudle, 1988; Vargas, 1968). We note that Hakim and Lifson found that in dog intestinal mucosa L_p measured under hydrostatic gradients was greater than the osmotic L_p by a factor of about 10^4.

A number of proposals have been put forward to explain these effects. Possibly unstirred layers at the osmotic barrier in the tissue, or an asymmetrical osmotic barrier, could account both for the smaller driving force under osmotic gradients and for polarity effects under some conditions but it is difficult to envisage how such mechanisms could operate in hypodermal sleeves, which are essentially dead material, in pure water. Steudle (1988) concluded that under hydrostatic gradients water moved primarily in the apoplast whereas osmotic gradients brought about larger movement from cell to cell through the symplast and cell vacuoles. Mechanical effects have also been proposed (Spyropoulos, 1983) and possibly the suberin lamellae could play a role in blocking pores in the hypodermal tissue to varying extents depending on the magnitude and direction of applied pressure. If a Casparian band is present in this tissue, one might speculate that the endodermis and hypodermis have similar hydraulic properties, and our experiments on applying axial pressure or suction to the stele seem to support this hypothesis. Nevertheless, suberin lamellae develop in both tissues with increasing age of the root. The polarity of water movement in the hypodermis may have adaptive significance in so far as it limits water loss by the cortex to dry soil while allowing some absorption of water and nutrients under more favourable environmental conditions.

Acknowledgements

We thank Ann Flood, Helen Ponting and John Sanderson for experimental assistance.

References

Clarkson D T, Robards A W, Sanderson J and Peterson C A 1978 Permeability studies on epidermal-hypodermal sleeves isolated from roots of *Allium cepa* (onion). Can. J. Bot. 56, 1526–1532.

Clarkson D T, Robards A W, Stephens J E and Stark M 1987 Suberin lamellae in the hypodermis of maize (*Zea mays*) roots: Development and factors affecting the permeability of hypodermal layers. Pl. Cell Env. 10, 83–93.

Hakim A A and Lifson N 1969 Effects of pressure on water and solute transport by dog intestinal mucosa *in vitro*. Am. J. Physiol. 216, 276–284.

Peterson C A 1988 Significance of the exodermis in root function *In* Structural and Functional Aspects of Transport in Roots. Eds. B C Loughman *et al.* pp. 35–40. Kluwer Academic Publishers. Dordrecht, The Netherlands.

Shone M G T, Bartlett B O and Flood A V 1984 Permeability of maize nodal roots to tritium-labelled water in relation to root anatomy. A F R C Letcombe Laboratory Annual Report for 1983, 75–77.

Spyropoulos C S 1983 Hydraulic conductivity of Nitella cells using the intracellular perfusion technique. J. Memb. Biol. 76, 17–26.

Steudle E 1988 Water transport in roots. *In* Structural and Functional Aspects of Transport in Roots. Eds. B C Loughman *et al.* pp. 139–145. Kluwer Academic Publishers, Dordrecht, The Netherlands.

Vargas F F 1968 Filtration coefficient of the (squid) axon membrane as measured with hydrostatic and osmotic methods. J. Gen. Physiol. 51, 13–27.

Weatherley P E 1982 Water uptake and flow in roots. *In* Encyclopaedia of Plant Physiology. New Series Vol. 12B, Physiological Plant Ecology II. Water Relations and Carbon Assimilation. Eds. O L Lange *et al.* pp 79–109. Springer-Verlag.

B. C. Loughman et al. (Eds.), Structural and functional aspects of transport in roots, 165–168.
© 1989 by Kluwer Academic Publishers.

Patterns of long-distance movement of water in roots

I. MISTRÍKOVÁ and V. KOZINKA
Institute of Experimental Biology and Ecólogy, CBES, Slovak Academy of Sciences, CS-814 34 Bratislava, Dúbravská 14, Czechoslovakia

Key words: functional vessels, individual roots, root system, *Zea mays*

Introduction

Determining the relationship between xylem anatomy and water flow is essential to our understanding of its development and function in the plant. Studies on the conducting capacity of xylem vascular tissues enables us to examine the potential ability of water flow in the plants. For a complete description of long distance water flow it is important to identify the functional vessels. The pattern of vasculature has been analyzed mostly in shoots of arborescent monocotyledons (Zimmermann, 1983; Zimmermann and Tomlinson, 1968), dicotyledons (Chaney and Kozlowski, 1977; Newbanks *et al.*, 1983; Zimmermann, 1978) and conifers (Ewers, 1985; Petty, 1970). Very few investigations of patterns of water movement in conducting elements of roots of herbaceous plants have been carried out.

Experiments reported here attempt the investigation of the conduction of water in individual roots and in the root system of intact *Zea mays* L. plants. For this aim a periodic acid Schiff's reagent staining technique (Chaney and Kozlowski, 1977) was used, that might possibly reveal the existence of the functional character of individual xylem vessels.

Materials and methods

Zea mays L. cv. CE-330 were cultivated in nutrient solution and in the field. The first group of plants was grown in a growth chamber in a continually aerated Knop nutrient solution. During the 14 h dark period the temperature was 25°C, the RH about 50·70%. The second group of maize plants was grown under field conditions. The root system of these plants was washed clean of soil before staining.

The PAS technique was used for staining. The 2/3 of the apical parts of the roots of intact plants were immersed in 0.1% solutions of periodic acid at a pH of 5.5 for 1 h. They were then rinsed in distilled water and placed in Schiff's reagent until the dye was visible in the uppermost leaves (usually after 1–2 h). All dye ascents were conducted in the growth chamber with supplementary lighting and circulating fans to increase transpiration. 5-mm long segments of the basal part of roots were fixed with 50% acetone for 30 minutes, then they were dehydrated in 100% acetone for 2×60 minutes and transferred through acetone–benzene (1:1) for 60 minutes and 100% benzene for 2×30 minutes into paraffin. Transverse sections, $20\,\mu m$ thick, were cut with a Cryocut microtome and the sections were transferred $3 \times$ in xylene, followed by mounting with Canada balsam.

Schiff's reagent, a leuco-compound formed when basic fuchsin dye is reduced with a sulphite, is used for the detection of aldehydes. The periodic acid has the capacity to break the carbon bonds between 1,2–glycol groups (CHOH-CHOH) and produce dialdehydes which are then available to react with Schiff's reagent (Pearse, 1962).

Such 1,2–glycol groups exist in the glucose moieties of cellulose in cell walls. When periodic acid is transported through the conducting elements of a root, the 1,2–glycol groups in the cell walls are converted to dialdehydes. These react with Schiff's reagent that is subsequently transported and give a violet colour. 'The stain is fixed in the tissue because the oxidized Schiff's reagent dialdehyde complex becomes an integral part of the cell wall structure. The stain is not removed during cutting of thin sections or during dehydration for the preparation of permanent slides' (Chaney and Kozlowski, 1977).

Under the microscope only vessel walls can be seen and the method does not show the presence of liquid in the vessel lumen. The wall staining takes place only as a result of the liquid flowing through the vessel lumen.

Results and discussion

Patterns of water movement in individual roots

8-day-old maize plants from the growing chamber were used for experiments to observe the staining sequence of individual xylem vessels. The roots of intact plants were immersed in the solution of periodic acid for 1 h. After washing they were transferred to Schiff's reagent for 30, 40, 50 and 60 minutes. The segments were derived from basal parts of primary seminal roots. Only the vessels of protoxylem and early metaxylem were stained by a 30 minute treatment of Schiff's reagent (Fig. 1a). The vessel walls of protoxylem, early metaxylem and parts of late metaxylem vessels neighbouring the early metaxylem were stained after 40 minutes (Fig. 1b). The parts of walls adjacent to early metaxylem of all late metaxylem were stained after 50 minutes of staining (Fig. 1c). All the vessels and walls of parenchymatous cells surrounded by the late metaxylem vessels were stained after 60 minutes treatment with Schiff's reagent (Fig. 1d). This phenomenon explains 'contact staining' (Zimmermann, 1983). Wall staining can take place

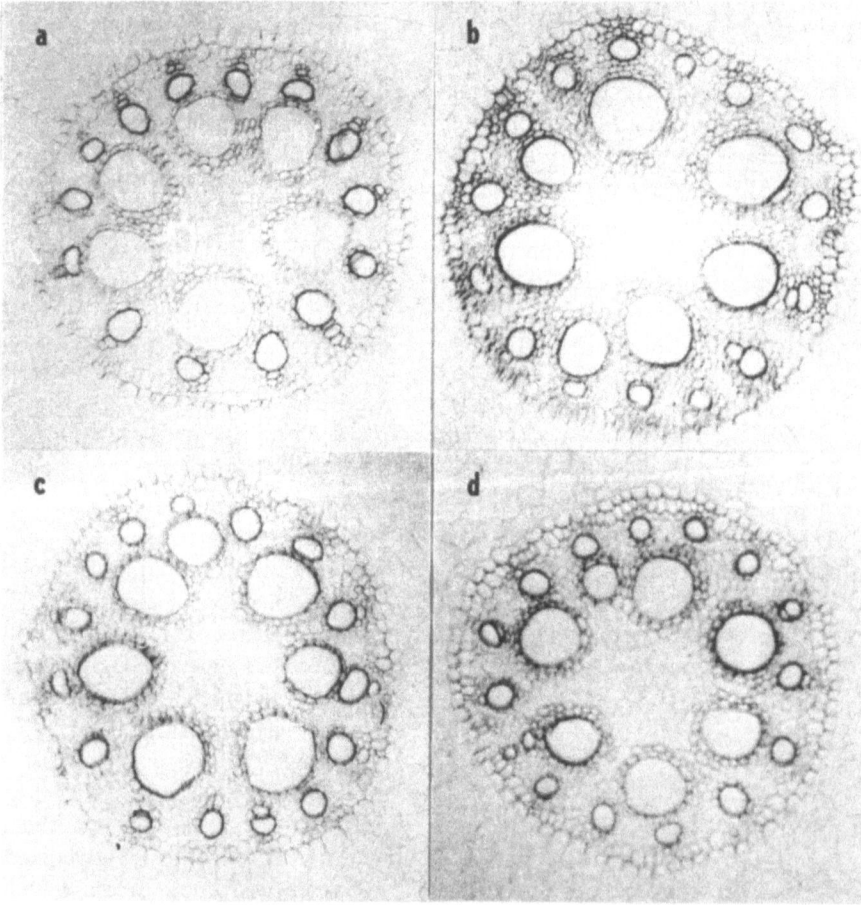

Fig. 1. Transverse section of 8-day-old maize root section from the basal part of a primary seminal root showing vascular elements stained after treatment of periodic acid for 60 minutes and Schiff's reagent for a = 30 minutes, b = 40 minutes, c = 50 minutes, d = 60 minutes.

not only by the liquid flowing through the vessel lumen, but also by the liquid seeping into the wall of an unstained cell from a neighbouring conducting element (stained vessel).

Patterns of water movement in the root system

Plants from the field trials were used for this experiment. In young maize plants (up to 30 days old), water movement, as indicated by staining, occurred primarily in the vessels of primary seminal and seminal adventitious roots. The participation of nodal roots in water conduction increased with the age of the plant.

Experiments with 62-days-old maize plants have shown that in primary seminal roots (Fig. 2a) and in seminal adventitious roots (Fig. 2b) all the vessels were stained. In nodal roots of the 1st node

(Fig. 2c) and of the 2nd node (Fig. 2d) all the vessels were also stained. In nodal roots of the 3rd node only the walls of protoxylem and early metaxylem were stained, the late metaxylem vessels not being open, *i.e.* have not yet started to accomplish their function. In nodal roots of the 4th node no vessels were stained.

The most satisfactory method for detection of vascular tissues available for water transport is by staining (Talboys, 1955). In the present work the modified PAS staining method was used with maize roots and it was necessary to ascertain the correct concentration of the periodic acid solution. In connection with the possibility of contact staining it was important to determine the optimal time of the reaction of Schiff's reagent. Our results have shown that the method used is suitable for determining functional tissues in maize roots.

Results of the experiments with individual roots

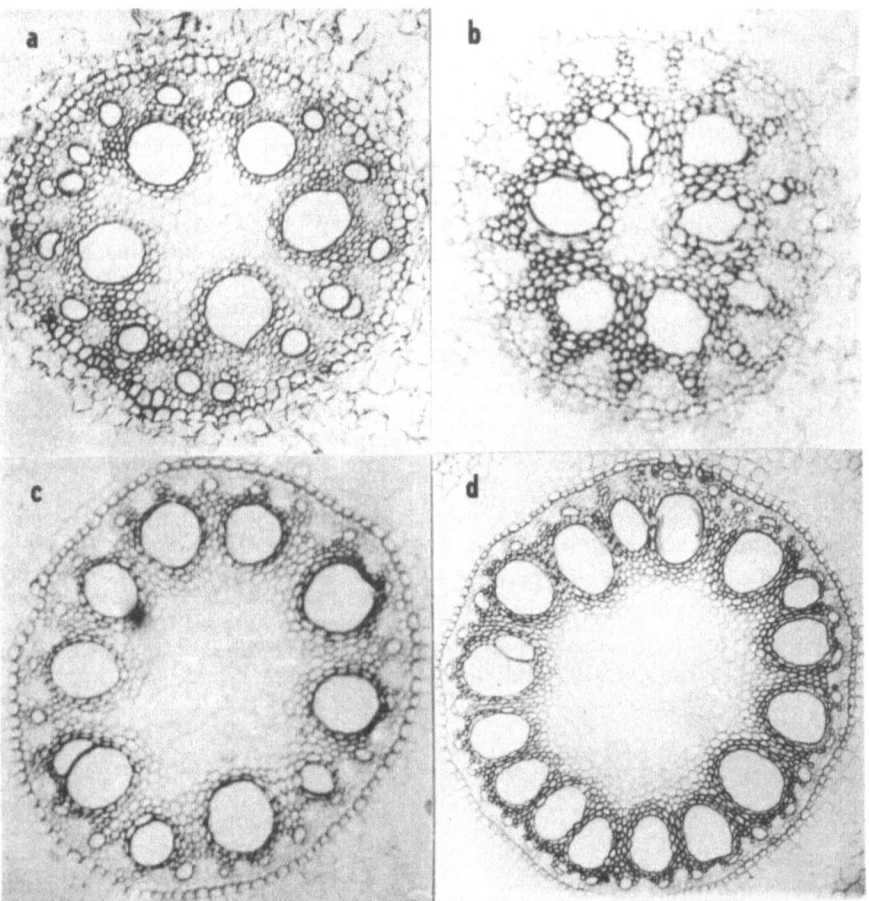

Fig. 2. Transverse section of a 62-day-old maize plant root system showing vascular elements stained after conduction of periodic acid Schiff's reagent. a = primary seminal root, b = adventitious root, c = nodal root of the 1st node, d = nodal root of the 2nd node.

are in accordance with results of studies on ^{32}P transport (Burley *et al.*, 1970) and $^{35}SO_4^{2-}$ (Weigl and Lüttge, 1962) through maize roots. The results of studies on ^{32}P transport and dye flow experiments suggest that only the outer, early differentiated metaxylem vessels and the protoxylem are pathways of upward transport to the shoot (Burley *et al.*, 1970). Weigl and Lüttge (1962) showed in microautoradiographic experiments on $^{35}SO_4^{2-}$ transport through maize roots that the outer metaxylem vessels were the most heavily labelled of the vessels. According to Postlethwait and Nelson (1957) the protoxylem serves for water conduction initially and during elongation, whereas metaxylem assumes this function immediately at its maturation. In maize roots there is a gradual, slow development of metaxylem vessels along the length of the root, the outer, early metaxylem vessels differentiate much more rapidly than the inner, late ones (Burley *et al.*, 1970). Therefore, maize root segments, through which transport occurs, may contain both mature and still differentiating metaxylem vessels. Hence, it is difficult to relate the transport function to a particular developmental stage of the metaxylem. In maize the physiological evidence suggests that transport occurs in mature fully differentiated vessels (Läuchli *et al.*, 1974). Anatomical evidence in support of this phenomenon was provided by Aubin *et al.* (1986); they observed that living elements persist in the late metaxylem up to 20–30 cm from the tip. Our results in maize nodal roots confirm these observations.

Conclusions

Patterns of long-distance water movement in maize roots were identified by staining with periodic acid Schiff's reagent. In individual roots the water movement occurs primarily in the outer, early differentiated metaxylem vessels and in

protoxylem. The vessels of the late metaxylem also participated in water flow. In young maize plants, water movement occurs primarily in conducting vessels of primary seminal roots and seminal adventitious roots. The participation of nodal roots in water conduction increases with the age of the plant.

References

Aubin G, Canny M J and McCully M E 1986 Living vessel elements in the late metaxylem of sheathed maize roots. Ann. Bot. 58, 577–588.

Burley J W A, Nwoke F I O, Leister G L and Popham R A 1970 The relationship of xylem maturation to the absorption and translocation of ^{32}P. Am. J. Bot. 57, 504–511.

Chaney W R and Kozlowski T T 1977 Patterns of water movement in intact and excised stems of *Fraxinus americana* and *Acer. sacharum* seedlings. Ann. Bot. 41, 1093–1100.

Ewers F W 1985 Xylem structure and water conduction in conifer trees and lianas. IAWA Bull. 6, 309–317.

Läuchli A, Kramer D, Pitman M G and Lüttge U 1974 Ultrastructure of xylem parenchyma cells of barley roots in relation to ion transport to the xylem. Planta 119, 85–99.

Newbanks D, Bosch A and Zimmermann M H 1983 Evidence for oxylem dysfunction by embolization in Dutch elm disease. Phytopathol. 73, 1060–1063.

Pearse A G E 1962 Histochemistry. Moscow, p 207–224.

Petty J A 1970 Permeability and structure of the wood of Sitka spruce. Proc. Roy. Soc. Lond. B 175, 149–166.

Posttlethwait S N and Nelson O E 1957 A chronically wilted mutant of maize. Am. J. Bot. 44, 628–633.

Talboys P W 1955 Detection of vascular tissues for water transport in the hop by colourless derivates of basic dyes. Nature (London) 175, 510.

Weigl J and Lüttge U 1962 Mikroautoradiographishe Untersuchungen über die Aufnahme von $^{35}SO_4^{2-}$ durch Wurzeln von *Zea mays* L.: Die Funktion der primären Endodermis. Planta (Berl.) 59, 15–28.

Zimmermann M H 1978 Hydraulic architecture of some diffuse porous trees. Can. J. Bot. 56, 2286–2295.

Zimmermann M H 1983 Xylem Structure and the Ascent of Sap. Springer-Verlag, Berlin, Heidelberg, New York, Tokyo.

Zimmermann M H and Tomlinson P B 1968 Vascular constriction and development in the aerial stem of Prionium (Juncaceae). Am. J. Bot. 55, 1100–1109.

B. C. Loughman et al. (Eds.), Structural and functional aspects of transport in roots, 169–173.
© 1989 by Kluwer Academic Publishers.

Water uptake in the root system of Gramineae

V. KOZINKA

Institute of Experimental Biology and Ecology CBES, Slovak Academy of Sciences, CS-814 34 Bratislava, Dúbravská 14, Czechoslovakia

Key words: nodal roots, seminal roots, water uptake, *Zea mays*

Introduction

Recent interest in modelling the water relations of plants has revealed gaps in the experimental information needed for a full description of water uptake by a complete root system. Calculating rates of water uptake by root systems requires not only knowledge of the demands placed on the root system by the shoot together with the magnitude of

the water source term, but also accurate assessment of the water transfer function of the root system (Sanderson, 1983).

The capacity of the root system to absorb and transfer water does not rise in direct proportion to the increasing length or area of individual roots, since new roots are being added while older roots are maturing and becoming less permeable. Obviously, roots varying so much in structure, must

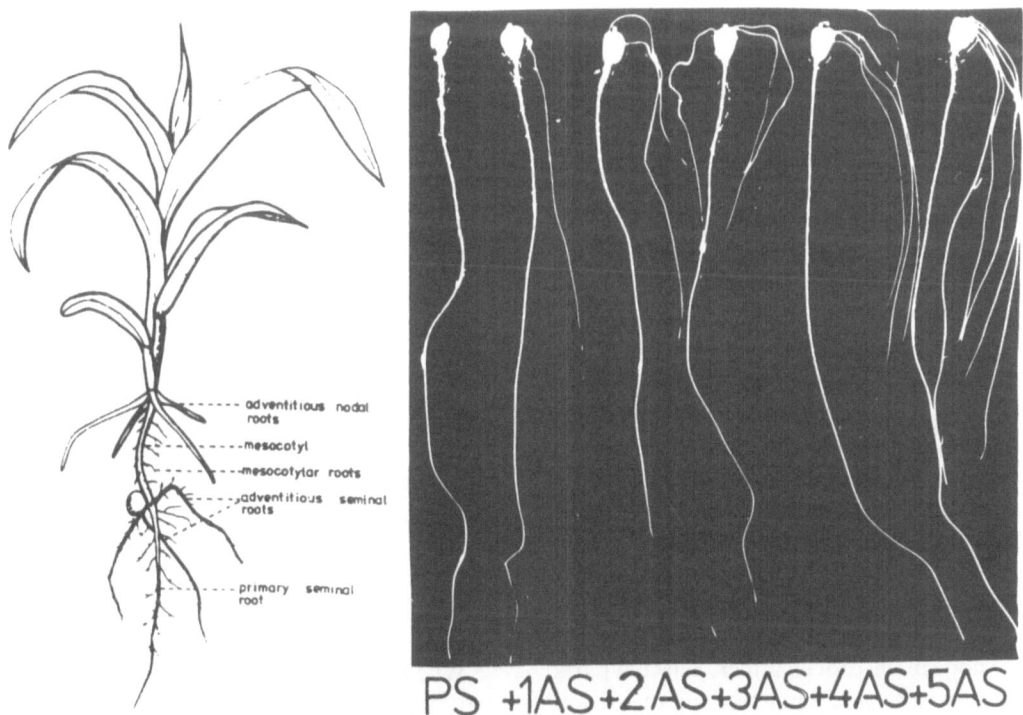

adventitious nodal roots

mesocotyl

mesocotylar roots

adventitious seminal roots

primary seminal root

PS +1AS+2AS+3AS+4AS+5AS

Fig. 1. Schematic illustration of *Zea mays* seedling (*on the right*) and seminal root system of the 91 hour old plants cv. CE-330 (*on the left*).

169

vary widely in permeability to water (Fiscus and Markhart, 1979). The effectiveness of a deep, wide spreading root system depends partially on its longitudinal or axial conductance. This is usually assumed to be high. On the other hand some authors found high resistance to axial water movement in certain grasses (Passioura, 1980).

A complicated development of the actual absorptive and transfer capacity of the root system is apparent in annual plants of the family Poaceae. The root system grows into a complex form where the individual roots are of different origin. One part of them originates as early as in the embryo (seminal roots). The others grow on the base of the stem internodes (adventitious stem or nodal roots). It has been postulated that seminal roots function only at the beginning of ontogenesis then wither, their function being taken up by the nodal roots of the 1st up to n-th node.

The root system of *Zea mays* is formed by the seminal primary root, by none or only some adventitious seminal roots and by adventitious stem roots. Furthermore, adventitious roots can grow on the mesocotyl (Fig. 1). In the present paper the rate of elongation of the youngest expanding leaf of intact maize plants was measured as a function of water uptake. A linear variable transducer (LVDT) was used to measure the elongation rate after changing the water potential of the solution bathing different parts of the root system.

Materials and methods

The results reported here were obtained with *Zea mays* cv. CE-330 plants grown in the growth chamber in a continually aerated Knop nutrient solution. During the 14 h dark period the temperature was 25°C, the RH about 50% and 70% respectively. During the light period the elongation rate of the intact youngest expanding leaf was continually measured by an electromechanical linear variable differential transducer. The nondestructive nature of this measurement enables plant growth response to environment changes to be immediately recorded. The intact maize plant was clamped at the coleoptilar node in the root chamber. Selected parts of the root system were immersed in the Knop nutrient solution and a low external water potential was induced by rapidly exchanging the pretreat-

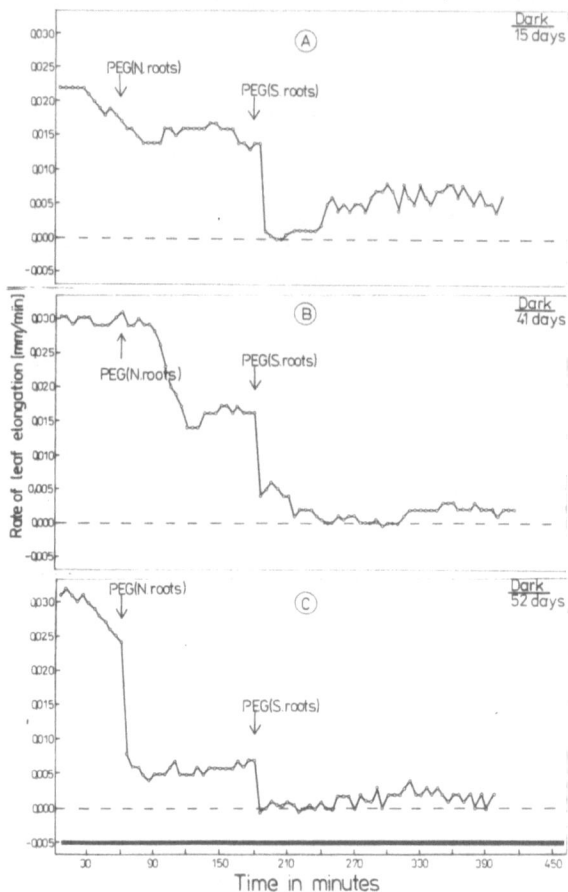

Fig. 2A, B, C. Effects of sudden change in water potential of the medium surrounding nodal (*the first arrow*) and seminal (*the second arrow*) roots on elongation rate of the youngest expanding leaf of *Zea mays* cv. CE-330. Right top age of plant in days. Experiment in dark.

ment nutrient solution for the low water potential nutrient solution with PEG-6000 (Serva). One concentration was used with a water potential of the nutrient solution of -0.23 MPa and all solutions were continually oxygenated prior to use. Oxygenation of PEG-solutions prior to use was necessary since the partial pressure of O_2 was low in nonoxygenated solutions.

Results

When the roots of intact maize plants are subjected to a sudden reduced external water potential a rapid decrease in the elongation rate of the youngest expanding leaf is observed (Fig. 2). Changes in the elongation rate are more dynamic at a sudden change in the water potential of the medium bathing the seminal roots and these changes are seen in seedlings and young plants but less so in older plants. A lowered water potential of the root medium should increase the oscillatory tendency of the water regulatory system (Fig. 2).

If we consider the changes in the elongation rate of the youngest expanding leaf as a function of the changes in the water uptake rate our experimental results support the view that seminal roots of maize are very effective in water absorption and translocation.

The results provide additional evidence supporting and extending previous conclusions that leaf elongation in maize responds immediately and is extremely sensitive to changes in water status of the root medium. The rapidity of this response is in agreement with the idea of water in the xylem acting as a hydraulic unit to bring about immediate changes in the leaf water status when the water potential of the root medium is changed (Kozinka, 1985). The comparison of changes in the elongation rates of the youngest expanding leaf reveals that a specific part of the root system rather than the age of the plant seems to be a dominant factor controlling water uptake characteristics in the root system of Gramineae.

Discussion

We have shown that the primary seminal root and adventitious seminal roots are the permanent part of the root system grown in water culture.

They maintain the capacities of the organs responsible for the longitudinal water flow to the end of the life of the plant (Kozinka, 1977; 1978). The conducting efficiency of all maize roots for the longitudinal water flow is high on the basis of the measurement of the actual flow volumes through the detached segment of roots of the cultivar VIR-17 (Table 1; Kozinka, 1981; Luxová and Kiozinka, 1970; 1973).

When the aerial part of the maize plant becomes larger an increasing number of thicker nodal roots is formed on higher nodes thus increasing the absorptive and transfer capacity of the root system. The dimensions of each internode may be regarded as a quantitative expression of a set of physiological conditions present at the time of its formation.

The elongation of intact young maize leaves was found to be dynamically dependent on soil water supply. With adequate water the elongation was remarkably constant but slowed down when the water potential of the soil dropped from -0.01 to -0.02 MPa and stopped when this dropped to -0.25 MPa (Acevedo *et al.*, 1971; Kozinka, 1984; 1985). It is generally accepted that under short-term conditions changes in the external water potential mainly affect the physical properties of elongating tissues such as cell turgor, irreversible plastic cell wall extension and the concomitant uptake of water by expanding cells (Steudle, 1985).

Boyer (1985) noted that water transport through the whole plant is complicated by several flow types occurring simultaneously. A large volume of water is lost by transpiration during the day and smaller amounts are devoted to cell enlargement, cell metabolism and phloem transport. At night, transpira-

Table 1. Structure and conductivity characteristics of different types of roots in the root system of *Zea mays cv.* VIR-17 (Luxová and Kozinka, 1970, 1973)

Root type	Area of transection			Mean number		Flow volume[a]	
	Total root (mm²)	Central cylinder (mm²)	Conducting tissues %	Protox. poles no	Large m. vessels no	Exp. cond. cm³ h⁻¹	Relat. cond. cm³ cm⁻² h⁻¹
Prim. sem.	1.11	0.26	3.22	14.5	6.0	24.7	2602.0
Adv. sem.	0.53	0.13	2.67	10.5	5.4	18.3	1212.0
Nodal 1st	0.81	0.26	5.42	21.4	10.1	65.8	8481.0
2nd	1.47	0.43	3.38	25.1	12.0	87.5	6172.0
3rd	4.90	1.57	2.57	40.7	18.4	181.0	3871.0
4th	10.58	3.83	1.76	59.2	32.8	288.3	2783.0
5th	20.09	8.34	1.33	88.2	54.3	557.1	2757.0

[a] Segment length 5 mm; pressure 4.10^4 Pa; temperature of water 25°C.

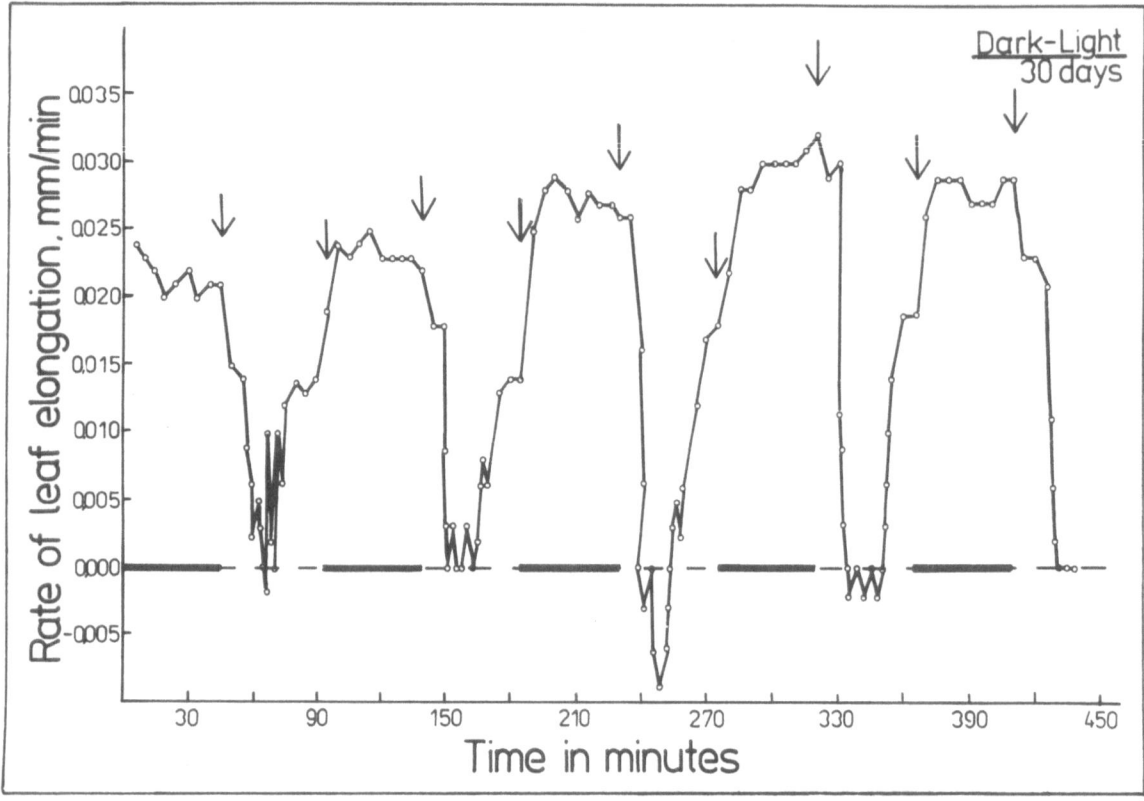

Fig. 3. The elongation rate of the youngest expanding leaf of *Zea mays* cv. CE-330 at periodical change of the dark and light period. *The solid bars* indicate the dark periods. Right top age of plant in days.

tion is diminished and small amounts of water may become the dominant flow components. It is possible to consider all the smaller flows together as 'growth' for which water uptake for enlargement is the major component.

During non-steady flow, one must also consider water movement for tissue rehydration and dehydration and this water can be considered to be stored in tissue capacitances. It is not a part of transpiration or of the growth components. Each day, as transpiration begins, the plant dehydrates (tissue capacitance loses water) and the water potential decreases until water enters the plant rapidly enough to prevent further dehydration. The reverse happens as the day ends so that the leaf water content and leaf water potential rise and fall diurnally. The conservation of mass requires that each of these flows are to be additive with due regard to whether water is entering, leaving or being stored by the plant. Accordingly:

$A + T = G + H$ where A and T are fluxes for absorption and transpiration (the total water gained and total water lost by the plant). G is always positive or zero and H can be positive or negative depending on whether the plant is hydrating or dehydrating and represents the storage fluxes for growth and hydration-dehydration, respectively. T is always negative since it represents the loss by transpiration.

In steady conditions the water status of the plant is constant, water potentials do not change with time and H is zero ($A + T = G$). Water absorption must exceed transpiration sufficiently to also supply water for growth which proceeds at a rate that is permitted by the water potentials necessary to bring this equation to a steady state that is to bring absorption to a level that replenishes the evaporating surfaces and supplies the water for G. Leaf water potentials decrease until this aim is achieved after which they become constant. When transpiration is rapid the water potentials within the plant may also be so low that G approaches zero (A = − T) as shown in Fig. 3.

Conclusions

The elongation of the youngest expanding leaf responds differently to a sudden change in water potential of the medium surrounding the seminal or nodal roots of *Zea mays* cv. CE-330. Experimental data support the view that seminal roots are very effective in water absorption and translocation.

References

Acevedo E, Hsiao T C and Hendersen D W 1971 Immediate and subsequent growth response of maize leaves to changes in water status. Plant Physiol. 48, 631–636.

Boyer J S 1985 Water transport. Annu. Rev. Plant Physiol. 36, 473–516.

Fiscus E L and Markhart A H 1979 Relationship between root system water transport properties and plant size in Phaseolus. Plant Physiol. 64, 770–773.

Kozinka V 1977 Primary seminal root a permanent part of the root system of *Zea mays* L. Biológia (Bratislava) 32, 779–786.

Kozinka V 1978 Seminal adventitious roots—a permanent integral part of the root system of the maize (*Zea mays* L.). Biológia (Bratislava) 33, 51–56.

Kozinka V 1981 Conducting efficiency of roots for the longitudinal flow of water. *In* Structure and Function of Plant Roots. Eds. R Brouwer *et al*. pp 165–169. Martinus Nijhoff/Dr W Junk Publishers, The Hague/Boston/London.

Kozinka V 1984 A method for study of solutions uptake by the root system. *In* Mineral Nutrition of Plants. Ed. T Kudrev. pp 307–310. Publishing House of the Central Cooperative Union, Sofia.

Kozinka V 1985 Water relations of the root system—a part of integrated plant water relations. *In* Regulation of Plant Integrity. Eds. S Procházka and J Hradilík. Acta Universitatis Agriculturae 32, No. 3, Brno.

Luxová M and Kozinka V 1970 Structure and conductivity of the corn root system. Biol. Plant. (Praha) 12, 47–57.

Luxová M and Kozinka V 1973 Study of the vascular flow in the root segments of *Zea mays* L. Biológia (Bratislava) 28, 227–234.

Passioura J B 1980 The transport of water from soil to shoot in wheat seedlings. J. Exp. Bot. 31, 333–345.

Sanderson J 1983 Water uptake by different regions of the barley root: Pathways of radial flow in relation to development of the endodermis. J. Exp. Bot. 34, 240–253.

Steudle E 1985 Water transport as a limiting factor in extension growth. *In* Control of Leaf Growth. Eds. N R Baker *et al*. pp 35–55. Cambridge University Press.

Session 4

Root-shoot relationships with
respect to transport processes

B. C. Loughman et al. (Eds.), Structural and functional aspects of transport in roots, 177–181.
© 1989 by Kluwer Academic Publishers.

Competition for nutrients between fruits and roots of tomato

Z. STARCK, E. STAHL and B. WITEK-CZUPRYŃSKA

Inst. of Plant Biology, Warsaw Agricultural University, PL-02-528 Warsaw, Rakowiecka 26/30, Poland

Key words: competition, fruits, growth regulators, ions, photosynthates, roots, source-sink interactions

Abbreviations: BA, benzyladenine; GR, growth regulators; NOA, β-naphthoxy-acetic acid; Z, zeatin

I. Nutrient allocation between fruits and roots in intact plants

In fruiting plants there is competition between sinks for photosynthates, ions and hormones. It is affected by sink size, sink activity and progress of senescence of individual organs, and modified by environmental factors. Therefore, fruit-root interaction with respect to partitioning of substances consists of a syndrome of effects, regulated by hormones. Generative organs tend to become dominant sinks exerting strong 'priority demand' both for photosynthates and ions. Developing fruits also monopolize hormones (Herzog, 1986), including cytokinins, major endogenous senescence retardants, creating even in some cases deficiency in the leaves (Nooden, 1985). Removal of fruits of grape vines leads to a drastically enhanced cytokinin content in the leaves retarding their senescence (Hoad *et al.*, 1977). Partial defoliation of tomato plants greatly increases the zeatin content of the fruits and even more in the seeds (Monseline *et al.*, 1978; Varga and Bruinsma, 1976). In the opinion of some authors, generative organs are an autonomous sink regulating many processes by endogenously produced hormones (Chalmers, 1985).

During generative development much of the dry matter is accumulated in the fruits, as shown in tomato plants by Widders and Lorenz (1982). Because of nutrient diversion to the generative organs, fruit demand for organic substances surpasses the supply of current photosynthates thus promoting mobilization of temporarily stored sub-

stances, at the cost of the growth of vegetative organs. Diversion of C-supply to the generative organs restricts C-transport to the roots and in consequence may reduce absorption and transport of ions to the shoot. Roots are not able to absorb ions fast enough to supply the shoot. Widders and Lorenz (1979) showed, that the concentration of K^+ in the xylem sap of tomato plants declines drastically during fruit development. Strong remobilization of K^+ from vegetative to generative organs in some cases may even cause self-imposed K^+-stress in the leaves (Lingle and Lorenz, 1969). The extent of remobilization of K^+ ions varied greatly in the individual genotypes with high or low nutrient efficiency (Widders and Lorenz, 1979), an important limiting factor in the tomato yield. Mohamed and Marshall (1979) demonstrated similar mobilization of phosphorus in developing wheat ears and nitrogen retranslocation to the filling pods in soybean (Nooden, 1985).

Much more dramatic is fruit-root competition for hormones in plants grown under stress conditions. Under heat stress (Itai and Benzioni, 1973) or low irradiation (Kinet and Leonard, 1983) the cytokinin level decreases.

Deficit of P in tomato plants decreased flower number as a consequence of restriction of cytokinin transport from the roots (Menary and van Staden, 1976). In wheat and barley grown at suboptimal or supraoptimal temperature, water stress or N-deficit provoke decrease of cytokinin synthesis in the roots, which may become critical for maintenance of leaf activity as reviewed by Herzog (1985). In turn, cytokinin levels may limit assimilate mobilization during early grain development (Patrick and Wareing, 1980). Herzog (1986) suggests that competition between organs for cytokinins seems to be most important in respect to duration of ear filling.

II. Fruit-root competition in plants with modulated sink-source relations

Application of growth regulators (GR) usually enhances assimilate transport to the treated organ. It is especially effective when the endogenous hormonal level limits sink growth. Stimulation of tomato fruit growth by NOA + GA₃ was observed at the expense of the vegetative organs (Starck, 1983; Starck and Stahl, 1986; Starck *et al.*, 1987) and in turn it may reduce cytokinin production. This supposition was checked in an experiment with rooted tomato cuttings (one cluster with four fruits, one leaf below and a fragment of stem). Inflorescences were sprayed twice a week with 50 ppm NOA + 30 ppm GA₃ or 50 ppm NOA + 30 ppm GA₃ + 10 ppm BA solution. In both series, treated with GR, strong stimulation of fruit growth was accompanied by drastic restriction of root growth (Fig. 1). The rates of assimilation of ¹⁴CO₂ and ¹⁴C-export from the blades, were lowest

in the control plants (Table 1), where roots were supplied with more ¹⁴C than fruits in contrast to both series treated with GR. Cytokinin altered the competition for photosynthates between fruits and roots and possibly also between fruits and leaves causing an improved supply of carbon to the root. In contrast BA increased competition within the cluster, between the distal and proximal fruits (Fig. 2), estimated both as their fresh matter and ¹⁴C-photosynthate distribution. Proximal fruits, developing first with high mobilizing power created locally by BA, intercepted most of the substances transported to the cluster.

Strong restriction of root size in tomato cuttings, may affect the shoot and fruits by negative feedback changing the ion supply. Therefore, absorp-

Table 1. Assimilation rate of ¹⁴CO₂ and ¹⁴C-distribution in rooted tomato cuttings cv. Ostona (details in Fig. 1). ¹⁴CO₂ assimilation — as 10^3 cpm./g fr.wt^{-1}

	¹⁴CO₂ assimilation	Total export % ¹⁴C	¹⁴C-distribution (%)		
			Fruits	Roots	Other organs
Control	702	34	21	27	52
NOA + GA₃	1087	43	49	11	40
NOA + GA₃ + BA	1165	43	39	17	44

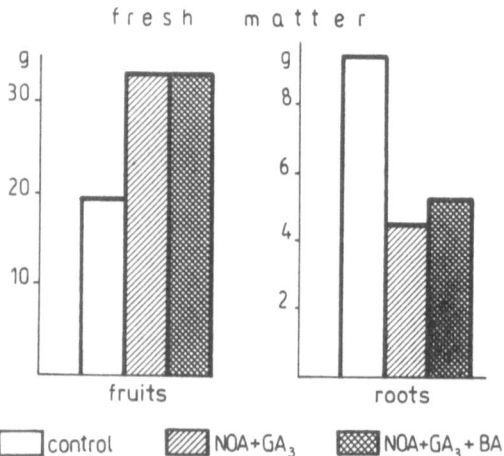

FRUITS-ROOTS COMPETITION IN TOMATO ROOTED CUTTINGS

Fig. 1. Fruit-root competition in rooted cuttings of tomato 5 weeks after beginning of flowering. Average of 3 replicates, cv. Ostona. Inflorescences of 116 day plants were sprayed with growth regulators twice a week.

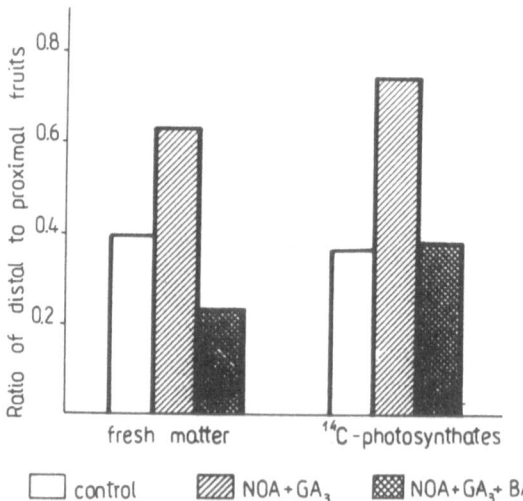

COMPETITION BETWEEN FRUITS IN THE CLUSTER.

Fig. 2. Competition between fruits within the cluster in rooted cuttings (details — as in Fig. 1). The cluster had 4 fruits with a leaf below on a fragment of stem.

Table 2. Effect of GR (NOA + GA₃) on the absorption and distribution of K, P and Ca in rooted tomato cuttings cv. Revermun, 3 weeks after beginning of flowering; average of 4 replicates. Figures marked with a or b are significantly different respectively at 5% or 1% level

	Amount (mg. organ^{-1})						Content (% d.m.)			
	Total cuttings		Fruits		Roots		Fruits		Roots	
	C	GR	C	GR	C	GR	C	GR	C	GR
K	189	208	23	74[b]	21	10[a]	4.9	5.5	6.3	4.7[a]
P	52	53	6	16[a]	9.4	6.5[a]	1.3	1.2	2.8	2.9
Ca	73	69	1.0	1.9	2.6	1.9	0.21	0.14	0.77	0.85

tion and distribution of K, P and Ca was studied in tomato cuttings of the control and NOA + GA₃ series (Table 2). The total amount of each element in the whole tomato cuttings was almost the same in both series, in spite of the strong disproportion in their root size. It indicates that the indirect negative effect of GR on root growth was fully compensated by an enhanced absorption rate, as shown by Davidson (1969) and Richards *et al.* (1979). In contrast to calcium, potassium content in the fruits of the NOA + GA₃ series was actually enhanced at the cost of the roots.

Fruit-root interactions were also examined in intact tomato plants grown in water culture with inflorescences treated with NOA + GA₃ as before or with zeatin (10 ppm) at the beginning of flower development and then with NOA + GA₃. In both series, treatment of flowers with GR strongly stimulated fruit growth partly at the expense of the roots (Fig. 3). More detailed analysis revealed that zeatin (Z) affected fruit-root relations as described for BA. In the control and NOA + GA₃ series fruit size negatively affected root growth in contrast to the positive relationship in the NOA + GA₃ + Z-series (Fig. 4).

The rate of ^{32}P absorption by excised roots from both series treated with GR exceeded that in control roots (Fig. 5) slightly compensating for their smaller size. A similar conclusion concerning the compensatory effect of roots, was drawn on the basis of higher K$^+$-concentration in the xylem sap exuded from the tomato stem, after shoot excision (Fig. 6) of plants treated with GR.

tribution and redirection of nutrient substances and hormones to developing fruits take place. Responsiveness to GR treatment depends on whether the plant system examined is source-limited or sink-limited and on the environmental conditions. If sink strength (fruits) increases it may cause:

1. Change of the pattern of distribution without effect on export from the donors (leaves).

2. Change of both export of nutrients and the pattern of their distribution; in some cases together with remobilization of C and N compounds and ions from vegetative organs.

3. Accompanying changes in activity of the leaf, fruit and root maintain balance in the whole organ-

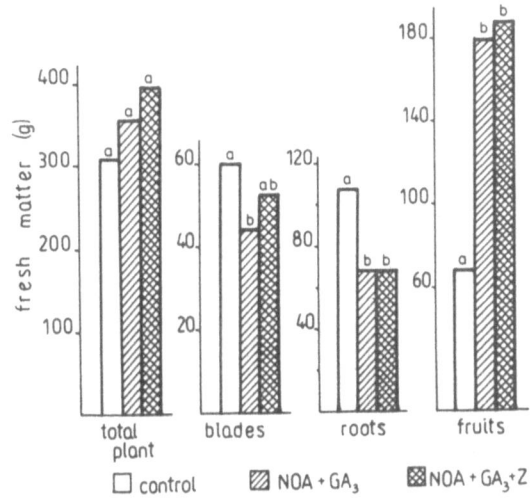

Fig. 3. Distribution of fresh matter in tomato plants grown in water culture; 4 weeks after onset of flowering, average of 6 replicates, cv. Remiz. Bars marked with the same letters — not significantly different at 5% level.

Conclusions

During generative development of plants, redis-

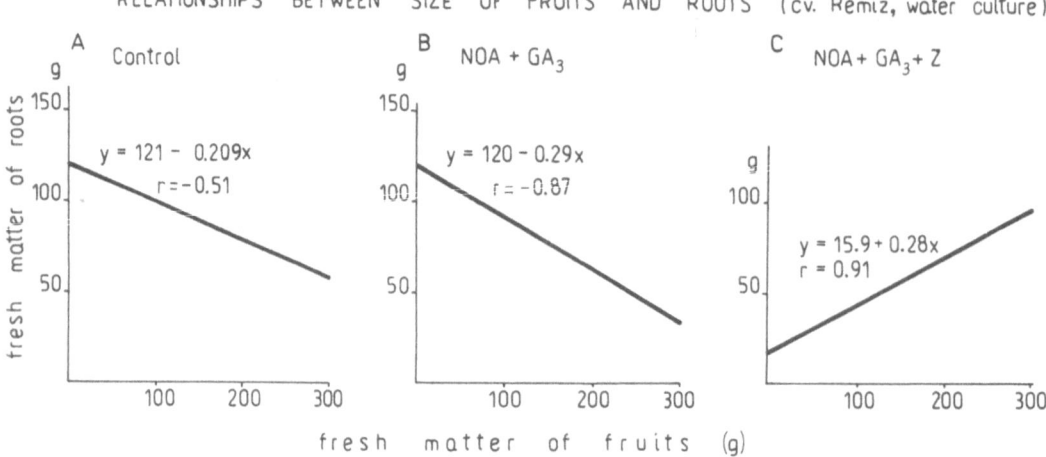

RELATIONSHIPS BETWEEN SIZE OF FRUITS AND ROOTS (cv. Remiz, water culture)

A   Control

$y = 121 - 0.209x$
$r = -0.51$

B   NOA + GA$_3$

$y = 120 - 0.29x$
$r = -0.87$

C   NOA + GA$_3$ + Z

$y = 15.9 + 0.28x$
$r = 0.91$

fresh matter of roots

fresh matter of fruits (g)

Fig. 4. Relationships between size of fruits and roots in tomato plants (details as in Fig. 3). **A**—control plants, **B**—treated with NOA + GA$_3$, **C**—treated with NOA + GA$_3$ + Z.

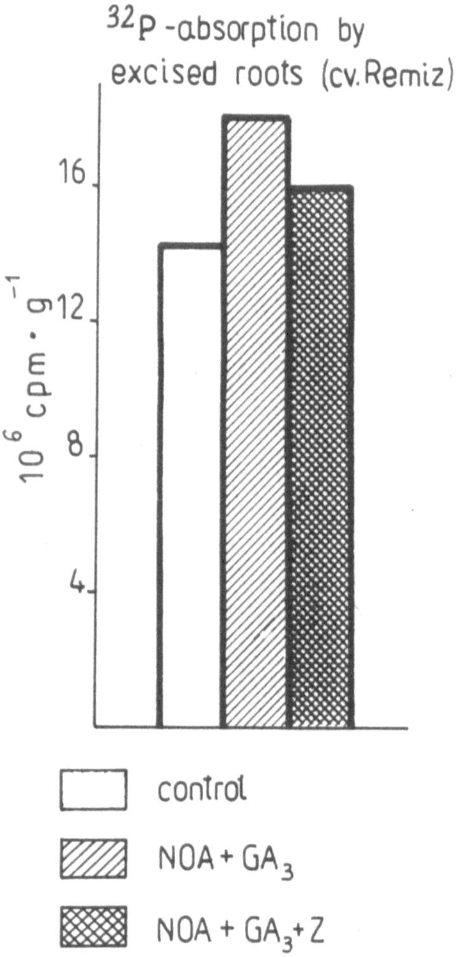

^{32}P -absorption by excised roots (cv. Remiz)

10^6 cpm · g^{-1}

control

NOA + GA$_3$

NOA + GA$_3$ + Z

K$^+$ concentration in xylem sap of tomato plants (cv. Robin. 1987)

mM

control

NOA + GA

NOA + GA

Fig. 5. Rate of ^{32}P-absorption by excised roots of tomato plants (details as in Fig. 3).

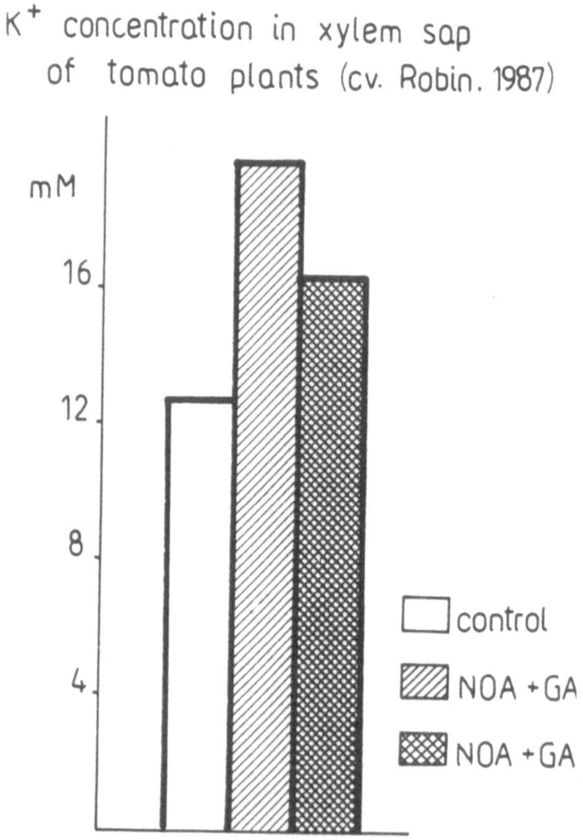

Fig. 6. Concentration of K$^+$ in xylem sap of tomato plants cv. Robin, 3 weeks after onset of flowering, average of 3 replicates.

ism, but if further remobilization and compensation are impossible some damage to the integrity of the plant is observed.

Acknowledgements

Some experiments were partially supported by the Institute of Vegetative Crops, in Skierniewice. The authors wish to thank Mrs E Ciesla for her excellent technical assistance.

References

Chalmers D J 1985 Position as a factor in growth and development effects. *In* Encycl. Plant Physiol. New Ser. Hormonal regulation of development, vol. 11, Role of Environmental Factors. Eds. R P Pharis and D M Reid. pp 169–192, Springer Verlag.

Davidson R L 1969 Effect of root/leaf temperature differentials on root/shoot ratios in some pasture grasses and clover. Ann. Bot. 33, 561–569.

Herzog H 1986 Source and sink during the reproductive period of wheat, Paul Parey Scientific Publ. Berlin, Hamburg.

Hoad G V, Loveys B R and Skene K G M 1977 The effect of fruit-removal on cytokinins and gibberellin-like substances in grape leaves. Planta 136, 25–30.

Itai C and Benzioni A 1973 Correlative changes in endogenous hormone levels and shoot growth induced by short-heat treatment to the root. Physiol. Plant. 27, 355–360.

Kinet J M and Leonard M 1983 The role of cytokinin and gibberellin in controlling inflorescence development in tomato. Acta Hortic. 134, 117–124.

Lingle J C and Lorenz O A 1969 Potassium nutrition in tomatoes. J. Am. Soc. Hort. Sci. 94, 679–683.

Marschner H 1983 Introduction and historical resume. *In* Encycl. Plant Physiol. New. Ser., General introduction to the mineral nutrition of plants, 15A. Ed. A Läuchli and R L Bieleski. pp 5–60, Springer Verlag.

Menary R C and Staden Van J 1976 Effect of phosphorus nutrition and cytokinins on flowering in the tomato *Lycopersicon esculentum*. Mill. Austr. J. Pl. Physiol. 3, 201–203.

Mohamed G E S and Marshall C 1979 The pattern of distribution of phosphorus and dry matter with time in spring wheat. Ann. Bot. 44, 721–730.

Monselise S P, Varga A, Knegt E, Bruinsma J 1978 Course of the zeatin content in tomato fruits and seeds developing on intact or partially defoliated plants. Z. Pflanzenernaehr. 90, 451–460.

Nooden L D 1986 Regulation of soybean senescence. *In* World Soybean Research Conference III: Proceedings. Ed. R Shibbles. pp 891–900. Westview Press Inc. Boulder and London.

Patrick J and Wareing P F 1980 Hormonal control of assimilate movement and distribution. *In* Aspects and prospects of plant growth regulators. Monograph 6, Joint DPGRS and BPGRG Symp. pp 65–84.

Richards D, Goubran F H and Collins K E 1979 Root-shoot equilibria in fruiting tomato plants. Ann. Bot. 43, 401–404.

Starck Z 1983 Photosynthesis and endogenous regulation of the source-sink relation in tomato plants. Photosynthetica 17, 1–11.

Starck Z and Stahl E 1986 Effect of fruit growth on the photosynthesis and remobilization of assimilates. *In* Symp. 'Regulation of photosynthesis efficiency in fruit trees' Bonn. Bad Godesberg, Wissenschaftszentrum, 8–10 Sept. 1982. Ed. A Lakso and F Lenz. pp 92–97, NY State Agr. Exp. Sta, Geneva.

Starck Z, Stahl E and Witek-Czuprynska B 1987 Responsiveness of tomato plants to growth regulators depends on light and temperature conditions. J. Plant. Physiol. 128, 121–131.

Varga A and Bruinsma J 1974 The growth and ripening of tomato fruits at different levels of endogenous cytokinins. J. Hort. Sci. 49, 135–142.

Widders I E and Lorenz O A 1979 Tomato root development as related to potassium nutrition. J. Am. Soc. Hort. Sci. 104, 216–220.

Widders I E and Lorenz O A 1982 Ontogenetic changes in potassium transport in xylem of tomato. Physiol. Plant. 56, 458–464.

B. C. Loughman et al. (Eds.), Structural and functional aspects of transport in roots, 183–188.
© 1989 by Kluwer Academic Publishers.

Effects of internal and external cytokinin concentrations on root growth and shoot to root ratio of *Plantago major* ssp *pleiosperma* at different nutrient conditions

DAAN KUIPER[1], JACQUELINE SCHUIT and PIETER J.C. KUIPER
Dept. of Plant Physiology, Univ. of Groningen, PO Box 14, 9750 AA Haren, The Netherlands, [1]Present address: Research Station for Floriculture, Linnaeuslaan 2A, 1431 JV Aalsmeer, The Netherlands

Key words: cytokinin, growth regulation, root growth, shoot to root ratio

Abbreviations. BA, benzyladenine; RGR, relative growth rate; S/R, shoot to root ratio

Introduction

In *Plantago major* L. an enormous genetic variability is present for a large number of plant traits (Van Dijk and Van Delden, 1981; Kuiper, 1982, 1983; Kuiper and Smid, 1985). Differences in relative growth rates (RGR) for shoot and roots were observed among inbred lines of *P. major*. Inbred lines also differed largely in the rate of growth responses, when plants were transferred from a nutrient-rich growth solution (100%) to a 50 times diluted nutrient solution (2%) and *vice versa*. Quick growth responses in plants of an inbred line belonging to subspecies *pleiosperma* (Pilger) transferred from a 100% to a 2% solution (100 — 2% plants) were accompanied by a rapid decrease in shoot to root ratio (S/R). The rapidity of the responses in *P. major* ssp *pleiosperma* and preliminary determinations of the contents of several ions raised the question about the decisive role of the availability of minerals in growth regulation. Therefore we hypothesized a role for plant growth substances in growth regulation. The experiments were set up with two aims: 1) to repeat the previous results by a non-destructive growth analysis, and 2) to study the role of plant hormones in growth responses by addition of plant growth substances to the nutrient solution.

Material and methods

Plants of inbred line Z_2 (Kuiper, 1982) of *Plantago major* ssp *pleiosperma* were grown as described in Kuiper and Staal (1987). The composition of the 100% solution was (KNO_3, 1.25 mM; $CA(NO_3)_2$, 1.25 mM; $MgSO_4$, 0.50 mM; KH_2PO_4, 0.25 mM; H_3BO_3, 11.5 μmM; $MnCl_2$, 2.3 μmM; $ZnSO_4$, 0.20 μmM; $CuSO_4$, 0.08 μmM; Na_2MoO_4, 0.13 μmM. Iron was added as Ferro-rexenol: 22.5 μmM. Plants grown on a 2% solution were exposed to a 50 times diluted 100% nutrient solution. Extended experiments (Kuiper and Staal, 1987) in which several plant growth substances were applied, clearly showed that in some way cytokinins are involved in growth regulation. For this reason we added 10^{-8} M benzyladenine (BA: a synthetic cytokinin) to the nutrient solution of a part of the plants, which were transferred from one to the other nutrient condition (100 — 2% and 2 — 100%).

Plant growth was followed by a non-destructive root volume measurement and expressed as fresh weight increment (Kuiper and Staal, 1987). Internal cytokinin contents of shoot and roots were determined according to Vonk *et al.* (1985).

Nitrate and phosphate were determined by a HPLC anion exchange method (Maas *et al.*, 1986). Potassium, magnesium and calcium were determined by atomic absorption spectrophotometry.

Results

The shoot to root ratio (S/R) of 100 — 2% plants decreased quickly after transfer and was significantly ($P \leqslant 0.05$) lower than that of the 100%

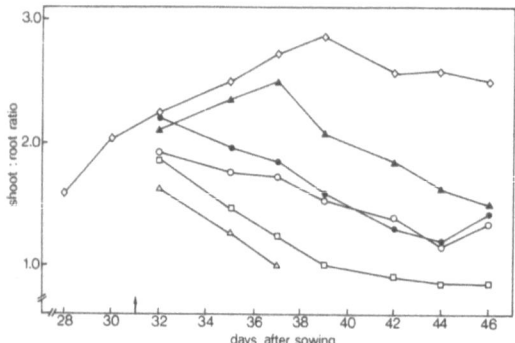

Fig. 1. Shoot to root ratios during the experiment with externally applied BA. ◇, 100%; ●, 2%; ○, 2% + $10^{-9}M$ BA; ▲, 2% + $10^{-8}M$ BA; □, 2% + $10^{-6}M$ BA; △, 2% + $10^{-5}M$ BA; *arrow* time of transfer.

plants. BA addition to the 2% nutrient solution retarded the effects of the 2% solution (Fig. 1). A short-term experiment produced similar data and 100—2% plants already had a significantly lower S/R and shoot growth rate after two days (Figs. 2 and 3). The S/R and shoot growth of plants trans-

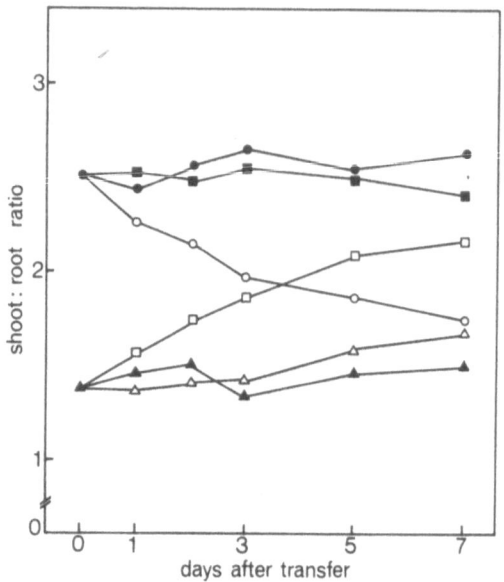

Fig. 2. Shoot to root ratios during the experiment with $10^{-8}M$ supplied BA. ●, 100%; ▲, 2%; ○, 100 → 2%; △, 2 → 100%; □, 2 → 100% + BA; ■, 100 → 2% + BA.

Table 1. Contents of some ions in shoot and root tissue of *P. major* ssp *pleiosperma* as affected by mineral nutrition

Treatment meq. (kg dry matter)$^{-1}$	t	Control[a] 100% S/R	Transfer 100 → 2% S/R	Transfer 2 → 100% S/R	Control 2% S/R
K$^+$	0	1230/1650	–	–	720/1130
	1	1350/1590	1280/1520	920/1390	780/1060
	2	1290/1510	1190/1530	1210/1450	740/1040
	4	1190/1500	1110/1510	1200/1560	640/920
	6	1210/1500	930/470	1300/1510	720/930
H$_2$PO$_4^-$	0	180/270	–	–	120/150
	1	190/290	190/290	160/230	110/140
	2	190/280	190/270	190/280	110/120
	4	200/270	180/270	190/270	100/90
	6	170/250	150/220	180/250	80/100
NO$_3^-$	0	1670/1180	–	–	1000/330
	1	1690/1025	1960/970	1500/830	710/430
	2	2020/1030	1840/1210	1900/1130	690/430
	4	1970/1230	1690/1230	2160/1400	600/370
	6	2190/1360	1270/740	1910/1160	660/490
Ca^{2+}	0	1100/210	–	–	810/200
	1	1140/190	1150/180	1000/200	830/200
	2	1190/190	1170/180	1130/190	830/210
	4	1210/170	1170/150	1190/190	820/180
	6	1190/190	980/180	1170/180	800/160
Mg^{2+}	0	280/870	–	–	270/870
	1	280/900	280/910	280/910	250/860
	2	285/890	270/905	290/915	265/850
	4	270/860	230/870	280/890	240/820
	6	280/820	240/850	260/840	250/830

[a] Control plants had been on nutrient solution for 14 days at t = 0. S, shoot; R, root; t, days after transfer. n = 10; SD ≤ 10%.

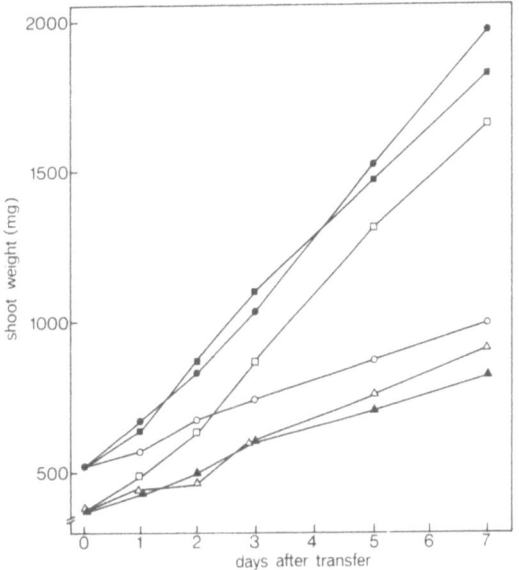

Fig. 3. Shoot growth (fresh weight in mg) during the experiment with $10^{-8}\,M$ supplied BA. Otherwise as for Fig. 2.

on mineral contents of shoot and root of the control plants (2% and 100%) and the transferred plants (100 — 2% and 2 — 100%). Data on BA treated plants are not yet available.

Another series of growth experiments was started to find out whether the effects induced by the 2% solution should be ascribed to the effects of only one compound of the 2% solution or more. Plants, previously grown on a 100% solution, were transferred to the '2%' solution and to 2% N, 2% P, 2% K and 2% Ca conditions, so that the supply of N, P, K and Ca, respectively, was reduced to the 2% level, whereas the remainder of the minerals was kept at the 100% level (using chlorides and sulphates instead of nitrate and phosphate and sodium instead of potassium and calcium; in the latter case sodium was also added to the control plants). Shoot growth (Fig. 4A, B) and S/R (Fig. 5A, B) of 2% N plants resembled the growth and S/R figures of the 2% plants. Plants of the 2% P treatment responded as the 2% N plants, but the responses were much slower. After 14 days exposure to 2% K conditions, the first significant ($P \leq 0.05$) difference in growth and S/R was noted between 2% K plants and 100% plants. The

ferred from a 2%- to the 100%-solution did not respond during the first week after transfer, while the addition of $10^{-8}\,M$ BA induced a quick raise in growth rate and S/R. Table 1 summarizes the data

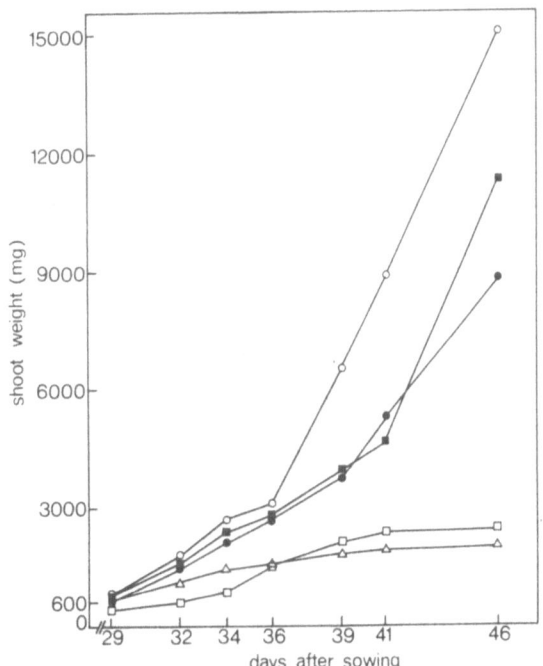
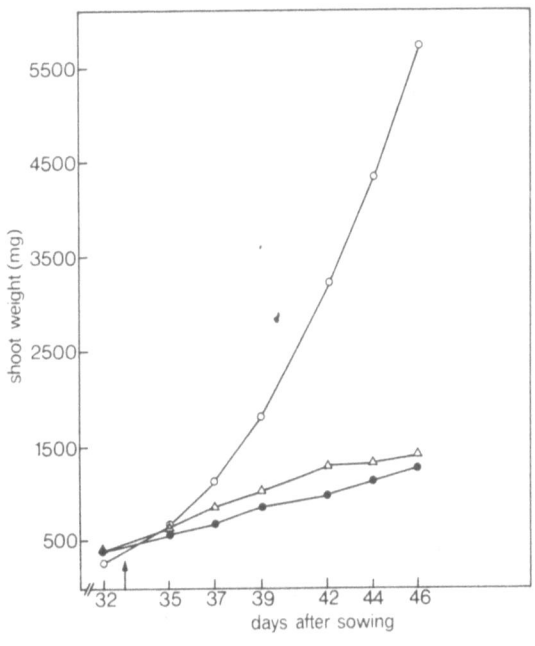

Fig. 4. Shoot growth (fresh weight in mg) during the experiment in which one component of the nutrient solution was lowered to the 2% level. A: ○, 100%; △, 100 → 2%; □, 100 → 2% N; ■, 100 → 2% P; ●, 100 → 2% K. B: ○, 100%; ●, 100 → 2%; △, 100 → 2% Ca. Time transfer is 29 days after sowing.

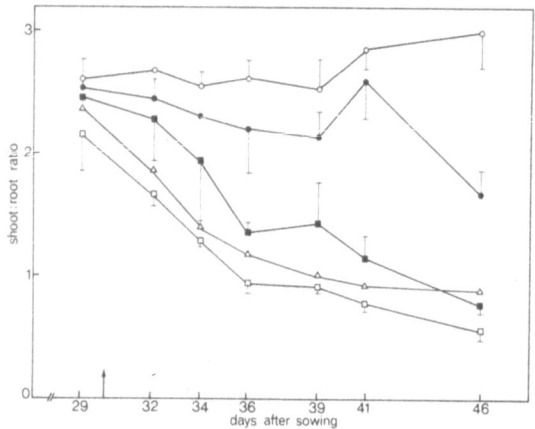 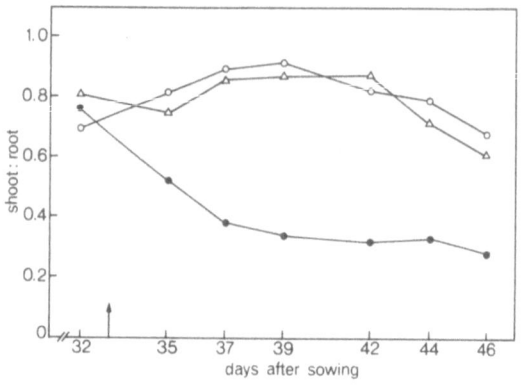

Fig. 5. Shoot to root ratios during the experiment in which one component of the nutrient solution was lowered to the 2% level. Otherwise as for Fig. 4.

growth response of 2% Ca plants was very fast, but no change in S/R was observed.

The internal cytokinin concentrations of shoot and root tissue were measured in several growth experiments (Tables 2 and 3). The cytokinin contents of 100—2% plants dropped to 50% in a 2-day period and they were on the 2% level after 5 days (Table 2). In the presence of BA in the 2% solution the cytokinin levels of the 100—2% plants were similar to the cytokinin contents of the 100% plants. The cytokinin contents in 2% N plants were very similar to those of the 2% plants (Table 3). The response in cytokinin contents was slightly less rapid in the 2% Ca plants, followed by the 2% P plants.

Discussion

Our research group is probably the first to substantiate a growth regulation model, in which the availability of minerals is not the very first factor in determining growth rate of plants of *Plantago major* ssp *pleiosperma*. On the one hand, the mineral contents of shoot and roots in 100—2% plants (Table 1) did not change during a period of 4 days after transfer, whereas growth rate and S/R decreased rapidly at the same time (Figs. 1, 2 and 3). On the other hand, mineral contents (Table 1) in 2—100% plants increased very quickly during the first 6-day period after transfer, in which no responses of growth and S/R were observed (Figs. 1, 2 and 3). In both situations growth responses were not accompanied by comparable responses in mi-

neral contents in plant tissue. However correlative responses were achieved when $10^{-8} M$ BA was added to the nutrient solution: the decrease in growth rate in 100—2% plants was postponed and the growth rate in 2—100% plants increased more quickly (Figs. 1 and 3). The S/R values were in agreement with the growth figures (Fig. 2). BA, a synthetic cytokinin, is a growth promoting agent which may temporarily counteract effects of plants exposed to a stress condition (Bradford, 1983; Itai *et al.*, 1973). The stimulatory effects of BA on the metabolism of the plant are due to increased protein synthesis and inhibited protein breakdown (Alvarez *et al.*, 1984; Letham and Palni, 1983). The cytokinins in roots and shoots were very responsive: the concentrations of 2% plants were much lower than that of 100% plants and 100—2% plants rapidly diminished their cytokinin contents (Table 2). When BA was present in the low nutrient solution of 100—2% plants, the cytokinin contents remained at the 100% level. Remarkably enough, BA promoted the growth in 100—2% plants by stabilization of the endogenous cytokinin contents at the 100% level. Preliminary results (not shown) on the cytokinin contents of 2—100% plants showed no response of the cytokinin contents unless BA was added. Thus, it seems that growth stimulation in 2—100% plants was also induced by an increase of the cytokinin contents.

In our model of growth regulation describing the influence of the availability of mineral nutrients in the rhizosphere on cytokinin production the main question was: Is one specific component of the nutrient solution responsible for the effects on the

Table 2. Cytokinin contents (pmol g^{-1} fresh weight) in shoots and roots of plants of *Plantago major* ssp *pleiosperma* grown at several treatments. Nucleotides and glycosides are excluded. Z, zeatine; ZR, zeatine riboside; t, days after sowing; T, days after the transfer. The data are mean values $\pm$ SD of six replicates

Treatment	t	T	Shoots-Z-Roots		Shoots-ZR-Roots	
100%	30	0	58 $\pm$ 5	76 $\pm$ 7	20 $\pm$ 2	29 $\pm$ 4
2%			14 $\pm$ 2	27 $\pm$ 2	7 $\pm$ <1	12 $\pm$ 1
100 $\rightarrow$ 2%	31	1	37 $\pm$ 5	55 $\pm$ 6	16 $\pm$ 2	24 $\pm$ 3
	32	2	22 $\pm$ 3	31 $\pm$ 3	12 $\pm$ 1	19 $\pm$ 1
	35	5	12 $\pm$ <1	24 $\pm$ 2	6 $\pm$ <1	13 $\pm$ 2
100 $\rightarrow$ 2% +	31	1	60 $\pm$ 7	78 $\pm$ 7	22 $\pm$ 2	31 $\pm$ 3
10^{-8} *M* BA	32	2	57 $\pm$ 6	81 $\pm$ 7	24 $\pm$ 3	29 $\pm$ 2
	35	5	55 $\pm$ 5	77 $\pm$ 6	21 $\pm$ 2	26 $\pm$ 3
100%	35	5	60 $\pm$ 5	79 $\pm$ 7	19 $\pm$ 2	31 $\pm$ 3
2%			11 $\pm$ 1	26 $\pm$ 3	8 $\pm$ <1	14 $\pm$ 1

cytokinin contents? In the literature nitrogen, phosphate and potassium are mentioned as the most likely effectors (Göring and Mardanov, 1976; Horgan and Wareing, 1980; Salama *et al.*, 1979). Since calcium is a phloem-immobile ion and therefore a likely limiting factor, we also included calcium. For this reason we studied the effects on the growth of plants which were transferred from a full strength solution (100%) to a 100% solution in which either nitrate or phosphate or potassium or calcium was reduced to the 2% level. We will refer

to these conditions as 2% N, 2% P, 2% K and 2% Ca, respectively.

The results of this series of growth experiments is summarized in Figs. 4 and 5. Growth and S/R ratio of 100 — 2% N plants behaved exactly as the 100 — 2% plants. Similar observations were made on 100 — 2% P plants but the responses were more slowly. The growth responses of 100 — 2% Ca plants were even faster than in 100 — 2% or 100 — 2% N plants but the S/R ratio was unaffected. The cytokinin contents (Tables 2 and 3) are in close agreement with the growth data. The 100 — 2% plants and 100 — 2% N plants decreased their cytokinin contents very rapidly, which were followed by a decrease in the 100 — 2% Ca plants at least 24 h later. After 5 days the contents of zeatine and zeatine riboside in 100 — 2% P plants were significantly lower, whereas no changes in the cytokinin contents of 100 — 2% K plants were observed during this 5-day period.

In Fig. 6 we summarize our results and propose the following hypothesis. Nitrate is the effector of the changes in cytokinin concentrations and they affect growth and S/R ratio in *P. major* ssp *pleiosperma*. Ammonium might be equally effective (Peterson and Miller, 1976). The slow responses of P-deprived plants suggests that P shortage will be effected by N limitation via lower energy availabil-

Table 3. Cytokinin contents (pmol g^{-1} fresh weight) in shoots and roots of plants of *Plantago major* ssp *pleiosperma* grown at several treatments. Nucleotides and glucosides are excluded. Z, zeatine; ZR, zeatine riboside; T, days after the transfer. The data are mean values of $\pm$ SD of six replicates

Treatment		Shoots - Z - Roots		Shoots - ZR - Roots	
T = 0	100%	52 $\pm$ 4	78 $\pm$ 5	13 $\pm$ 1	19 $\pm$ 1
T = 1	100%	55 $\pm$ 4	84 $\pm$ 5	15 $\pm$ 1	23 $\pm$ 1
	100 $\rightarrow$ 2% N	43 $\pm$ 2	59 $\pm$ 4	12 $\pm$ <1	17 $\pm$ 1
	100 $\rightarrow$ 2% K	57 $\pm$ 4	82 $\pm$ 6	14 $\pm$ 1	23 $\pm$ 1
	100 $\rightarrow$ 2% P	51 $\pm$ 4	85 $\pm$ 6	15 $\pm$ <1	21 $\pm$ 2
	100 $\rightarrow$ 2% Ca	58 $\pm$ 4	81 $\pm$ 5	16 $\pm$ 1	25 $\pm$ 1
T = 3	100%	65 $\pm$ 4	83 $\pm$ 5	18 $\pm$ 1	24 $\pm$ 2
	100 $\rightarrow$ 2% N	21 $\pm$ 1	38 $\pm$ 2	8 $\pm$ <1	9 $\pm$ <1
	100 $\rightarrow$ 2% K	60 $\pm$ 6	75 $\pm$ 5	16 $\pm$ 1	21 $\pm$ 1
	100 $\rightarrow$ 2% P	55 $\pm$ 5	84 $\pm$ 6	17 $\pm$ 1	25 $\pm$ 1
	100 $\rightarrow$ 2% Ca	31 $\pm$ 2	48 $\pm$ 3	8 $\pm$ <1	13 $\pm$ 1
T = 5	100%	60 $\pm$ 4	75 $\pm$ 5	20 $\pm$ 1	27 $\pm$ 2
	100 $\rightarrow$ 2% N	17 $\pm$ 4	29 $\pm$ 2	6 $\pm$ <1	7 $\pm$ <1
	100 $\rightarrow$ 2% K	62 $\pm$ 4	73 $\pm$ 4	19 $\pm$ 1	30 $\pm$ 2
	100 $\rightarrow$ 2% P	50 $\pm$ 3	61 $\pm$ 4	13 $\pm$ <1	20 $\pm$ 1
	100 $\rightarrow$ 2% Ca	12 $\pm$ <1	17 $\pm$ 1	2 $\pm$ <1	9 $\pm$ <1
	100 $\rightarrow$ 2%	15 $\pm$ 1	30 $\pm$ 1	8 $\pm$ <1	5 $\pm$ <1

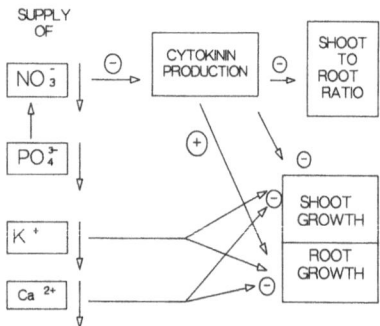

Fig. 6. Summary of the results. (−) and ↓, a decrease of; (+) and ↑, an increase of.

ity. An alternative may be a direct hampering effect of P *via* energy availability on growth rate of the root and via this on cytokinin production. Similarly, Ca deprivation inhibited growth immediately and damaged the root tips and as a result cytokinin production will be hampered. We postulate that in the case of calcium deprivation growth is limiting cytokinin production. The stable S/R ratio might form an indication for non-cytokinin-mediated growth inhibition. The action of K deprivation is not unambiguous. We tend to place potassium in the same category as calcium. K shortage will decrease growth on a longer term without an effect on S/R ratio (except on day 46). Similar experiments in combination with BA application may elucidate this problem.

Acknowledgements

These investigations were supported by the Foundation of Technical Sciences (STW) which is subsidized by the Netherlands Organization for the Advancement of Pure Research (ZWO).

References

Alvarez J, Guerra H and Corchete P 1984 Effect of benzyladenine and kinetin and the degradation of proteins in lentil protein bodies. In 4th Congress of the F.E.S.P.P., Strassbourg. Book of Abstracts, 166–167.

Bradford K J 1983 Involvement of plant growth substances in the alteration of leaf gas exchange of flooded tomato plants. Plant Physiol. 73, 480–483.

Göring H and Mardanov A A 1976 Influence of nitrogen deficiency on K/Ca ratio and cytokinin content of pumpkin seedlings. Biochem. Physiol. Plant. 170, 261–264.

Horgan J M and Wareing P F 1980 Cytokinins and the growth responses of seedlings of *Betula pendula* Roth. and *Acer pseudoplatanus* L. to nitrogen and phosphorus deficiency. J. Exp. Bot. 31, 525–532.

Itai C, Ben Zioni A and Ording L 1973 Correlative changes in endogenous hormone levels and shoot growth induced by short heat treatment to the root. Physiol. Plant. 29, 355–360.

Kuiper D 1982 Genetic differentiation in *Plantago major*: Ca^{2+}- and Mg^{2+}-stimulated ATPases from roots and their role in phenotypic adaptation. Physiol. Plant. 56, 436–443.

Kuiper D 1983 Genetic differentiation in *Plantago major*: Growth and root respiration and their role in phenotypic adaptation. Physiol. Plant. 57, 222–230.

Kuiper D and Smid A 1985 Genetic differentiation and phenotypic plasticity in *Plantago major* ssp *major*. I. The effect of differences in level of irradiance on growth, photosynthesis, respiration and chlorophyll content. Physiol. Plant. 65, 520–528.

Kuiper D and Staal M 1987 The effects of exogenously applied plant growth substances on the physiological plasticity in *Plantago major* ssp *pleiosperma*: Responses of growth, shoot to root ratio and respiration. Physiol. Plant. 69, 651–658.

Letham D S and Palni L M S 1983 The biosynthesis and metabolism of cytokinins. Annu. Rev. Plant Physiol. 43, 163–197.

Maas F M, Hoffmann I, Van Harmelen M J and De Kok L J 1986 Refractometric determination of sulphate and other anions in plants separated by High-Performance Liquid Chromatography. Plant and Soil 91, 129–132.

Peterson J B and Miller C O 1976 Cytokinin in *Vinca rosea* L. Crown Gall tumor tissue as influenced by components containing reduced nitrogen. Plant Physiol. 57, 393–399.

Salama A M S, El da and Wareing P F 1979 Effect of mineral nutrition on endogenous cytokinins in plants of sunflower (*Helianthus annuus* L.). J. Exp. Bot. 30, 971–981.

Smakman G and Hofstra J J 1982 Energy metabolism of *Plantago lanceolata* as affected by change in root temperature. Physiol. Plant. 56, 33–37.

Van Dijk H and Van Delden W 1981 Genetic variability in *Plantago* species in relation to their ecology. I. Genetic analysis of the allozyme variation in *P. major* subspecies. Theor. Appl. Genet. 60, 285–290.

Vonk C R, Davelaar E and Ribot S A 1986 The role of cytokinins in relation to flower-bud blasting in Iris cv Ideal: Cytokinin determination by an improved enzyme-linked immunosorbent assay. Plant Growth Regulation 4, 65–74.

B. C. Loughman et al. (Eds.), Structural and functional aspects of transport in roots, 189–193.
© 1989 by Kluwer Academic Publishers.

Root-shoot relationships in sorghum and maize plants with different numbers of seminal adventitious roots

TIMOTEJ JEŠKO

Institute of Experimental Biology and Ecology, CBES, Slovak Academy of Sciences, CS-814 34 Bratislava, Czechoslovakia

Key words: maize, root-shoot relationships, seminal adventitious roots, sorghum

Abbreviations: RPS, primary seminal root from latin Radix Primaria Seminalis; RAS, seminal adventitious roots; RAN, nodal adventitious roots

Introduction

It has been generally believed that seminal roots of cereal plants function only until nodal roots appear and then gradually die. However, according to Troughton (1962), Brower (1965), Kozinka (1977; 1978) and by our own findings (unpublished), seminal roots can be active throughout the life of the plant. Bloodworth *et al.* (1958) reported that the seminal roots penetrate earlier and deeper into the soil than most of the nodal roots, which occur more at the surface layer of the soil. Seminal roots are more branched and have a higher specific absorption capacity for water and this is important in drought periods.

The seminal or primary root system of maize usually forms two types of roots, the primary seminal root (RPS) and a varying number of seminal adventitious roots (RAS). The importance of RAS for shoot function during plant ontogeny is insufficiently understood. This work will summarize some results of experiments on the root-shoot relationships of sorghum and maize plants with different number of RAS.

Materials and methods

Zea mays L. cv. Microsperma (with violet grains) and *Sorghum saccharatum* L. Moench cv. Bucianski are experimental plants which do not produce RAS. *Zea mays* L. hybrids CE 380 and CE 330 are plants which produce RAS in different numbers. Experiments were carried out between 1964 and 1987 in controlled and field conditions. In controlled conditions plants were grown in growth chambers (original construction or the E8 Conviron, Canada) at 12 or 14 h photoperiods, photon flux density about $200\,\mu mol\,m^{-2}\,s^{-1}$, temperature $20°C \pm 0.5°C$, relative air humidity $65 \pm 5\%$, CO_2 concentration near to that of normal air in 5-l pots with Knop's nutrient solution. Plants were also grown in field conditions in the Bratislava region in 1986 and 1987. Net photosynthetic rate (P_N) was measured with an infrared gas analyzer CO_2 (Infralyt III, Juncalor, GDR). Plant growth was analysed according to Sestak *et al.* (1971). Further information about the materials and methods is described in our papers cited in the text.

Results and discussion

In plants which do not produce RAS, development of the RPS root system may not keep pace with the shoot demand. This may be the cause of very steep continuing increases in the shoot/root ratio (Fig. 1A, left from broken line). Towards the end of this period of plant growth temporary decreases are seen in shoot water loss (Fig. 1B), shoot photosynthesis (Fig. 1C) and chlorophyll *a* content (Fig. 1D). In maize plants this temporary decrease of chlorophyll content was found by Haspelová-Horvatovičová (1963), and decrease of water uptake by Navara (1987).

189

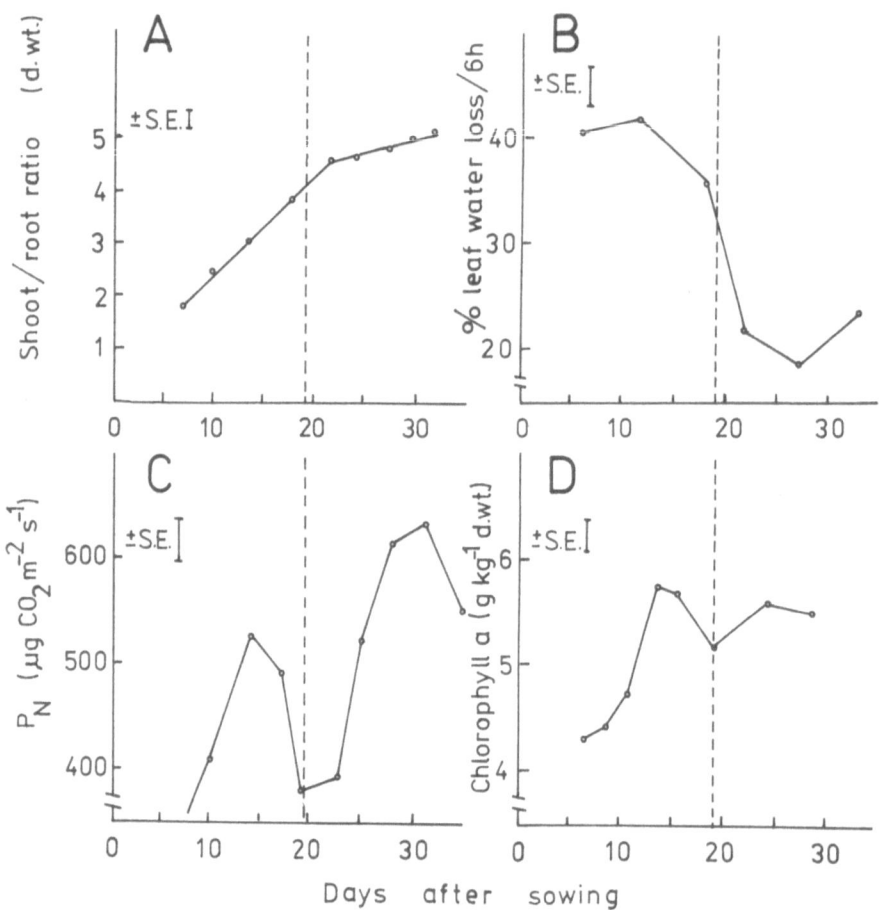

Fig. 1. Changes in shoot/root ratio (**A**), leaf water loss (**B**), net photosynthesis (**C**) and chlorophyll *a* content (**D**) in young *Sorghum saccharatum* L. Moench cv. Buciansky plants in growth phase with only one primary seminal root (RPS — *left from broken line*) and after emergence of nodal adventitious roots near coleoptile node (RAN — *right from broken line*). Characteristic transitory decreases of shoot functions occur just before the first nodal root emerge. Plants were grown in controlled conditions at 14 h photoperiod, photon flux density about 200 μmol m^{-2} s^{-1}, temperature 20°C $\pm$ 0.5°C, relative air humidity 65 $\pm$ 5%, CO_2 concentration near to the normal air in 5-l pots with Knop's solution. Leaf water loss was determined by weighing the detached shoot. Data are the same means from 20 plants in four replications at each measurement.

These decreases of shoot functions will stop when the first nodal root emerges (near the coleoptile node) and so development of the secondary (vegetative) root system starts (Fig. 1, right from broken lines). Due to the production of RAN the transport of cytokinins to the leaves increases and also their P_N (Ješko and Vizárová, 1980). These fluctuations in P_N which are connected with nodal root production in individual whorls was first described for Sorghum plants (Ješko, 1968; Ješko *et al.*, 1971), and then for *Zea mays* plants (Ješko, 1971; 1981).

Decrease of P_N just before the first nodal root emerged was observed in Sorghum and Zea plants which do not produce RAS, but only RPS (Fig. 1C, Fig. 2A). In Zea plants which produce RAS, the decrease of shoot functions was not observed (Fig. 2B). It is important to recall that RAS emerge at the stage of plant development between the RPS and RAN emergence and the results suggest that the functions of RAS enable the plant to overcome a critical phase in its development.

The number of developing RAS is also important. The number of RAS on a plant is mainly genetically determined but its realization also depends on a range of external and other internal conditions, *e.g.*, temperature or initial dry weight of grains as described by Kozinka (1985). Results

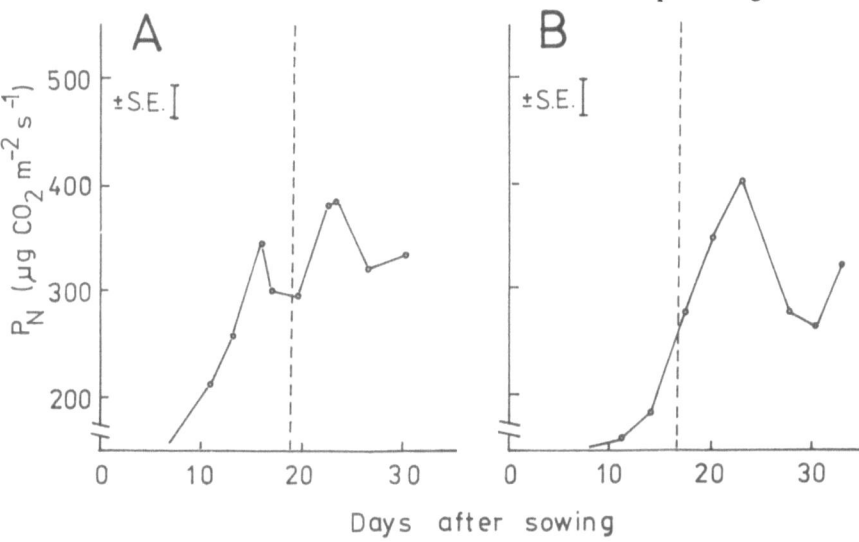

Fig. 2. Changes in the course of net photosynthetic rate (P_N) of *Zea mays* L. cv. Microsperma plants which do not produce seminal adventitious roots (RAS) but only RPS (**A**), and *Zea mays* L. hybrid CE 380 plants with production of RAS (**B**). The decrease of shoot P_N in plants (**A**) just before the first nodal root emerged was not observed or was not so marked in plants (1B) which produced RAS between the RPS and RAN production. Other as in Fig. 1.

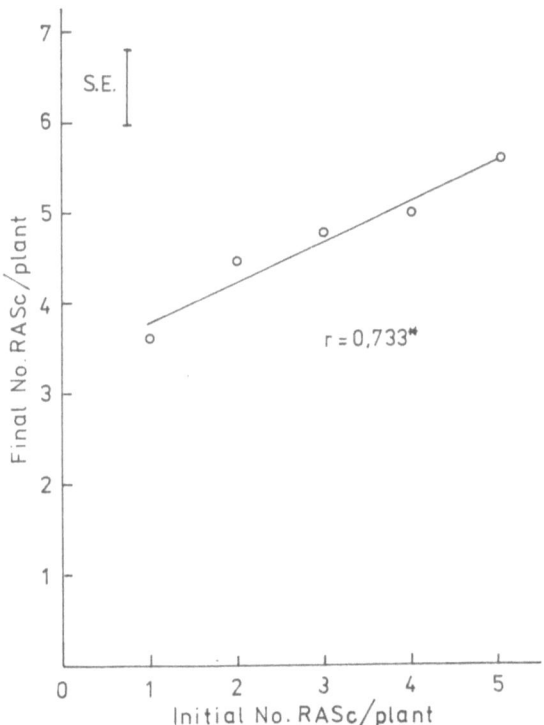

Fig. 3. Difference between the numbers of RAS or scutellar node roots (RASc) which were counted in the same plants first when coleoptile appears on the soil surface and secondly at the end of the vegetative period. It is supposed that RASc can have seminal as well as vegetative origin. The experimental plant was *Zea mays* L. hybrid CE 330 growing in field.

from the the field experiments suggest that not all the roots originating near the scutellar node are based seminally during embryogenesis and their production can still continue in the vegetative phase of plant development (Fig. 3). An increasing number of RAS on the plant changes the distribution of dry matter and extension growth within the different organs. When the number of RAS is smaller and so competition for assimilates is less the individual seminal roots (RPS, RAS) grew more intensively and *vice versa* (Fig. 4). However, the plant with the larger number of RAS produces

Table 1. Correlation coefficients for different numbers of scutellar node roots (RASc) at the emergence and mature plant stages, and some yield characteristics at harvest time of *Zea mays* L. hybrid CE 330 plants grown under field conditions. It is assumed that initial No. RASc determined at the emergence are the roots of seminal origin, and final No. RASc are the roots of seminal and also vegetative origin

Characteristics	Correlation coefficients for	
	No. RASc initial	No. RASc final
Total shoot d.wt (g)	0.2642	0.3683
Stem + leaves (g)	0.1932	0.5626[+]
Ears/plant (g)	0.3693	0.3395
Grain yield/plant (g)	0.3714	0.3395
1000 grains (g)	− 0.5403[+a]	− 0.6898[+]
No. grains/plant	0.8673[++]	0.9454[++]

[a] Significant at $P = 0.05$[+] and $P = 0.01$[++]

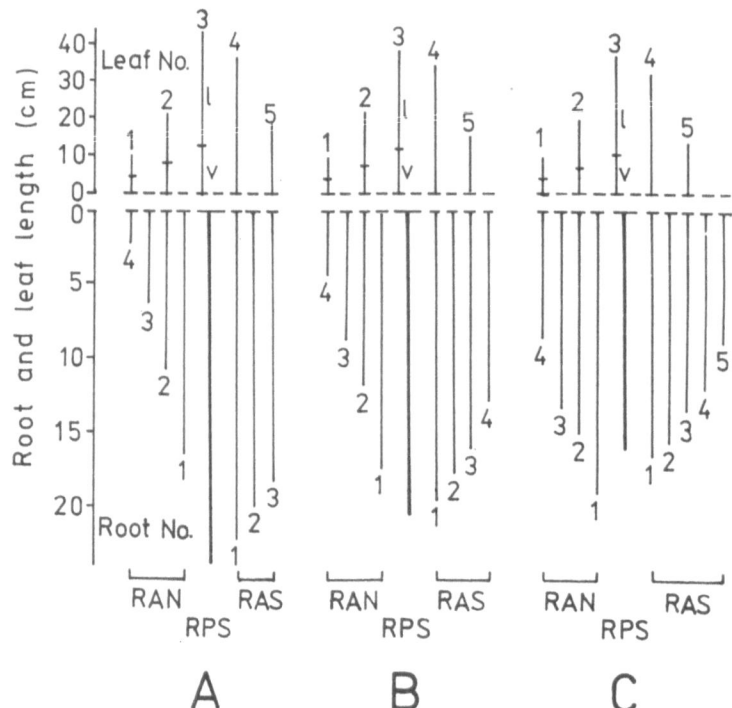

Fig. 4. Distribution of extension growth within the organs of young *Zea mays* L. hybrid CE 330 plants with different numbers of 3RAS (**A**), 4RAS (**B**) and 5RAS (**C**). All plants were in the same growth phase of three fully developed leaves. Growth conditions as in Fig. 1. l = leaf blade, v = l. sheath

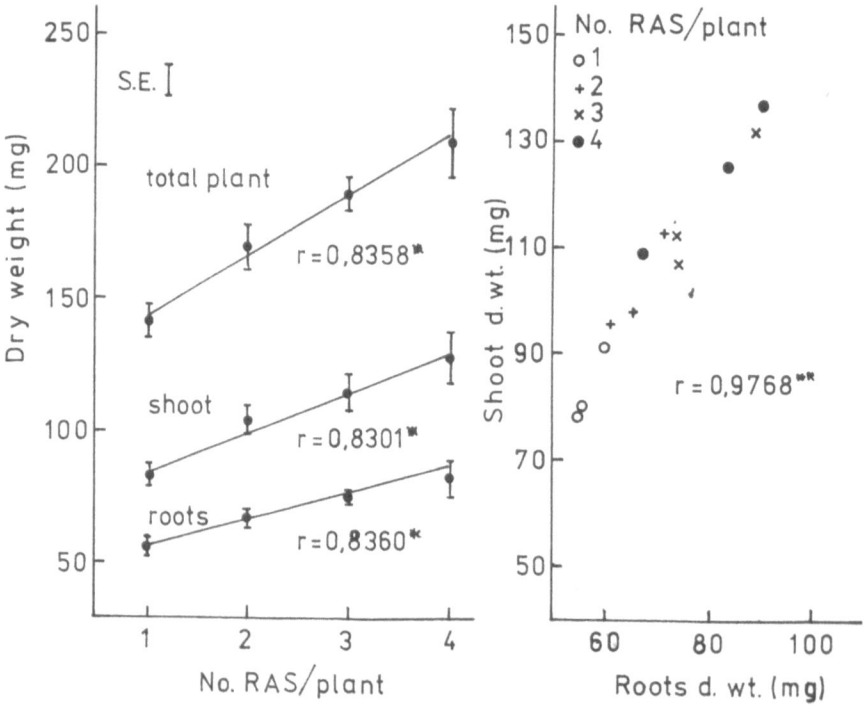

Fig. 5. Increase of total dry matter of the plant, its shoot and root proportionally with increasing number of RAS developed on young *Zea mays* L. hybrid CE 330 plants. Growth conditions as in Fig. 1.

more dry matter which is proportionally distributed between the shoot and roots (Fig. 5). This was repeatedly found for young maize plants particularly hybrids CE 380 and CE 330.

If the root system of the young plant is more vigorous then this will be true of the older plant and the findings concerning young *Zea mays* plants have practical meaning for improved yield. As an example, in Table 1 some results from field experiments are presented which indicate the positive correlation between the number of RAS or scutellar node roots and total grain yield per plant.

References

Bloodworth M E, Burleson C A and Cowley W R 1958 Root distribution of some irrigated crops using undisrupted soil cores. Agron. J. 50, 317–320.

Brouwer R 1965 Root growth of grasses and cereals. *In* The Growth of Cereals and Grasses. pp. 153–166. University of Nottingham.

Haspelová-Horvatovičová A 1963 Beitrag zur Physiologie letaler Pflanzen von *Zea mays* mit Chlorophylldefekt. IV. Veränderungen der grünen Blattfarbstoffkomponenten unter dem Einfluß künstlicher Ernährung. Biológia (Bratislava) 18, 334–347.

Ješko T 1968 Growth and Developmental Changes in Photosynthetic Activity in *Sorghum saccharatum* (L.) Moench, 198.

CSc Thesis, Inst. Experim. Bot. and Ecol., CBES, Bratislava.

Ješko T 1971 Relationships between root development and photosynthetic activity at Sorghum and Zea plants, 102. Final report of Inst. Experim. Bot. and Ecol., CBES, Bratislava.

Ješko T 1981 Inter-organ control of photosynthesis mediated by emerging nodal roots in young maize plants. *In* Structure and Function of Plant Roots. Eds. R Brouwer *et al.* pp 367–371. Martinus Nijhoff/Dr Junk W Publishers, The Hague.

Ješko T, Heinrichová K and Lukačovič A 1971 Increase in photosynthetic activity during the formation of the first node roots and first tiller in *Sorghum saccharatum* (L.) Moench. Photosynthetics 5, 233–240.

Ješko T and Vizárová G 1980 Changes of free endogenous cytokinins during transitorily increased photosynthetic rate initiated by formation of the first two whorls of nodal roots in *Zea mays* L. Photosynthetica 14, 83–85.

Kozinka V 1977 Primary seminal root; a permanent part of the root system of *Zea mays* L. Biológia (Bratislava) 32, 779–786.

Kozinka V 1978 Seminal adventitious roots — a permanent integral part of the root system of the maize (*Zea mays* L.) Biológia (Bratislava) 33, 51–56.

Kozinka V *et al.* 1985 Heterogeneity of the root system — structural and functional, 210. Final report of Inst. Experim. Biol. and Ecol., CBES, Bratislava.

Navara J 1987 Participation of individual root types in water uptake by maize seedlings. Biológia (Bratislava) 42, 17–26.

Šesták Z, Čatský J and Jarvis P F 1971 Plant Photosynthetic Production. Manual of Methods, 818. Dr. W Junk N V Publishers, The Hague.

Troughton A 1962 The roots of temperate cereals (wheat, barley, oats and rye). Mimeo Publ. 2, Commonwealth Bureau of Pasture and Field Crops, Hurley, Berkshire.

B. C. Loughman et al. (Eds.), Structural and functional aspects of transport in roots, 195–198.
© 1989 by Kluwer Academic Publishers.

Shoot/root ratio during the early heterotrophic growth of barley as influenced by mineral nutrition

LUBOMÍR NÁTR

Department of Plant Physiology, Charles University, Viničná 5, CS-128 44 Praha 2, Czechoslovakia

Key words: barley, growth efficiency, mineral nutrients, respiration

Introduction

Seed germination and the early stages of seedling growth have been extensively studied (Hänsel, 1962; Mayer and Shain, 1974). However, little attention has been paid to the efficiency of utilization of reserve substances from the kernel for seedling growth.

Growth efficiency was defined as a ratio between dry weight increment and the total amount of substrate utilized (Lambers, 1985; Tanaka, 1977). According to Tanaka (1977), the growth efficiency varies between 0.60 and 0.65 for many crops. Contreras and Gaudillère (1987) found values of 0.67 and 0.60 for wheat seedlings 9- and 13-days old, respectively.

Most of the research on growth efficiency analyzed the utilization of products of concomitant photosynthesis, but the early growth of most plants depends on reserve substances. The efficiency of their utilization is of prime importance for transition from heterotrophic to autotrophic growth and for the onset of photosynthesis.

Material and methods

Spring barley (*Hordeum vulgare* L., cv. Korál) seeds were germinated and seedlings cultured in the dark for a period of 19 days in one of the following three nutrient solutions: H_2O — distilled water, N — calcium and potassium nitrate, H3 — a complete Hoagland 3 nutrient solution. Dry matter of the shoot, root and kernels was measured at daily intervals from day 5 to 9 and on days 13, 15 and 19 from the start of the experiment. At each sampling date,

4 replicates, each containing 5 plants, were taken. The total amount of nonstructural substrate of the kernel was obtained as the difference between the initial and final kernel dry weight. The amount of respired substrate was calculated as the difference between the initial kernel dry weight and the sum of dry weight of the shoot, root and kernel at any sampling date. A comparison was made of (i) the total amount of respired substrate as derived from data at the end of the experiment and (ii) the rate of respiration calculated from the first five samplings. A linear regression was fitted separately to the values of the first five and the last three samplings.

Results and discussion

The growing medium significantly affected both the dry matter of shoot + root and the shoot/root ratio (Fig. 1). By the end of the experiment, the total plant (shoot + root) dry weight and the shoot/root ratio of the H3 treatment amounted to 132 and 188 per cent of the H_2O plants, respectively (Table 1).

Differences in dark respiration between nutrient treatments, derived from the total amount of substrate lost by the end of the experiment, resembled the differences calculated from the slope of the change in the total seedling dry weight during the first nine days of growth (Table 2).

The effects of mineral nutrients on heterotrophic growth are still poorly understood. To my knowledge, Kincl and Kolková (1984) and M R Sarič (personal communication) are the only investigators who observed that the final dry weight of

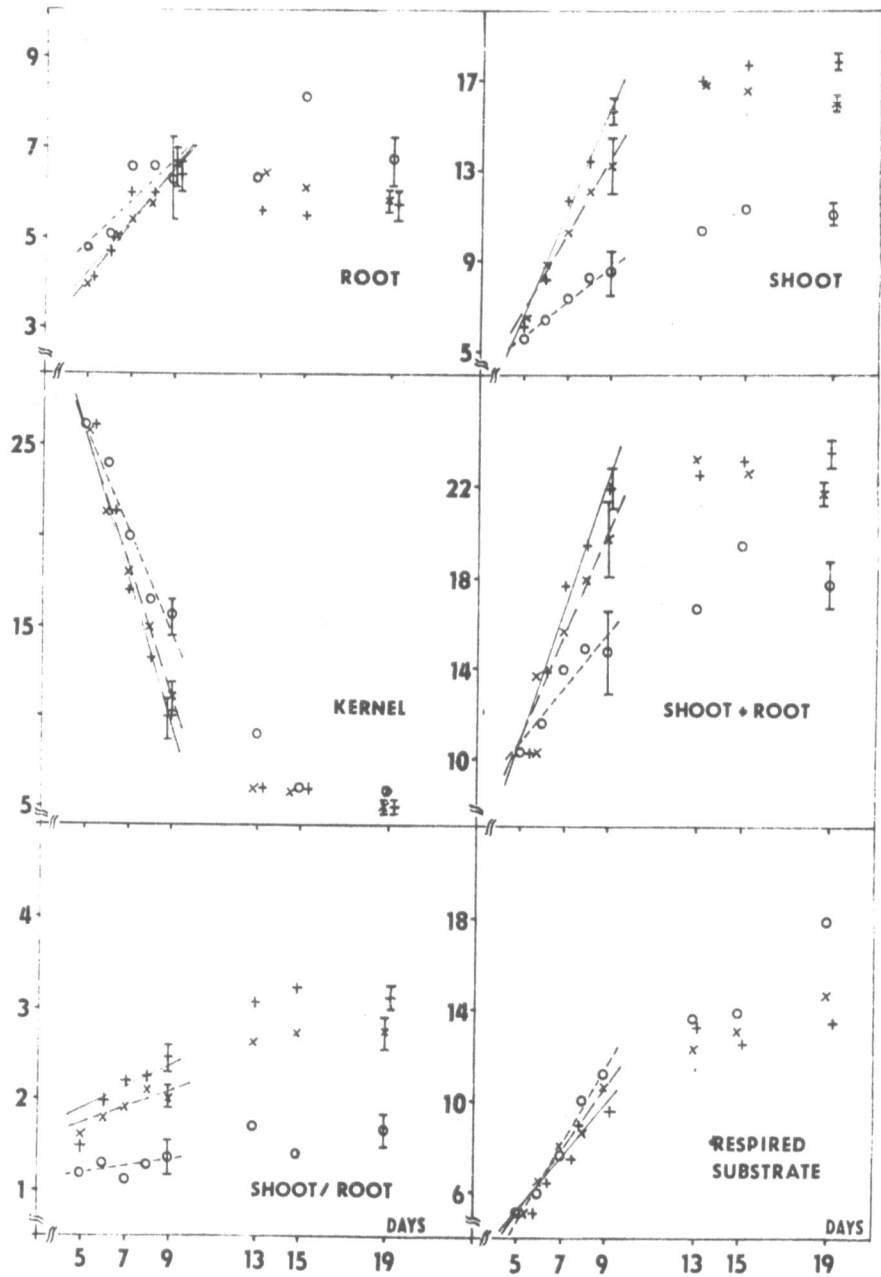

Fig. 1. The time course of dry weight (mg × plant⁻¹) of the kernel, root, shoot, shoot + root, shoot/root ratio and the amount of respired substrate of spring barley seedlings cultivated for 19 days in the dark in distilled water (O – – – O), in nitrate solution (× – – – ×) or in full Hoagland 3 nutrient solution (+ —— +). *Linear regression* was calculated using values of the first five samplings. *Vertical bars* indicate the ± t₀.₀₅ · sₓ̄ interval. The regression coefficients are given in Table 2.

wheat plants cultivated in continuous darkness is influenced by the nutrient composition of the root medium.

The cultivars of spring and winter wheat grown in the dark utilized 45 to 64 per cent of non-

structural dry matter of the kernel for shoot and root structures (Kincl and Kolková, 1984). The conversion efficiency was lowest in plants cultivated in pure water and highest in plants grown in soil rich in nutrients. Our results with barley agree

Table 1. Dry weight (x $\pm$ $s_{\bar{x}}$; mg $\times$ plant^{-1}) of shoot, root and kernel and shoot/root ratio of plants cultivated for 19 days in the dark in one of the following solutions: H_2O — distilled water, N — nitrate solution, H3 — full Hoagland 3 nutrient solution. Values followed by different letters indicate statistically significant differences among nutrient treatments (*P* 0.05)

Nutrient solution	Shoot	Root	Kernel	Shoot/root
H$_2$O	11.07 $\pm$ 0.18 a	6.74 $\pm$ 0.24 a	5.92 $\pm$ 0.07 a	1.65 $\pm$ 0.07 a
N	15.97 $\pm$ 0.12 b	5.15 $\pm$ 0.10 b	5.85 $\pm$ 0.14 a	2.73 $\pm$ 0.06 b
H3	17.83 $\pm$ 0.12 c	5.74 $\pm$ 0.13 c	5.14 $\pm$ 0.16 b	3.11 $\pm$ 0.05 c

very well with those of the above authors. Varietal differences as described by Kincl and Kolková (1984) for wheat were also confirmed for barley (Nátr, 1988, in press).

The effect of mineral nutrition (Table 1, Fig. 1) was unexpected. The higher shoot and root dry weight of plants in treatments N and H3 may to a small extent be due to the weight of the nutrients absorbed from the solution. Although no chemical analyses of plants were carried out, the contribution of weight of nutrients cannot fully explain the higher values of the H3 plants. Furthermore, the effects of nutrients manifested themselves in the shoot/root ratio (Fig. 1) and the habit of plants.

The dry matter lost in respiration can be explained by the energy requirements for growth and maintenance (McCree, 1970). The growth respiration usually amounts to about one third of the total daily photosynthesis (Szaniawski, 1983) although considerable variations of the values have been reported (Lambers, 1985). In the present experiments, the rate of photosynthesis should be replaced by the flow rate of reserve substances from the kernel, *i.e.* by the rate of decrease in kernel dry weight. Expressed as a portion of the amount of the

Table 2. Regression coefficients of the daily changes in the shoot, root and kernel dry weight and respired substrate (mg dry matter $\times$ day^{-1} $\times$ plant^{-1}), and of the shoot/root ratio calculated for the first five samplings as indicated in Fig. 1. Values followed by different letters indicate statistically significant differences between nutrient treatments (*P* 0.05)

	H$_2$O	N	H3
Kernel	−2.85 a	−3.64 b	−4.04 b
Shoot	0.76 a	1.70 b	2.33 c
Root	0.46 a	0.61 a	0.58 a
Respired substrate	1.63 a	1.34 b	1.13 c
Shoot/root	0.035 a	0.085 b	0.120 c

substrate transported out of the kernel, the amount of dry matter respired totalled 57.2 per cent (H$_2$O, 36.8 per cent (N) and 28.0 per cent (H3).

The respiration rate in the second part of the experiment, (between days 13 and 19), can be considered as maintenance respiration, as no further plant growth occurred, due to exhaustion of kernel reserves. The very low values of the change in dry weight did not permit us to record them precisely by the technique used. It was therefore not possible to determine the differences between the three treatments. However, the average amount of substrate respired per day, irrespective of nutrient treatment used, corresponded to about 2 per cent of the shoot and root dry weight. Hence, the value obtained was very similar to photoautotrophically grown plants (McCree, 1982).

The results obtained reveal interesting aspects of studies of the effect of mineral nutrients on growth efficiency and respiration components. In addition, there is also an agronomically interesting point. An ample supply of nutrients from the very beginning of seed germination accelerates transition to the photoautotrophic growth by enhancing the shoot growth.

In this connection, attention is called to the analogy of the results reported in this paper with the effects of fertilizers in crop production. An increase in the supply of fertilizers does not bring about an increase in the total biomass but rather increases the proportion of economically valuable parts (kernels, tubers, and so on). Hence, the most pronounced effects of fertilizers are on dry matter distribution (Gent and Kiyomoto, 1985). Surprisingly, the same effect was observed in the microsystem consisting of seedlings grown in darkness.

References

Contreras M P and Gaudillère J-P 1987 Efficacité de la croissance du blé lors du passage à l'autotrophie. Plant Physiol. Biochem. 25, 35–42.

Gent M P N and Kiyomoto R K 1985 Comparison of canopy and flag leaf net carbon dioxide exchange of 1920 and 1977 New York winter wheats. Crop Sci. 25, 81–86.

Hänsel H 1962 Untersuchungen über die Entleerung des Endosperms bei Samen von Winterweizen (*Triticum aestivum*, L.). Die Bodenkultur A 13, 294–309.

Kincl M and Kolková 1984 Růst pšenice z rezervních látek endospermu. Acta Fac. paedag. (Ostrava) 89, 5–13.

Lambers H 1985 Respiration in intact plants and tissues: Its regulation and dependence on environmental factors, metabolism and invaded organisms. *In* Encyclopedia of Plant Physiology. Eds. R Douce and D A Day. Vol. 18, pp 418–473. Springer-Verlag, Berlin.

Mayer A M and Shain T 1974 Control of seed germination. Annu. Rev. Plant Physiol. 25, 167–193.

McCree K J 1970 An equation for the rate of respiration of white clover plants grown under controlled conditions. *In* Prediction and Measurement of Photosynthetic Productivity. Ed. J. Setlik. pp 221–229. PUDOC, Wageningen.

McCree K J 1982 The role of respiration in crop production. Iowa State J. Res. 56, 291–306.

Nátr L 1988 Vliv genotypu a minerálních živin na heterotrofní růst mladých rostlin ječmene. Rostlinná Výroba 33. *In press*.

Szaniawski R K 1983 Adaptation and functional balance between shoot and root activity of sunflower plants grown at different root temperatures. Ann. Bot. 51, 453–459.

Tanaka A 1977 Photosynthesis and respiration in relation to productivity of crops. *In* Biological Solar Energy Conversion. Eds. A Mitsui *et al.* pp 213–229. Academic Press, London.

B. C. Loughman et al. (Eds.), Structural and functional aspects of transport in roots, 199–202.
© 1989 by Kluwer Academic Publishers.

Nitrogen and carbon utilization in shoots and roots of nitrogen-limited Pisum

PAULO DUARTE, PETTER OSCARSON, JAN-ERIC TILLBERG and CARL-MAGNUS LARSSON
Department of Botany, University of Stockholm, S-106 91 Stockholm, Sweden

Key words: growth, nitrate, photosynthesis, Pisum, respiration

Abbreviations: DW, dry weight; FW, fresh weight; RGR, relative growth rate; R_N, relative nitrogen addition rate

Introduction

Models of partitioning and utilization of C and N can be constructed by analysing xylem and phloem bleeding saps, as has been demonstrated *e.g.* in white lupin (Pate *et al.*, 1979), and in wheat grown with a split root system (Lambers *et al.*, 1982). When studying these parameters in relation to N availability, it is of major importance to characterize the relationship between N supply and growth. In fact, the N (or nutrient) supply can be used to control growth effectively, by allowing the plants to adapt to a chosen relative rate of N addition, which eventually leads to a linear relationship between the relative nitrogen (or nutrient) addition rate and the relative growth rate (reviewed by Ingestad 1982; Ingestad and Lund 1986). The applicability of this method in terms of control of dry matter increment and other growth parameters has been demonstrated for several species (Ericsson 1981a; b; Ingemarsson *et al.*, 1984; Ingestad 1980, 1981; Ingestad and Lund, 1979; Linder and Rook, 1983; Linder *et al.*, 1981) including *Pisum sativum* (Oscarson and Larsson, 1986).

This communication concerns growth, and carbon and nitrogen utilization in whole Pisum plants, growing with different relative rates of nitrogen addition. The aim is to form a basis for further application of this growth technique in detailed studies of fluxes, partitioning, and utilization of carbon in relation to long-term nitrogen nutrition.

Materials and methods

Plant material

Pisum sativum L. cv. Marma was grown as described elsewhere (Oscarson and Larsson, 1986).

Growth of the plants was monitored as dry weight increase in the interval 25 to 29 days after sowing. The relative growth rate (RGR) was calculated from the equation

$$W_t = W_0 e^{RGR\, t}$$

where t is the time in days, and W_0 and W_t the dry weights initially and after t days, respectively.

The seedlings were initially N-starved; however, from day 16 onwards, NO_3^- was added once daily in doses calculated to sustain relative rates of culture N increments (R_N; calculated analogous to RGR) of 0.06 day^{-1}, 0.10 day^{-1}, 0.12 day^{-1}, and 0.14 day^{-1}, respectively. The initial NO_3^- concentration of the medium, immediately after addition of the daily dose, ranged from 40 to 87 μM in the R_N 0.06 day^{-1} culture, and from 97 to 601 μM in the R_N 0.14 day^{-1} culture, depending on plant age and culture density. All NO_3^- added was assumed to be taken up during the 24 h between NO_3^--additions (see Oscarson and Larsson, 1986). The plants were cultured under continuous light, at a photon flux density of approximately 200 $\mu mol\, m^{-2}\, s^{-1}$.

Net CO_2-fixation and root respiration

Gas exchange measurements were performed using a system schematically presented in Fig. 1. It consists of a two-compartment perspex box enclosing the whole plant. The upper (shoot) compartment was connected to an infrared gas analyser (Series 225 Gas Analyser, Analytical Development Co. Ltd, Hoddesdon, UK). Net CO_2-fixation was measured at the same light conditions as mentioned above. The CO_2 concentration was 330–370 ppm

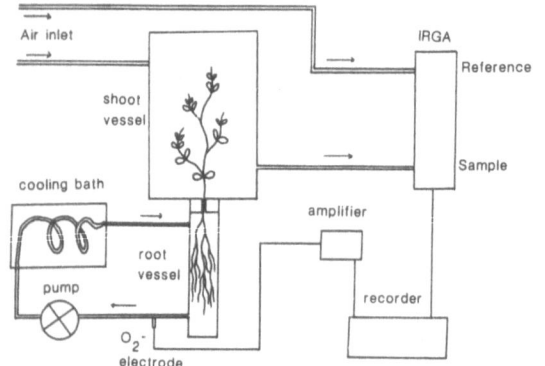

Fig. 1. Schematic representation of the system used for the measurement of net photosynthesis and root respiration.

depending on the CO_2 concentration of the outside air which was used for the measurements. The lower compartment was completely filled with the daily air saturated NO_3^- solution at the beginning of the experiment. The two compartments were separated by an air tight rubber stopper which also held the plant in a fixed position. A pump circulated the nutrient solution over a Clark-type oxygen electrode mounted in a flow-cell, and subse-

quently through a coil immersed in a thermostated water bath. Temperature of both compartments was maintained at 20°C. The gas exchange was measured for between 30 and 60 min.

Chemical analysis

Total C in the dried material was measured with a Carbo Erba elemental analyser.

Results and discussion

Growth

The growth pattern of Pisum at different relative nitrogen addition rates is shown in Fig. 2. Over-all growth of the plants during the experimental period correlated well with R_N (Fig. 2A). This observation is in agreement with previous observations on *Pisum* (Oscarson and Larsson, 1986; see also Oscarson *et al.*, 1988). The root:shoot ratio in

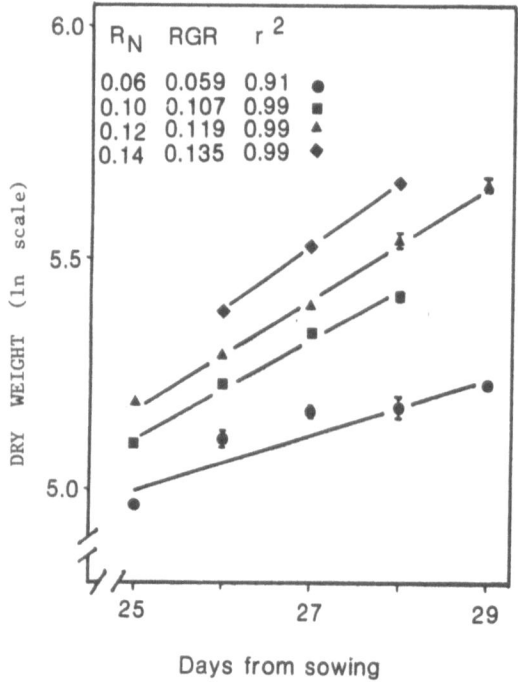

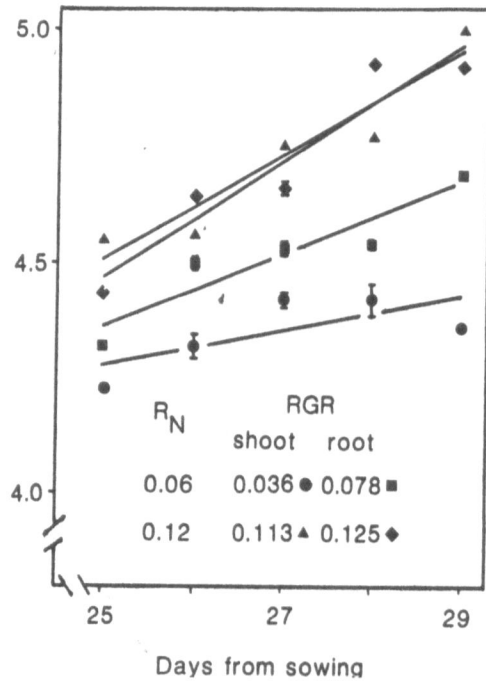

Fig. 2. Growth of *Pisum sativum* at R_N 0.06 day^{-1}, 0.10 day^{-1}, 0.12 day^{-1} and 0.14 day^{-1} during the time period 25 to 29 days. RGR values and square correlation coefficients are indicated. Each point represents mean value $\pm$ SE for twelve plants. **A**: intact plants. **B**: roots and shoots.

Table 1. Net photosynthesis and root respiration in Pisum grown at different R_N, expressed on a whole plant dry matter basis

R_N	Net photosynthesis $mg\,C\,g^{-1}\,DW\,day^{-1}$	Root respiration $mg\,O_2\,g^{-1}\,DW\,day^{-1}$
0.06	32.2 ± 16.3	16.5 ± 4.3
0.12	72.6 ± 3.5	48.3 ± 12

27-day old plants was 1.08 in the R_N 0.06 day^{-1} culture, but only 0.91 in the R_N 0.12 day^{-1} culture. This difference is explained both by uneven growth of root and shoot in the stage of adaptation to the chosen R_N, and by the slightly differing growth rate observed also in the adapted stage, where the root grew consistently faster than the shoot (Fig. 2B). Nevertheless, for the purpose of measuring the daily processing of nitrogen and carbon, plants harvested in the interval 25 to 29 days after sowing could be considered adapted with respect to their growth rate to the chosen R_N.

Gas exchange and over-all carbon-nitrogen budget

Gas exchange measurements performed on roots and shoots of plants cultured at R_N 0.06 day^{-1} and 0.12 day^{-1}, respectively, are presented in Table 1. Both photosynthesis and root respiration of the plants adapted to different R_N were practically constant on a weight basis throughout the experimental period (data not shown); not surprisingly, however, the rates depended clearly on R_N.

The data for the R_N 0.12 day^{-1} culture were, together with previous data on N assimilation and

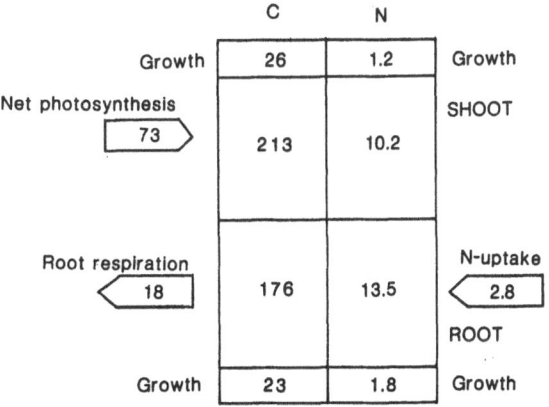

Fig. 3. Model for the assimilation of C and N and their growth related net distribution to shoot and root on day 27 after sowing.

partitioning (Oscarson and Larsson, 1986), used to construct a model for assimilation of C and N, and growth-related net distribution of these elements to root and shoot on day 27 after sowing. The model contains the following elements: the observed initial C and N levels in roots and shoots on day 27; the expected partitioning of C and N to growth in the respective organ, assuming constant C and N levels and based on the observed RGR values; the actual amount of N received by the plant on day 27; the observed rates of C assimilation and O_2 consumption. The N addition (as taken from the preset N addition scheme) to the culture corresponded to 2.8 mg N g^{-1} DW plant. Taking the observed nitrogen content of the plant into account, this would result in an observed R_N of 0.112 day^{-1}, which correlates well to the expected values. Similar considerations can be made concerning C assimilation. If net photosynthesis is compensated for root respiration (assuming a 1:1 ratio between O_2 consumed and CO_2 evolved), the relative rate of carbon increment would be 0.141 day^{-1}, which compares with the predicted value of 0.12 day^{-1}.

The scheme of Fig. 3 is deficient in certain important aspects, *e.g.*, the C and N exchange between root and shoot, and in determination of actual respiratory quotients in root tissue. However, the obtained data demonstrate adaption of the C assimilation and growth processes to the chosen R_N level. The method presented appears to be a promising tool for further investigations on carbon utilization under N limitation; two advantages that can be identified are the possibility to predict and control growth, and the possibility to relate the energy metabolism of the root to the processing of a known and controllable amount of NO_3^-.

Acknowledgement

This work was supported by the Swedish Natural Science Research Council.

References

Ericsson T .1981a Growth and nutrition of three Salix clones in low conductivity solutions. Physiol. Plant. 52, 239–244.
Ericsson T 1981b Effects of varied nitrogen stress on growth and nutrition in three Salix clones. Plant Physiol. 51, 423–429.
Ingemarsson B, Johansson L and Larsson C-M 1984 Photosyn-

thesis and nitrogen utilization in exponentially growing nitrogen-limited cultures of *Lemna gibba*. Physiol. Plant. 62, 363–369.

Ingestad T 1980 Growth, nutrition and nitrogen fixation in grey alder at varied rate of nitrogen nutrition. Physiol. Plant. 50, 353–364.

Ingestad T 1981 Nutrition and growth of birch and grey alder seedlings in low conductivity solutions and at varied relative rate of nutrient addition. Physiol. Plant. 52, 454–466.

Ingestad T 1982 Relative addition rate and external concentration; driving variables used in plant nutrition research. Plant Cell Environ. 5, 443–453.

Ingestad T and Lund A-B 1979 Nitrogen stress in birch seedlings. I. growth technique and growth. Physiol. Plant. 45, 137–148.

Ingestad T and Lund A-B 1986 Theory and techniques for steady state mineral nutrition and growth of plants. Scand. J. For. Res. 1, 439–453.

Lambers H, Simpson R J, Beilharz V C and Dalling M J 1982 Growth and translocation of C and N in wheat (*Triticum aestivum*) grown with a split root system. Physiol. Plant. 56, 421–429.

Linder S, McDonald J and Lohammar T 1981 Effect of nitrogen status and irradiance during cultivation on photosynthesis and respiration in birch seedlings. Energy Forestry Project Technical Report, Vol 12, Swedish University of Agricultural Sciences, Uppsala, Sweden.

Linder S and Rook D A 1983 Effects of mineral nutrition on carbon dioxide exchange and partitioning of carbon in trees. *In* Nutrition of Forest Trees in Plantations. Eds. E D Bower and E K S Nambiar. Academic Press, London.

Oscarson P, Ingemarsson B, Ugglas M and Larsson C M 1988 Characteristics of NO_3^- uptake in Leinna and Pisum. Plant and Soil 111, 203–205.

Oscarson P and Larsson C-M 1986 Relations between uptake and utilization of NO_3^- in Pisum growing exponentially under nitrogen limitation. Physiol. Plant. 67, 109–117.

Pate J S, Layzell D B and McNeil D L 1979 Modeling the transport and utilization of carbon and nitrogen in nodulated legume. Plant Physiol. 63, 730–737.

B. C. Loughman et al. (Eds.), Structural and functional aspects of transport in roots, 203–206.
© 1989 by Kluwer Academic Publishers.

The effect of shoot-root ratio and temperature on K^+ influx in rye

PHILIP J. WHITE[1], MICHAEL J. EARNSHAW[2] and DAVID T. CLARKSON[3]
[1]*Department of Botany, University of Edinburgh, The King's Buildings, Mayfield Road, Edinburgh EH9 3JH, UK,* [2]*Department of Cell and Structural Biology, University of Manchester, Williamson Building, Oxford Road, Manchester M13 9PL, UK and* [3]*Long Ashton Research Station, University of Bristol, Long Ashton, Bristol BS18 9AF, UK*

Key words: acclimation (K^+ influx), low temperature (root), plasma membrane ATPase, *Secale cereale* (K^+ influx), shoot/root ratio

Abbreviations: DT, differential temperature pretreatment; Ea, activation energy; $[K^+]$cyt, cytoplasmic potassium concentration; $[K^+]$ext, potassium concentration of nutrient medium; $[K^+]$int, root potassium concentration; WG, warm grown pretreatment

Abstract. The temperature sensitivity of K^+ influx into rye roots and root plasma membrane ATPase activity were compared in plants grown at different temperatures. It was shown that ATPase activity obeyed the Arrhenius relationship with temperature, whereas K^+ influx into intact plants was linearly related to temperature and markedly influenced by shoot/root ratio. A model for acclimation of K^+ influx to low temperatures based on the regulation of the K^+ carrier mechanism by plant 'demand' for K^+ is described.

Introduction

Extended exposure of cereal root systems to low temperatures often results in an increase in their K^+ influx capacity (Clarkson, 1976; Deane-Drummond and Glass, 1983; Siddiqi *et al.*, 1984). Since this response serves to compensate for the low temperature constraint on K^+ influx it may be termed 'acclimatory'. Such enhancement of K^+ influx capacity has either been attributed to an increase in the number of plasma membrane K^+ porters *per se* (Glass, 1983) or, alternatively, to the development of a temperature-insensitive, regulatory component to K^+ influx, which is enhanced by plant K^+ demand (White *et al.*, 1987).

The influx of potassium across the root plasma membrane may be mediated by a primary active K,H-ATPase at low $[K^+]$ext ($< 1\,mM$) and by facilitated diffusion at higher $[K^+]$ext (Briskin, 1986). Under the conditions employed for K^+ influx studies detailed in this paper (external medium containing $0.6\,mM$ K^+, $0.15\,mM$ Ca^{2+}, pH 6.5), it is considered unlikely that passive K^+ channels are

operative since (i) the K^+ channels of the characean plasma membrane, presumed to be similar to those of higher plants, are closed at low $[K^+]$ext in the presence of calcium (Keifer and Lucas, 1982; Smith and Kerr, 1987) and (ii) passive K^+ influx appears to be thermodynamically unfavourable. It is thought, therefore, that potassium influx in these experiments was mediated by an active K,H-ATPase.

In this paper we shall describe the *in vitro* temperature dependence of the plasma membrane ATPase of rye roots isolated from plants grown under differing temperature regimes and contrast this with corresponding data, reported by White *et al.* (1987), for K^+ influx into intact plants.

Materials and methods

Plants were grown hydroponically in a full nutrient medium, with 16 h daily illumination at $250\,\mu mol$ photons $m^{-2} s^{-1}$ (Clarkson, 1976). Plants were either maintained at 20°C (warm grown pretreatment: WG) or were transferred to a temperature regime of 20°C shoot/8°C root three days prior to experimentation (differential temperature pretreatment: DT). Experiments were performed 14 d (WG) or 15 d (DT) post germination.

K^+ influx into intact plants was determined from a nutrient solution identical to that in which the plants were grown, isotopically labelled with $3\,MBq\,l^{-1}\,^{86}Rb^+$. Incubation and desorption times at the assay temperature were 20 min and 10 min respectively. Shoot temperature was maintained at 20°C. $^{86}Rb^+$ content was determined by Cerenkov radiation techniques following acid digestion of the whole plant.

Plasma membrane was collected between 31–35% sucrose following sucrose density gradient centrifugation of a root microsomal fraction (DuPont and Hurkman, 1985). ATPase activity was determined by Pi release from $MgATP^{2-}$ using the method of Ames (1966). The reaction medium contained $3\,mM\,MgATP^{2-}$, $50\,mM$ KCl, $0.5\,mM$ sodium azide and $100\,\mu M$ sodium molybdate in $5\,mM$ PIPES/NaOH buffer, pH 6.5. Protein determinations followed the method of Bradford (1976).

Results

The plasma membrane ATPase activity increased exponentially with increasing assay temperature (Fig. 1a). Arrhenius transformations of the data were linear. The activation energy (Ea) of the ATPase was 60.5 ± 2.8 and $54.1 \pm 4.5\,kJ\,mol^{-1}$ for WG and DT plants respectively. Plant growth temperature pretreatment had no effect on either the Ea or specific activity of the plasma membrane ATPase.

In contrast K^+ influx into intact plants showed a linear dependence upon assay temperature (Fig. 1b) implying a decrease in the apparent Ea of K^+ influx with increasing temperature. In addition, there was an apparent increase in the K^+ influx capacity of DT plants compared to WG plants. This latter observation is misleading, however, since the absolute rate of K^+ influx into intact plants was proportional to shoot/root ratio (Fig. 2) and the enhancement of K^+ influx in DT plants in Fig. 1 may be attributed to an increased shoot/root ratio in this treatment.

The temperature sensitivity of K^+ influx de-

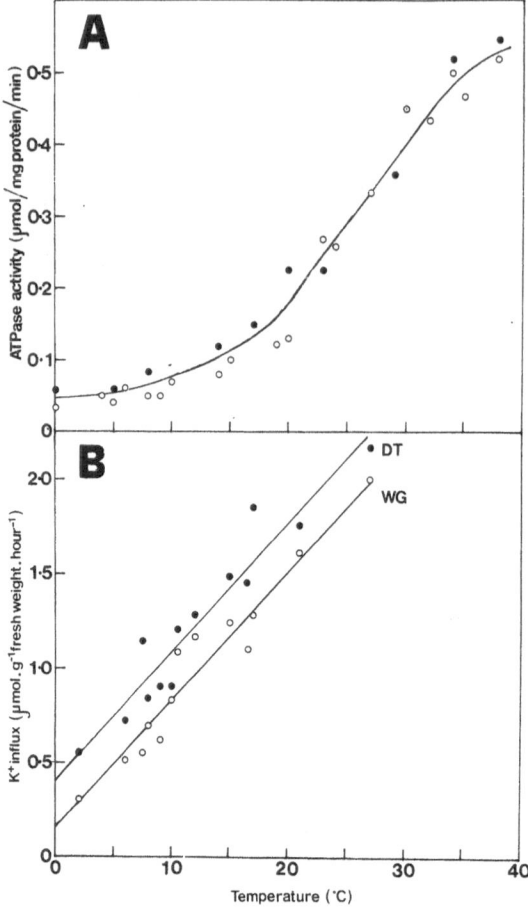

Fig. 1. The effect of temperature on *in vitro* plasma membrane K^+-stimulated, Mg^{2+}-dependent ATPase activity (**A**) and K^+ influx into intact plants (**B**). The shoot/root ratios (fresh weight) of WG (○) and DT (●) plants were 1.76 and 2.25 respectively.

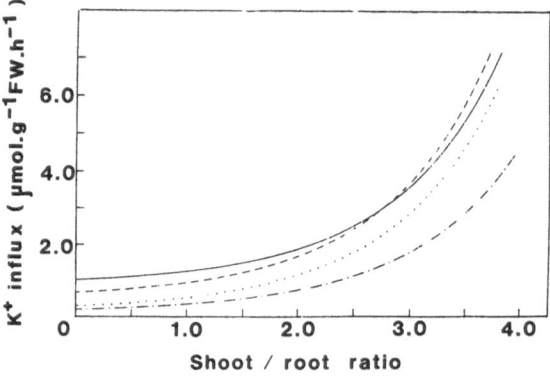

Fig. 2. Relationship between shoot/root ratio and K^+ influx into WG plants at 8 (–·–·–) and 20°C (– – –) and DT plants at 8 (·····) and 20°C (———). *Regression lines* were calculated from data detailed in White *et al.* (1987).

creased with increasing shoot/root ratio. The ratio of K$^+$ influx at 20°C/K$^+$ influx at 8°C for WG plants approached a minimum value of 1.9, whereas that for DT plants approached unity (calculated from data presented in Fig. 2). This implies that upon acclimation K$^+$ influx becomes less temperature dependent.

Discussion

Rye root plasma membrane ATPase activity showed a classical biochemical dependence on assay temperature (Fig. 1), similar to that for barley (Caldwell and Haug, 1981). Plant growth temperature pretreatment had no significant effect on either the specific activity or the Ea of the ATPase, implying that the acclimation of K$^+$ influx to low temperatures in rye did not involve modulation of plasma membrane ATPase characteristics. This contrasts with studies of microsomal fractions from bean, wheat and oat roots (Kähr and Møller, 1976; Kuiper, 1972; Lundborg *et al.*, 1983), which suggest an increase in ATPase activity after roots have been cooled for several days.

A comparison of the effect of temperature upon ATPase activity and K$^+$ influx clearly demonstrated that *in vitro* ATPase activity did not reflect the *in vivo* activity of the K$^+$ porter. There are several explanations for this observation:

(i) not all the ATPase activity can be attributed to the K,H-ATPase, since ATPase activity is clearly electrogenic (Sze, 1985).

(ii) alternative K$^+$ porter mechanisms may operate as well as the putative, K,H-ATPase (Hager *et al.*, 1986).

(iii) the K$^+$ carriers at the plasma membrane appear to be allosterically regulated by [K$^+$]cyt (Glass, 1983) and, since influx in intact plants is well correlated with both shoot growth (Pitman, 1972) and shoot/root ratio (Fig. 2), K$^+$ carrier activity might be determined by depletion of [K$^+$]cyt by the removal of K$^+$ to the shoot.

Potassium influx appears to be enhanced at low temperatures following the development of a temperature insensitive component (Fig. 2), rather than by a general increase in carrier number (*cf.* Fig. 1a). The magnitude of this component increases in response to shoot growth. We have previously shown that, upon exposure to low root

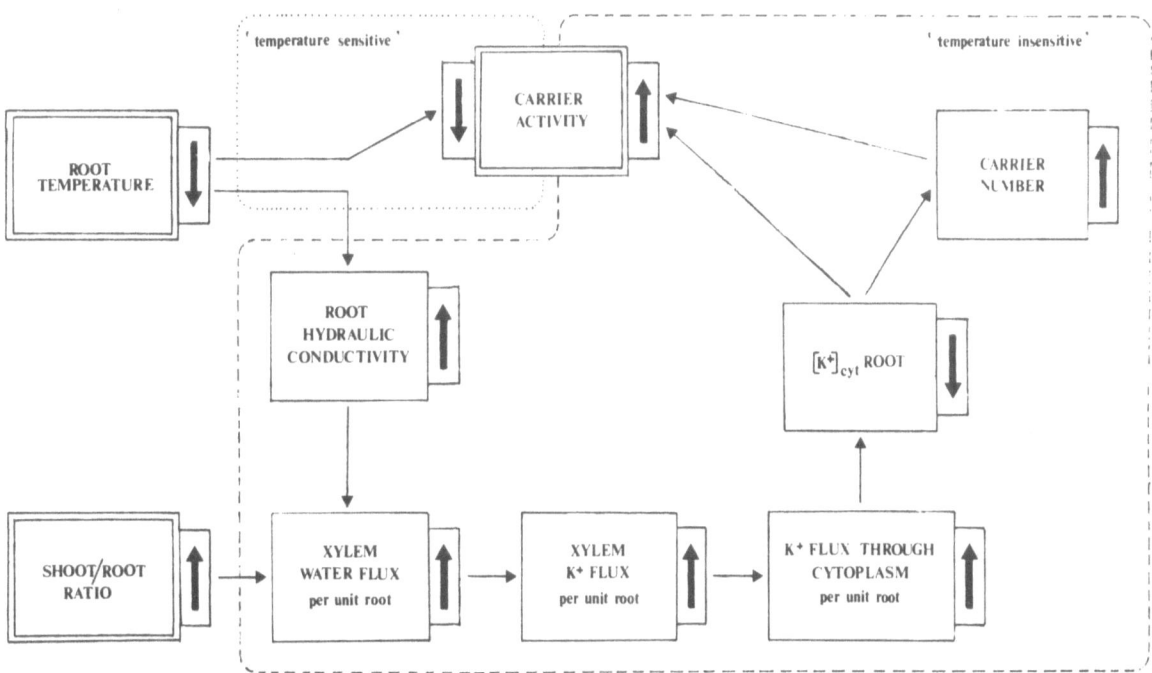

Fig. 3. Schematic diagram of the effect of reduced root temperature on K$^+$ carrier activity. The 'temperature sensitive' component of K$^+$ influx is manifest by the biochemical temperature dependence of the K$^+$ carrier. The proposed 'temperature insensitive' mechanism is, in essence, regulatory: it allows enhanced K$^+$ influx upon acclimation, which is modulated by plant shoot/root ratio, *i.e.* shoot 'demand'.

temperatures, changes in root membrane composition occur (Clarkson *et al.*, 1980) which, although not affecting plasma membrane ATPase activity (Fig. 1a), serve to increase the hydraulic conductivity of the root system (Clarkson, 1976). An increase in water flux through the root would enhance K^+ removal to the xylem at low temperatures, thereby releasing the K^+ carrier from feedback inhibition by $[K^+]$cyt and stimulating K^+ carrier synthesis (Glass, 1983). A model incorporating these aspects concerning the regulation of K^+ influx in cooled roots in response to shoot growth is shown in Fig. 3.

References

Ames B M 1966 Assay of inorganic phosphate, total phosphate and phosphatases. Methods Enzymol. 8, 115–118.

Bradford M 1976 A rapid and sensitive method for the quantitation of microgram quantities of protein utilising the principle of protein dye binding. Analyt. Biochem. 72, 248–254.

Briskin D P 1986 Plasma membrane ATPase: Role in potassium ion transport? Physiol. Plant. 68, 159–163.

Caldwell C R and Haug A 1981 Temperature dependence of the barley root plasma membrane bound Ca^{2+}- and Mg^{2+}-dependent ATPase. Physiol. Plant. 53, 117–124.

Clarkson D T 1976 The influence of temperature on the exudation of xylem sap from detached roots of rye (*Secale cereale*) and barley (*Hordeum vulgare*). Planta 132, 297–304.

Clarkson D T, Hall K C and Roberts J K M 1980 Phospholipid composition and fatty acid desaturation in the roots of rye during acclimatisation to low temperatures. Planta 149, 464–471.

Deane-Drummond C E and Glass A D M 1983 Compensatory changes in ion fluxes in barley (*Hordeum vulgare* L. cv Betzes) seedlings in response to differential root/shoot growth temperature. J. Exp. Bot. 34, 1711–1719.

DuPont F M and Hurkman W J 1985 Separation of the Mg^{2+}-ATPases from the Ca^{2+}-phosphatase activity of microsomal membranes prepared from barley roots. Plant Physiol. 77, 857–862.

Glass A D M 1983 Regulation of ion transport. Annu. Rev. Plant Physiol. 34, 311–326.

Hager A, Berthold W, Biber W, Edel W-C, Lanz Ch and Schiebel G 1986 Primary and secondary energised ion translocating systems on membranes of plant cells. Ber. Deutsch. Bot. Ges. Bd. 99, 284–295.

Kähr M and Møller I M 1976 Temperature response and effect of Ca^{2+} and Mg^{2+} on ATPases from roots of oats and wheat as influenced by growth temperature and nutritional status. Physiol. Plant 38, 153–158.

Keifer D W and Lucas W J 1982 Potassium channels in *Chara corallina*. Control and interaction with the electrogenic H^+ pump. Plant Physiol. 69, 781–788.

Kuiper P J C 1972 Temperature response of adenosine triphosphatase of bean roots as related to growth temperature and to lipid requirements of the adenosine triphosphatase. Physiol. Plant. 26, 200–205.

Lundborg T, Jensen P and Kylin A 1983 The relationship between Rb^+ influx and K^+-stimulated Mg^{2+}-ATPase activity of oat roots of different K^+ status — flexible coupling? Physiol. Plant 59, 277–284.

Pitman M G 1972 Uptake and transport of ions in barley seedlings. III. Correlation between transport to the shoot and relative growth rate. Aust. J. Biol. Sci. 25, 905–919.

Siddiqi M Y, Memon A R and Glass A D M 1984 Regulation of K^+ influx in barley. Effects of low temperature. Plant Physiol. 74, 730–734.

Smith J R and Kerr R J 1987 Potassium transport across the membranes of Chara. IV. Interactions with other cations. J. Exp. Bot. 38, 788–799.

Sze H 1985 H^+-translocating ATPases: Advances using membrane vesicles. Annu. Rev. Plant Physiol. 36, 175–208.

White P J, Clarkson D T and Earnshaw M J 1987 Acclimation of potassium influx in rye (*Secale cereale*) to low root temperatures. Planta 171, 377–385.

Session 5

Effects of stress on function of roots

B. C. Loughman et al. (Eds.), Structural and functional aspects of transport in roots, 209–213.
© 1989 by Kluwer Academic Publishers.

Root functioning under stress conditions: An introduction

PIETER J.C. KUIPER, DAAN KUIPER[1] and JACQUELINE SCHUIT
Department of Plant Physiology, University of Groningen, POB 14, 9750 AA Haren, The Netherlands.
[1]*Research Station for Horticulture, Linnaeuslaan 2A, 1431 JV Aalsmeer, The Netherlands*

Key words: adaptation, degradation, electrogenic pumps, genotype, hormones, mycorrhiza, nutrient stress, salt stress

Abbreviations: BA, benzyladenine; RGR, relative growth rate

Stress physiology of plants is ordinary plant physiology

The only difference from the usual approach in plant sciences is that the experimental plants are exposed to suboptimal environmental conditions or even severe conditions, the so-called stress conditions.

Environmental conditions may induce various responses in plants. These responses are expressed as variation in morphology, physiology and metabolism and they may represent: 1) changes in adaptive value; 2) changes in morphology and metabolism without further significance for the fitness of the plant; 3) stress-induced degradation processes.

Interpretation of experimental data in the above-mentioned three terms often is very difficult. When growth of lupin roots is hampered by soil compaction (Atwell, this volume) and the percentage of root growth inhibition matches the percentage of increase of osmotic compounds of the same roots, the second possibility seems straightforward: the increased osmotic potential is a secondary effect of the compaction-induced growth inhibition and a possible adaptive significance remains to be demonstrated.

Salinity stress and root function

Within the plant kingdom numerous strategies exist in order to cope with salinity stress. As an example, many halophytes are Na^+ includers and many glycophytes are Na^+ excluders (or K^+ in-cluders). At very low NaCl concentration, 1 or 10 mM, *Plantago maritima* absorbs and transports Na^+ to the shoot (De Boer 1985; Tànczos *et al.*, 1981). The halophyte preferentially absorbs Na^+ and transports this ion to the xylem. In contrast, at such low NaCl concentrations the glycophyte *P. media* restricts to a large extent Na^+ absorption and transport to the shoot: the glycophyte pumps Na^+ ions from the cortex cells to the root environment. The presence of an exodermis should be essential for functioning of roots of this glycophyte under salt stress (Peterson, this volume). In addition, Na^+ may be readsorbed from the xylem and transported outwards via the root cortical cells to the environment.

Many species follow the above strategies for halophytes and glycophytes, but variations do exist: *Spergularia marina* is a Na^+ includer at low salt levels and a Na^+ excluder at high salt (Cheeseman, 1984). Clearly, the distinction between glycophytes and halophytes is a relative one.

The mechanism of ion transport from the root cortex to the xylem is important for research in salt tolerance and mineral nutrition. De Boer *et al.* (1983) were the first to demonstrate the involvement of two, spatially separated, electrogenic pumps in plant roots: the first one operating at the plasmalemma of the cortical cells/root environment interface, and the other at the plasmalemma/xylem vessel interface. The two pumps operate in opposite directions, pumping protons to the root environment and to the xylem vessels, respectively. Under limiting oxygen supply in the root environment the inner pump will be the first to depolarize.

Another important fact is the hormonal regulation of the ion pump located at the xylem apoplast. In experiments in which roots were perfused by solutions, it was shown that indoleacetic acid (De Boer *et al.*, 1985), abscisic acid (De Boer, 1985) and fusicoccin (De Boer *et al.*, 1985; De Boer and Prins, 1985) activated the pump at the cortex/xylem interface, pumping protons into the xylem and (re)absorbing K^+ into the cortex.

Abscisic acid strongly activates xylem flow (Glinka, 1980) and the activation was ascribed to an increased K^+ flux to the xylem. Whether such an efflux is due to release of K^+ from late metaxylem elements (Canny, this volume), in parallel to the situation with stomata, remains uncertain.

Recently Van Steveninck *et al.* (1988) showed that abscisic acid stimulated xylem exudation in lupin roots could be reversed by cycloheximide, indicating utilization of K^+ in the root tissue. One may speculate whether drought-induced accumulation of abscisic acid in roots may facilitate accumulation of K^+ in the root cortex, stimulating root growth into unexploited soil; McCully (1987) provided evidence for improved lateral root growth in corn under drought stress.

Mycorrhiza and nutrient stress

Mycorrhizal hyphae may serve as an extension of the root system. For an element such as P the soil supply parameters are at least as important for P uptake by the plant as the root parameters. A high efficiency of mycorrhiza to utilize soil P can be explained by the extension of the root system, or more precisely by extended length of the hyphae and reduced root radius (Silberbush and Barber, 1983). For further analytical studies a special set up for hydroculture was developed by Kähr and Arveby (1986); in normal water culture tree roots are unable to develop into mycorrhizal associations. In the set up the roots are lying on slopes, with the nutrient solution flowing over the roots.

Mycorrhizal associations may result in reduced growth (Bethlenfalvay *et al.*, 1983, soybean with *Phlomus fasciculatus*), unaffected growth (Kähr and Arveby, 1986; pine tree with *Suillus bovinus*) and stimulated growth (Kamminga-van Wijk and Prins, 1988; Douglas fir with *Laccaria laccata*). In

the last example growth stimulation was observed under nutrient limitation of growth; RGR was about one sixth of the maximal RGR.

In Douglas fir, at very low limiting amounts of nutrients, mycorrhizal trees absorbed more N, P, K, Ca and Mg than non-mycorrhizal trees (Kamminga-van Wijk and Prins, 1988). In short term experiments P uptake, on a basis of weight of roots (+mycorrhiza), was equal for both types of trees, indicating that the mycorrhiza served as a surface extension of the root system as far as P was concerned. For ammonia and potassium a larger uptake by the mycorrhizal plants was observed, indicating that the mycorrhizal partner exhibited improved absorption parameters for these ions such as a higher affinity for NH_4^+ and K^+ and increased branching of the root/mycorrhiza system.

Cytokinin-mediated growth responses in barley and wheat to salt stress

In this volume D Kuiper *et al.* (1988) report on growth responses of *Plantago major* to nutrient stress as mediated by exogenous cytokinins. This section provides evidence that in crops like wheat and barley this group of hormones is involved in responses to salt stress. The concept that reduction of shoot growth by salt stress may, at least partly, be due to decreased cytokinin production by the roots is not new: Itai and Vaadia (1965) already showed that NaCl caused a decreased level of cytokinin in the xylem exudate of sunflower roots. Kinetin application reversed negative effects of salt stress such as accelerated ageing of the leaves.

The growth of several varieties of barley and wheat which differ in salt tolerance was followed in a non-destructive way in culture solutions of 25% of a Hoagland solution, adjusted to pH 6. NaCl (65 mM) was added, when the seedlings were 13 days old. A salt-sensitive barley variety, Antares, continued shoot growth in a logarithmic fashion, undisturbed by the salt stress, up to 9 days after NaCl application (Fig. 1). A salt-tolerant variety, Featherstone, reduced its shoot growth within a week after salt addition. A similar observation could be made for the effect of salt stress on the shoot/root ratio: the salt-sensitive Antares showed an unaffected shoot/root ratio which continued to increase with time, while the tolerant Featherstone

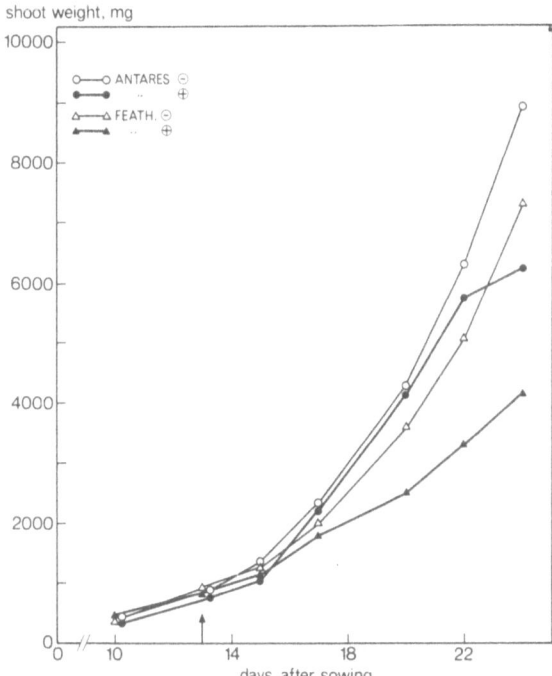

Fig. 1. Time course of shoot fresh weight (mg) of a salt-tolerant barley variety, Featherstone (*triangles*) and a salt-sensitive variety, Antares (*circles*), as affected by salt addition (*closed symbols*) on day 13; control plants without salt addition are presented as *open symbols*.

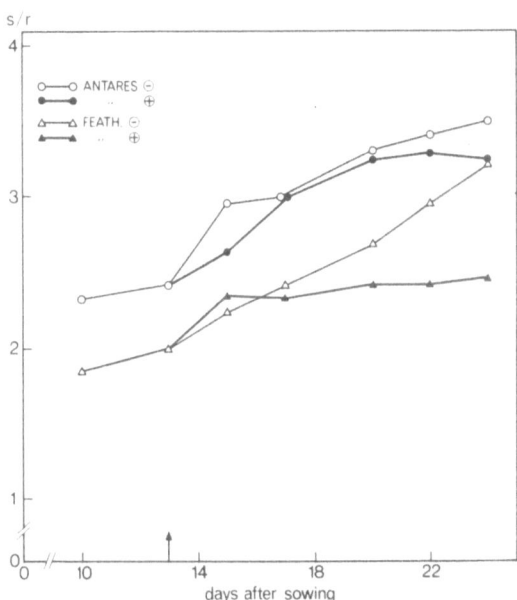

Fig. 2. Time course of the shoot to root ratio on a fresh weight basis of a salt-tolerant barley variety, Featherstone (*triangles*) and a salt-sensitive variety, Antares (*circles*), as affected by salt addition (*closed symbols*) on day 13; control plants without salt addition are presented as *open symbols*.

quickly adjusted the shoot/root ratio after salt addition and the shoot/root ratio no longer increased with time (Fig. 2).

A survey of barley and wheat varieties which differ in salt tolerance (judged from agricultural experience) revealed that the four tolerant barley and two tolerant wheat varities showed a quick and significant reduction of RGR of the shoot during the first 5 days after salt application (Table 1). The 2 sensitive barley varieties and the sensitive wheat variety maintained the RGR of the shoot during this period. During the next 5 days all barley and wheat varieties showed salt-induced reduction of shoot growth. The results on RGR of roots pointed in the same direction: 3 out of 4 tolerant barley varieties showed a quick reduction in RGR upon salt addition. The shoot/root ratio in general increased with time but in the tolerant barley and wheat varieties the shoot/root ratio levelled already 8 days after salt addition, in contrast to the sensitive barley and wheat varieties (Table 2). Clearly,

Table 1. Relative growth rate (RGR) in $mg\,g^{-1}$ fresh weight. day^{-1} of shoot and roots of salt-tolerant and -sensitive barley and wheat varieties as affected by salinity. Salt was added on day 13 after sowing. RGR was determined for 2 conductive periods of 5 days after salt addition. ., significance at the $P = 0.05$ level for differences between salt treatment and control

Crop and variety	Tolerance	NaCl (65 mM)	RGR (root)		RGR (shoot)	
			Days after addition of salt			
			0–5	5–10	0–5	5–10
Barley						
Featherstone	+	−	155	163	192	210
		+	126.	121.	163.	172.
Calif. Mariout	+	−	168	171	170	181
		+	142.	140.	153.	150.
White Gatami	+	−	200	181	204	189
		+	213	215.	148.	174.
Club Mariout	+	−	175	184	183	195
		+	149.	168.	141.	171.
Antares	−	−	180	185	228	233
		+	171	95.	215	141.
Arivat	−	−	175	186	202	210
		+	179	111.	206	133.
Wheat						
Kharchia	+	−	196	172	209	181
		+	212.	168	164.	160.
Candeal	+	−	210	198	180	184
		+	215	189	148.	159.
Anza	−	−	187	162	205	191
		+	188	131.	191	135.

Table 2. Time course of shoot to root ratio of salt-tolerant and salt-sensitive varieties of barley and wheat as affected by salinity. Salt was added at day 13 after sowing. ., significance at the $P = 0.05$ level for the differences between salt treatment and control

Crop and variety	Tolerance	NaCl (65 mM)	Days after NaCl addition			
			0	4	9	11
Barley						
Featherstone	+	−	1.99	2.26	2.95	3.21
		+		2.36	2.41.	2.45.
Calif. Mariout	+	−	1.85	2.11	2.65	2.86
		+		2.03	2.15.	2.17.
White Gatami	+	−	1.45	1.31	1.34	1.66
		+		1.43	0.94.	0.96.
Club Mariout	+	−	1.67	1.58	1.68	1.82
		+		1.53	1.29.	1.18.
Antares	−	−	2.42	2.65	3.40	3.48
		+		2.96	3.28	3.23
Arivat	−	−	2.19	2.28	2.92	3.08
		+		2.35	2.80	2.86
Wheat						
Kharchia	+	−	1.54	1.65	1.95	1.98
		+		1.52	1.36.	1.26.
Candeal	+	−	0.98	1.07	1.08	1.12
		+		0.98	0.44.	0.58.
Anza	−	−	1.84	1.95	1.91	2.01
		+		1.98	1.84	1.89

quickly reduced shoot growth and a quickly established constant shoot/root ratio have adaptive value for salt tolerance in wheat and barley.

Addition of benzyladenine together with salt could modulate the salt-affected RGR of the shoot of the tolerant barley Featherstone (Table 3) and its

Table 3. Effect of addition of benzyladenine on day 13 on the relative growth rate (RGR on a fresh weight basis in mg.g^{-1}.day) of salt-affected barley of the tolerant variety Featherstone. Salt was added on day 13 after sowing. ., significance at the $P = 0.05$ level for differences between controls (+ or − salt) and BA addition (control, + salt). RGR was determined for 2 consecutive periods of 5 days after salt addition

Treatment		RGR (root)		RGR (shoot)	
NaCl (65 mM)	BA (M)	Days after addition of salt and/or BA			
		0–5	5–10	0–5	5–10
−	−	185.	183.	198.	210.
+	−	146	151	163	172
+	10^{-9}	171.	159	183.	180
+	10^{-8}	189.	123.	201.	158
+	10^{-7}	151	121.	128	100.
+	10^{-6}	123	86.	102	74.

Table 4. Effect of addition of benzyladenine (BA) on day 13 on the shoot/root ratio of salt-affected barley of the tolerant variety Featherstone. Salt was added on day 13 after sowing. ., significant difference between controls (+ NaCl, − NaCl) and BA addition (control, + NaCl) at the $P = 0.05$ level

Treatment		Days after addition of salt and/or BA			
NaCl (65 mM)	BA (M)	0	4	9	11
−	−	2.01	1.96.	2.35.	2.54.
+	−		1.66	1.73	1.71
+	10^{-9}		1.73	2.11.	2.06.
+	10^{-8}		2.05.	2.10.	1.98.
+	10^{-7}		1.51	1.42.	1.19.
+	10^{-6}		1.45.	1.18.	1.02.

shoot/root ratio (Table 4). Application of 10^{-9} to 10^{-8} M BA resulted in increased shoot growth and the effect of salt addition, like in sensitive cultivars, was absent. Growth reduction by mineral stress of Plantago (Kuiper and Staal, 1987) could be temporarily alleviated by exogenous BA and the same conclusion may be reached for salt-stressed barley. Higher concentration of BA, 10^{-7} and 10^{-6} M, should give negative effects; ageing of new leaves was accelerated and the salt stress was intensified.

In conclusion, in barley and wheat adaptive responses to salt stress occur *via* a quickly reduced shoot growth; experiments with added benzyladenine to a salt-tolerant variety indicate that the salt stress induced growth reduction was due to decreased internal cytokinin production.

Acknowledgements

The investigations on barley and wheat were supported by the Foundation of Technical Sciences (STW) which is subsidized by the Netherlands Organization for Scientific Research (NWO). Seeds were supplied by Dr S Ceccalli, JCARDA, Aleppo, Syria.

References

Bethlenfalvay G J, Pacovsky R S, Bayne H G and Stafford A E 1982 Introduction between nitrogen fixation, mycorrhizal colonization, and host-plant growth in the *Phaseolus-Rhizobium-Phlomus* symbiosis. Plant Physiol. 70, 446–450.
Cheeseman J M 1984 Sodium and potassium transport in *Spergularia marina*, a coastal halophyte. Plant Physiol. 75, S 1023.
De Boer A H 1985 Xylem/symplast Ion Exchange: Mechanism

and Function in Salt Tolerance and Growth. PhD dissertation, University of Groningen, the Netherlands, 112 pp.

De Boer A H, Prins H B A and Zanstra P E 1983 Biphasic composition of trans-root electrical potentials in roots of Plantago species: Involvement of spatially separated electrogenic pumps. Planta 157, 259–266.

De Boer A H, Katou K, Mizumo A, Kojima H and Okomoto H 1985 The role of electrogenic xylem pumps in K$^+$ absorption from the xylem of *Vigna unguiculata*: The effects of aeration and fusicoccin. Plant, Cell and Environment 8, 579–586.

De Boer A H and Prins H B A 1985 Xylem perfusion of tap root segments of *Plantago maritima*: The physiological significance of electrogenic pumps. Plant, Cell and Environment 8, 587–594.

Glinka Z 1980 Abscisic acid promotes both volume flow and ion release to the xylem in sunflower roots. Plant Physiol. 65, 537–540.

Itai C and Vaadia Y 1965 Kinetin-like activity in root exudate of water-stressed sunflower plants. Physiol. Plant. 18, 941–946.

Kähr M and Arveby A S 1986 A method for establishing ectomycorrhiza on conifer seedlings in steady-state conditions of nutrition. Physiol. Plant. 67, 332–339.

Kamminga-van Wijk C and Prins H B A 1988 Ectomycorrhizal effects on the physiology and growth of *Pseudotsuga menziesii* grown on hydroculture. Proc. Workshop on Ectomycorrhiza and Acid Rain. Air pollution res. rep. 12, Comm. Europ. Communities, pp. 153–168. Ed. A. Jansen *et al.*

Kuiper D and Staal M 1987 The effects of exogenously applied plant growth substrates on the physiological plasticity of *Plantago major* ssp pleiosperma: Responses of growth, shoot to root ratio and respiration. Physiol. Plant. 69, 651–658.

McCully M E 1987 Selected aspects of the structure and development of field-grown roots with special reference to maize. *In* Root development and function. Eds. P I Gregory *et al.* Soc. Exp. Biol. 30, 53–70.

Silberbush M and Barber S A 1983 Sensitivity of simulated phosphoric uptake to parameters used by a mechanistic-mathematical model. Plant and Soil 74, 93–100.

Tànczos O G, Erdei L and Snijder J 1981 Uptake and translocation in salt-sensitive and salt-tolerant *Plantago* species. *In* Structure and Function of Plant Roots. Eds. R Brouwer *et al.* pp 193–198. Martinus Nijhoff Publishers, The Hague.

Van Steveninck R F M, Van Steveninck M E and Läuchli A 1988 The effect of abscisic acid and K$^+$ on xylem exudation from excised roots of *Lupinus luteus*. Physiol. Plant. 72, 1–7.

B. C. Loughman et al. (Eds.), Structural and functional aspects of transport in roots, 215–218.
© 1989 by Kluwer Academic Publishers.

Anatomy and gas exchange of the roots of wheat seedlings following root anaerobiosis

ERNST MANFRED WIEDENROTH and BETTINA ERDMANN
Department of Biology, Div. of General Botany, Humboldt-University Berlin, Invalidenstr. 43, DDR-1040 Berlin, GDR

Key words: aerenchyma, hypoxia, microabsorption method, oxygen concentration, roots, *Triticum aestivum* L.

Introduction

Growth and development of plants are dependent upon a functional equilibrium between their autotrophic and heterotrophic parts. In accordance with this point of view, compensatory reactions of plants, following oxygen deficiency of the root system, are important in order to cope with stress (Wiedenroth, 1981).

Adaptation of cereals to hypoxia in the rhizosphere include changes in morphology, anatomy and physiology, as well as in biochemical and biophysical capacities. This paper will focus on aspects of morphology and anatomy as well as the gas exchange of wheat seedlings (*Triticum aestivum* L. cv. Hatri) following 7 days of severe hypoxia in the rooting medium, caused by nitrogen flushing of the nutrient solution in water culture or by flooding of sand in sand culture.

Results and discussion

Growth retardation

Oxygen deficiency was started at the 3rd day after swelling of the caryopses. After 7 days the overall morphology of seedlings was strongly affected compared with the control (Wiedenroth and Erdmann, 1985): The elongation of the first three roots was inhibited almost completely, roots number 4 and 5 were retarded, while the later ones were stimulated (Fig. 1). The first pair of adventitious roots (number 9 and 10) clearly differed from all

others in shape and colour; they emerged up to day 10 only in the oxygen deficient medium. The morphological structure of the whole root system changed from a hierarchic one (with root lengths corresponding to the insertion level) to a higher number of roots having similar lengths.

These changes were accompanied by a retardation of biomass accumulation amounting to 70 per cent of the control in roots and shoots; the R/S ratio remained unchanged. Because of the higher number of roots, the number of apical meristems per plant increased, which may be important for cytokinin production.

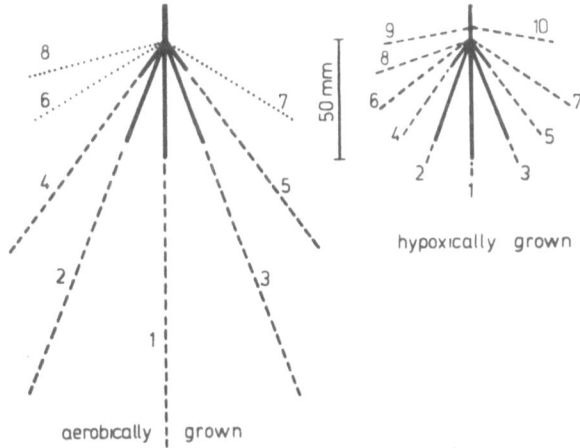

Fig. 1. Scheme of the roots (with numbers) of wheat seedlings following different cultivation regimes. *Solid lines* indicate root lengths at the beginning of differing oxygen supply (day 3), *broken lines* those at day 10. *Dotted lines* indicate roots developed in less than 50 per cent of the plants.

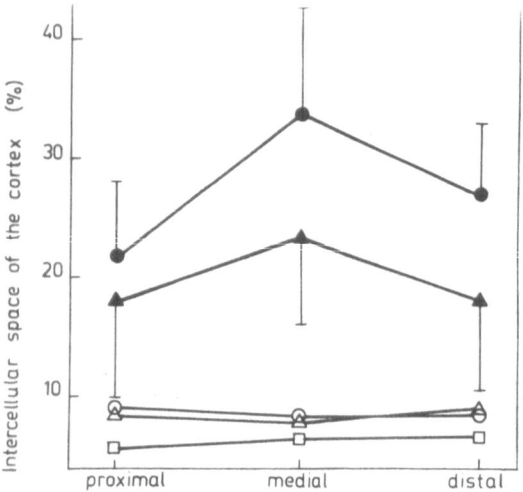

Fig. 2. Intercellular space of the root cortex (average values of all roots) of 3 (□) and 10 day old wheat seedlings following different cultivation regimes in their proximal, medial, and distal region. n = 20. △ nutrient solution, aerated; ▲ nutrient solution, N_2-flushed; ○ sand, three times watered per day; ● sand, flooded.

Aerenchyma formation

Following hypoxia the intercellular space of the root cortex increased from about 6 per cent in the aerated control to 25–30 per cent. This was due to radial elongation of cortical cells, particularly those of the 2nd and 3rd layer, as well as to lysis of cells (Erdmann *et al.*, 1986). Aerenchyma was particularly pronounced in the medial region of all roots (Fig. 2). Furthermore, an enhanced lignification of some cortical and stelar cells could be seen, possibly compensating for the decrease in mechanical strength caused by the loosening of tissues by aerenchyma formation.

Gas exchange

During the early stages of ontogenesis the gas exchange rate of roots (measured as O_2-uptake per hour and g root dry matter) decreased with increasing dry matter, to about $5\,mg\,O_2\,h^{-1}\,g^{-1}$, a nearly constant level, at day 9. At this time the respiration rate was similar in all roots of aerobically grown plants. Following oxygen deficiency, the later emerging roots from number 4 upwards showed a significantly higher respiration rate than the control roots (Erdmann *et al.*, 1986). With respect to

the respiratory activity (expressed per hour and per plant) which depends to a large extent on the biomass of the respiring organ, it should be noted that roots number 4 and higher are responsible for 50 per cent of the total respiration following hypoxia, compared with only 20 per cent in aerated plants.

Oxygen concentration within the roots

Root respiration was measured as O_2-uptake from a closed volume of previously aerated nutrient solution by a Clark type electrode, at an external oxygen concentration between 6–$8\,mg\,l^{-1}$ (equivalent to a concentration of 16–20 per cent in the gas phase). A limited volume of nutrient solution (75 ml per roots of 4 plants) was almost completely deprived of oxygen by root respiration within 90 minutes but roots survived such conditions at least for 36 hours.

This may be explained by oxygen supply from the shoot system to the roots or by changes in the metabolic pathway, from respiration to fermentation, combined with a cessation of respiratory overflow mechanisms (Lambers and Steingröver, 1987; Wiedenroth und Hoffmann, 1984). To analyze these 2 possibilities the composition of gas bubbles extracted from detached roots under lowered pressure was investigated, using a micro-absorption method (Steinmann and Brändle, 1981).

Independent of the cultivation method and pretreatment (to minimize contamination with air during the transfer of roots into the extraction vessels) we found very high amounts of oxygen in the roots (Table 1). Even when detached roots of hypoxically grown plants were submerged into nitrogen flushed nutrient solution before gas extraction, the oxygen concentration within the roots remained at 10 per cent. A prolongation of this pretreatment up to 20 hours only reduced the oxygen content to 6 per cent, which corresponds to a calculated respiration rate much lower than that measured as O_2-uptake of roots from air-saturated ambient medium. The oxygen content in roots, adapted to hypoxia by aerenchyma formation, was always significantly lower than in roots with smaller intercellular spaces.

To explain these unexpected results we examined *Carex acutiformis* EHRH. and *Phalaris arundinacea* L. species known to have very expanded air

Table 1. Oxygen content (per cent) of the gas extracted from the roots of wheat seedlings following different cultivation and pretreatment procedures. Pretreatments lasted 3–5 hours or 20 hours*

Cultivation regime	n	Pretreatment		
		Aerated Intact plants	Nitrogen flushed Intact plants	Detached roots
Nutrient solution	20	19.0 ± 2.0	16.2 ± 2.1	
Flooded sand	7	18.4 ± 1.4	13.1 ± 1.3	10.2 ± 1.7 *6.0 ± 0.8

Table 2. Oxygen content (per cent) of the gas extracted from the roots of different species in relation to the volume of extractable gas, following a 3–5 hour pretreatment. n = 8–12

Taxon	*Triticum* *aestivum*	*Zea mays*	*Triticum* *aestivum*	*Zea mays*	*Carex* *acutiformis*
Cultivation regime	Aerated	Aerated	Flooded	Flooded	
Oxygen content (%)	16.8 ± 1.5	12.3 ± 2.0	10.5 ± 0.9	5.4 ± 1.3	3.3 ± 0.9
Extractable gas ($\mu l\,cm^{-3}$)	6	31	74	120	127

spaces in their subterranean parts. Furthermore, roots of *Zea mays* L. were included in these experiments (Erdmann and Wiedenroth, in preparation). The measuring procedures were the same as described above. A linear relationship was found between the oxygen concentration and the volume of gas extracted from a fixed volume of root material; this relationship may be used as a parameter of the size of total air spaces (Table 2).

Additional experiments with rhizomes of the two helophytes yielded similar results as described in the literature (Studer und Brändle, 1984) underlining the validity of the methods used.

We suppose that a considerable part of the oxygen within the roots is not available for respiratory demand. According to our findings that detached roots with larger gas spaces (*e.g.* Carex) contain less oxygen following 3–5 hours of external

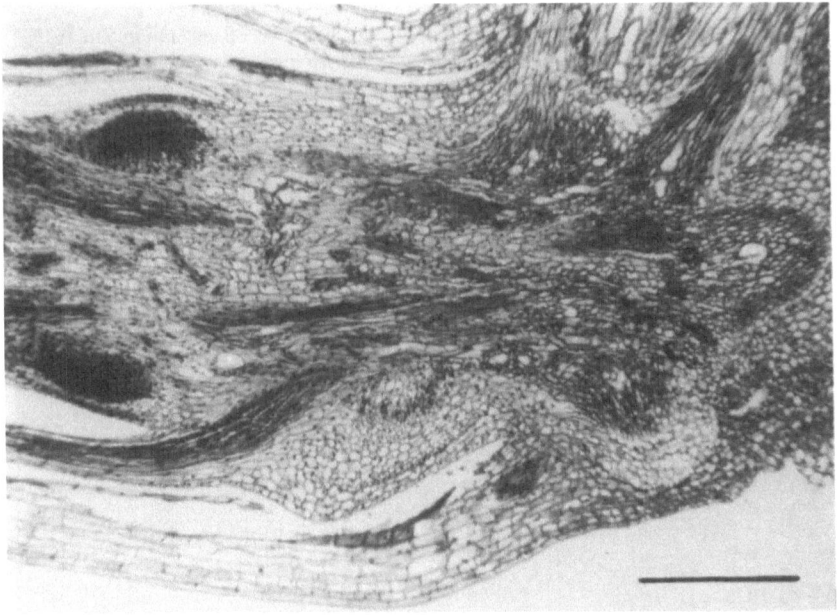

Fig. 3. Longitudinal section through the shoot-root-transition zone of a 10-day-old wheat seedling grown under oxygen deficiency. *Bar* indicates 0.5 mm.

anoxia we conclude that the unavailable oxygen may be bound *in vivo* at the surface of the intercellular spaces: The smaller the gas space, the greater their relative surface.

For extraction roots were submerged in a highly concentrated barrier solution to collect the small bubbles withdrawn from the aerenchyma. This treatment killed the root cells probably caused by an abrupt breakdown of adsorption capacity of the inner surface, and oxygen may be released into the bubbles. Of course, this preliminary explanation needs further investigation.

Internal oxygen transport

It was shown that independent of the general high values (Table 1), oxygen concentrations in submerged roots 3–5 hours after pruning of the shoots, were significantly lower than in those of intact plants. This together with previous findings (Wiedenroth and Poskuta, 1978) indicates the existence of an internal oxygen transport from the leaves into the roots even in mesophytic wheat seedlings used in our experiments. Oxygen diffusion through the intercellular system (Armstrong, 1979; Prioul and Guyot, 1985) is affected by the low porosity of the shoot-root-transition zone (Fig. 3, Erdmann and Wiedenroth, in preparation). We suppose that oxygen diffuses mainly through the air spaces between leaf sheaths. This would explain the enhanced growth of adventitious roots (*cf.* Fig. 1) under oxygen deficiency, because adventitious roots penetrate those interspaces which are in direct contact with air.

References

Armstrong W 1979 Aeration in higher plants. *In* Advances in Botanical Research, Vol. 7. Ed. H W Woolhouse. pp 226–332. Academic Press, London.

Erdmann B, Hoffmann P and Wiedenroth E M 1986 Changes in the root system of wheat seedlings following root anaerobiosis. I. Anatomy and respiration in *Triticum aestivum* L. Ann. Bot. 58, 597–605.

Lambers H and Steingröver E 1978 Growth respiration of a flood-tolerant and a flood-intolerant *Senecio* species: Correlation between calculated and experimental values. Physiol. Plant. 43, 219–224.

Prioul J L and Guyot C 1985 Role of oxygen transport and nitrate metabolism in the adaptation of wheat plants to root anaerobiosis. Physiologie Végétale 23, 175–185.

Steinmann F und Brändle R 1981 Die Überflutungstoleranz der Teichsimse (*Schoenoplectus lacustris* (L). Palla): III. Beziehungen zwischen der Sauerstoffversorgung und der 'Adenylate Energy Charge' der Rhizome in Abhängigkeit von der Sauerstoffkonzentration in der Umgebung. Flora 171, 307–314.

Studer C und Brändle R 1984 Sauerstoffkonsum und Versorgung der Rhizome vom *Acorus calamus* L., *Glyceria maxima* (Hartmann) Holmberg, *Menyanthes trifoliata* L., *Phalaris arundinacea* L., *Phragmites communis* TRIN. und *Typha latifolia* L. Bot. Helv. 94, 23–31.

Wiedenroth E M 1981 Relations between photosynthesis and root metabolism of cereal seedlings influenced by root anaerobiosis. Photosynthetica 15, 575–591.

Wiedenroth E M and Erdmann B 1985 Morphological changes in wheat seedlings (*Triticum aestivum* L.) following root anaerobiosis and partial pruning of the root system. Ann. Bot. 56, 307–316.

Wiedenroth E M und Hoffmann P 1984 Regulative Wechselbeziehungen zwischen den autotrophen und heterotrophen Teilen der Pflanze. Wiss. Z. Humboldt-Universität Berlin, Math.-Nat. R. XXXIII, 318–323.

Wiedenroth E M and Poskuta J 1978 Photosynthesis, photorespiration, respiration of shoots, and respiration of roots of wheat seedlings as influenced by oxygen concentration. Z. Pflanzenphysiol. 89, 217–225.

B. C. Loughman et al. (Eds.), Structural and functional aspects of transport in roots, 219–222.
© 1989 by Kluwer Academic Publishers.

Quantification of air-filled root porosity: A comparison of two methods

M. VAN NOORDWIJK and G. BROUWER
Institute for Soil Fertility, P.O. Box 30003, Haren (Gr.), The Netherlands

Key words: anaerobiosis, aerenchyma, methods root research, pycnometer, *Zea mays*

Introduction

Air-filled porosity of the root cortex is important for aeration of roots in situations where the external oxygen is insufficient. A quantitative theory predicting the depth a vertically growing root can penetrate into the soil is now available for simultaneous internal and external oxygen transport as a function of the air-filled porosities of soil and root (De Willigen and Van Noordwijk, 1987). Maximum depth of root penetration in the soil depends on root diameter, respiration rate, conductance of root epidermis-plus-exodermis for oxygen and air-filled porosity of both soil and root. Reliable methods for quantification of the air-filled porosity of roots or root segments are needed for practical applications of this theory. Two measurement techniques will be discussed here, direct measurements on microscopic sections and the pycnometer method as described by Jensen *et al.* (1969). To obtain root material with significant variation in porosity, maize plants were grown with and without aeration, including some factors stimulating or reducing the normal formation of air spaces via ethylene (Konings, 1983).

Materials and methods

Plant material

Maize plants were grown on a nutrient solution in a growth chamber (temperature 22°C, relative humidity about 40%) on a full-strength nutrient solution, replaced once a week; 3 plants were grown per 5-l pot from 2 weeks after sowing onwards. To obtain low root porosities, AgNO$_3$ was added to reduce ethylene formation; to obtain high porosi-

ties a low nitrogen supply was used in the presence of the ethylene precursor ACC (1-aminocyclo-propane-1-carboxylic acid). Following Konings (1983), four treatments were used, each in two variants: AA. Aerated, with AgNO$_3$ ($10^{-6} M$), A. Aerated, N. Nonaerated, NN. Nonaerated, with ACC ($10^{-5} M$), low N-supply (3.4 instead of 10.2 me l^{-1}) during pretreatment and without N in the last week before harvest.

In variant 1 the pretreatments were implemented at the start of the solution culture period; in variant 2 treatments were only implemented in the last week before the harvest and the plants were aerated in a full-strength nutrient solution beforehand. Plants were harvested 11 weeks after sowing. All treatments were in duplicate.

Measurements on microscopic sections

The surface area of root pores in microscopic sections was calculated from visual estimates (calibrated with computer measurements) of the part of the cortex tissue occupied by air spaces and from measured diameters of stele, cortex and exodermis-plus-epidermis. Photographs of selected sections were compared by Dr H Konings with his quantitative results.

Pycnometer measurements

The pycnometer method is based on a comparison of the density of intact root tissue, including air-filled pores, and that of root homogenate without air. Intact root samples (1 to 3 g fresh weight) are carefully cut into pieces of 5 cm length directly before they are placed in a pycnometer flask (50 ml). Root samples, pycnometer and water are allowed to come to room temperature beforehand. After it has been filled, the pycnometer is weighed. After removing the water adhering to the roots by centrifugation for 30 s in a household centrifuge, the fresh weight of the roots is determined. Air is removed from the roots by grinding them in a mortar, and the density of the root homogenate is measured in the pycnometer. Alternatively, air can be removed from intact roots in the pycnometer by applying vacuum (the two methods give equal results, the latter is more reproducible and easier). The temperature at the time of each pycnometer measurement is recorded.

Calculations for the pycnometer method

For calculations of air-filled root porosity the following quantities are used: ε_r = percentage air-filled porosity of the root (on a volume basis), F = (fresh) weight [g], V = volume [cm^3], X(T) = density of water at T°C [$g\,cm^{-3}$].

As subscripts are used: r = roots with air-filled pores, r* = roots without air-filled pores, a = air-filled pores, p = pycnometer filled with water, pr = pycnometer with roots and water.

As roots replace a volume of water equal to their own volume when put in the pycnometer, the following relation holds (at temperature T1):

$$F_{pr}(T1) = F_p(T1) - V_r \cdot X(T1) + F_r \quad (1)$$

Rearranging gives:

$$V_r = \{F_p(T1) - F_{pr}(T1) + F_r\}/X(T1) \quad (2)$$

After grinding we find for the root mass without air-filled porosity, at temperature T2:

$$F_{pr*}(T2) = F_{pr}(T2) + V_a X(T2) \quad (3)$$

By rearranging we obtain:

$$V_a = \{F_{pr*}(T2) - F_{pr}(T2)\}/X(T2) \quad (4)$$

From (2) and (4) we find for the air-filled root porosity, ε_r:

$$\varepsilon_r = \frac{100 \cdot V_a}{V_r}$$

$$= \frac{100\{F_{pr*}(T2) - F_{pr}(T2)\} \cdot X(T1)}{\{F_p(T1) - F_{pr}(T1) + F_r\} \cdot X(T2)} \quad (5)$$

If T1 equals T2 the formula can be simplified to the form presented by Jensen *et al.* (1969). Values for X(T) can be found by interpolating between the following values for the density of water from Weast (1975): 0.99862, 0.99823 and 0.99707 $g\,cm^{-3}$ for 18, 20 and 25°C, respectively.

If an error of 1°C is made in the correction for temperature X(T), the estimated root air-filled porosity ε_r may, for the values of F_r and F_p we use, deviate from the real value by about 1% porosity. In 25 measurements of F_p we found an average value of 49.9622 g and a standard deviation of 0.0057 g. Using this value we may expect for a porosity estimate ε_r of 5% that a 95% probability interval is 3.9–6.1%, under the conditions of our measurements. Duplicate samples usually differ by less than 0.5% in calculated ε_r.

Table 1. Air-filled root porosity (%) quantified by the pycnometer method and by direct observation; the first two columns indicate treatment; in the pycnometer method three categories of roots were measured, intact seminal, intact nodal roots and broken branch roots ('rest'); for the microscopic sections of intact nodal roots two diameter classes were used and a weighted average is given, based on the percentage of each class of roots in the root system as a whole; in the last two columns average root diameter (mm) and specific root length (m g^{-1}) are given

		Pycnometer method			Microscopic sections			Aver. Diam.	Spec. root length
		Seminal	Nodal	Rest	<0.5 mm	>0.5 mm	ave.	Diam.	Length
AA	1	–	3.3	–	0	0	0	0.67	123
	2	1.5	4.4	–	0.5	8.2	0.9	0.26	426
A	1	0.7	5.1	1.5	5.0	12.1	5.9	0.28	192
	2	1.7	4.5	2.8	3.4	5.4	3.5	0.30	504
N	1	5.1	15.0	7.5	15.4	23.5	15.6	0.32	387
	2	1.7	7.7	1.2	6.1	13.8	7.4	0.25	182
NN	1	6.8	18.8	11.0	15.3	21.6	15.9	0.33	305
	2	0	9.2	1.6	8.9	15.4	9.4	0.26	382

Results and discussion

Results are summarized in Table 1. The treatments of the maize roots induced the expected variation in air-filled root porosity. With both observation methods considerable differences were observed in air-filled root porosity between parts of a single root system. Comparison of the two methods therefore is only possible if representative samples are considered. In the pycnometer method, intact seminal roots always had a lower air-filled porosity than intact nodal roots. Broken branch roots, from both nodal and seminal roots, generally had an intermediate porosity.

For the microscope sections only intact nodal roots were considered. As porosity appeared to be related to root diameter, calculations were based on the average porosity of 10 to 20 measurements in two diameter classes (with an average standard error of the mean of 2.18% air-filled porosity) and on the percentage of roots in each diameter class found for the nodal root system as a whole.

When considered on this basis, the average porosities of nodal roots as estimated by the two methods agree well. The main exception is that in roots of treatment AA hardly any cavities in the cortex were observed, while the pycnometer method indicated an air-filled root porosity of 3 to 4%. As in the direct observations intercellular spaces in the intact root tissue are not measured, such a difference is not surprising. If 3.5% porosity is added to all microscopic measurements to account for such intercellular air cavities, the results are slightly higher than the pycnometer indicates, but within the range of experimental error.

Continuous presence of $AgNO_3$ in the solution had a pronounced effect on root morphology: branch roots were initiated, but hardly grew beyond the cortex of the main axes. This resulted in a high average root diameter (Table 1). Response of porosity to treatments was much more pronounced in variant 1, with continuous treatment differences, than in variant 2 where root conditions were only affected in the last week before harvest.

The results show agreement between the two methods of measuring air-filled root porosity. With the pycnometer a relatively rapid estimate of air-filled porosity of a well-defined part of the root system is possible. With the microscopic technique the relation between air cavities and root diameter

can be studied in more detail, but more effort is needed to obtain a representative average value. In the study by Konings (1983) where the physiological mechanism of cavity formation was of prime interest, the microscopic technique, employed at a well defined distance from the root tip of nodal roots was appropriate. To obtain average values for calculations on root systems in the soil, the pycnometer method is preferable.

The difference in air cavities between main and branch roots is interesting as such. Thin roots have a better oxygen supply to all root cells than thick roots and hence need a lower air-filled root porosity provided that the external pathway through the soil yields oxygen. If, however, the root depends completely on internal oxygen transport, thin roots will lose more oxygen to the environment per unit volume of root tissue and hence need a higher air-filled root porosity to attain the same root length (De Willigen and Van Noordwijk, 1987). From our observations on microscopic sections and from the relevant literature, it seems that no continuity exists between air channels in the cortex of a main axis and air channels in the cortex of a branch root. At present we are investigating this matter experimentally. A diffusion barrier between air channels in main and branch roots may be advantageous to the plant if oxygen transport along the main root is insufficient to sustain all branch roots during a period of anaerobiosis in the soil. Priority of survival of the main axis may have ecological value for the plant, as it allows a rapid recolonization of the soil by new branch roots. Some observations on the response to flooding in the field agree with this view. Plumbing aspects of air channels and differences in air-filled porosity of different parts of a root system (Kozinka, 1979) have to be incorporated in models of oxygen transport to make them more realistic.

References

De Willigen P and Van Noordwijk M 1987 Roots, Plant production and Nutrient Use Efficiency. PhD thesis Agricultural University, Wageningen.

Jensen C R, Luxmoore R J, Gundy S D and Stolzy L H 1969 Root air space measurements by a pycnometer method. Agron. J. 61, 474–475.

Konings H 1983 Formation of gas spaces (Aerenchyma) in

seedling roots of *Zea mays* under aerated and non-aerated conditions. *In* Wurzelökologie und ihre Nutzanwendung. Eds. W Bohm *et al.* pp 761–765. Int. Symp. Gumpenstein 1982, Irdning.

Kozinka V 1979 Conditions for 'internal aeration' in the seminal root system of *Zea mays* L. Biologia (Bratislava) 34, 531–539.

Weast R C (Ed.) 1975 Handbook of Chemistry and Physics, 56th ed. The chemical rubber Co., Cleveland.

B. C. Loughman et al. (Eds.), Structural and functional aspects of transport in roots, 223–230.
© 1989 by Kluwer Academic Publishers.

Influence of oxygen deficiency on growth and function of plant roots

B.W. VEEN

Centre for Agrobiological Research, P.O. Box 14, 6700 AA Wageningen, The Netherlands

Key words: bicarbonate efflux, ion uptake, nitrate reductase, oxygen stress, root respiration, water uptake

Introduction

An increasing amount of the fruit–vegetables tomato, cucumber and paprika and a number of ornamental flowers grown in glasshouses are cultivated on artificial substrates such as rockwool and peat. Water and nutrients are supplied continuously by dripping a nutrient solution at the base of each plant. To prevent water shortage and accumulation of salts in the rooting medium the supply of nutrient solution is as an average twice the daily water uptake. Because of the variation in plant size and rate of water supply this ratio is often exceeded. Under these wet conditions diffusion of oxygen to the root system is assumed to be the limiting factor for root respiration and hence for growth and development of plants. Different from field conditions where rain showers can periodically cause oxygen deficiency, plants grown on artificial substrates are permanently supplied with suboptimal amounts of oxygen.

There are, however, great differences in sensitivity between plant species to oxygen deficiency in the root medium (Yu *et al.*, 1969). Oxygen transport from shoot to root was substantiated by the observation of large gas-filled spaces, aerenchyma, in shoots and roots of wetland plants (Armstrong, 1979). In non-wetland plants aerenchyma is formed in adventitious roots growing under anaerobic conditions, by degeneration of cortical cells (Trought and Drew, 1980). The tolerance of plants to oxygen deficiency is improved by mineral nutrition, especially nitrate (Guyot and Prioul, 1985). Trought and Drew (1981), explain this as an alleviation of the negative effect of oxygen deficiency on the mineral nutrition of the plant. Another explanation is that nitrate serves as an alternative electron acceptor to free oxygen in roots (Garcia-Novo and Crawford, 1973), thus enabling root respiration and root functions to continue. In line with this conception nitrate reductase activity is found to be enhanced in roots under anaerobic conditions (Lambers *et al.*, 1978). The aim of the experiments was to determine the change in root activity in relation to a reduction of root respiration of three species, maize, cucumber and tomato which are known to be differently sensitive to oxygen stress and to reveal the possible cause of these differences.

Material and methods

Measuring system

The measuring system for root growth and root activity is derived from a system described before (Veen, 1977). Only one plant is grown on the system with the roots situated in the root vessel, a transparent Perspex box measuring $20 \times 15 \times 15$ cm. The stem base is fixed water tight with a closed-cell rubber stopper into an opening in the top of the root vessel (Fig. 1). With a small rotary pump the nutrient solution is pumped at a speed of $1 \, l \, min^{-1}$ through the measuring chamber in which electrodes can be mounted. The measuring chamber is situated at the highest level in the system and can be moved up and down to keep the pressure inside close to atmospheric pressure which is necessary for a good functioning of the reference electrode. After passing the measuring chamber the nutrient solution flows back to the root vessel. Because the root vessel and the measuring chamber are closed, the volume of the nutrient solution in the system is displayed in the aeration vessel (A)

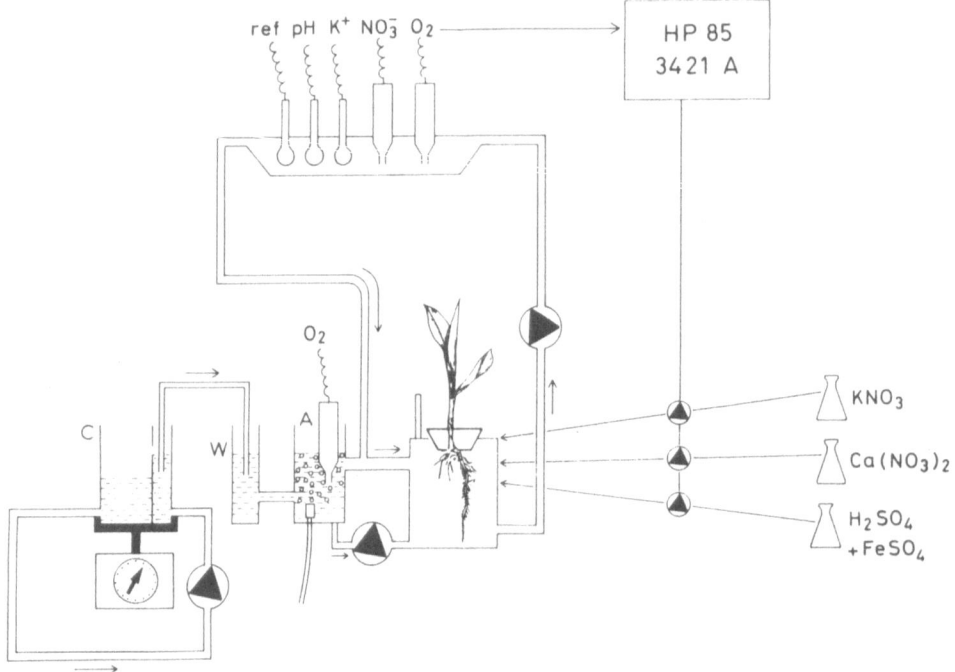

Fig. 1. Measuring system for simultaneous measurements of uptake of water, oxygen, potassium and nitrate, and the production of bicarbonate by a plant root system. For further details see text.

and vessel W, which have open connections with the main circulation stream.

Next to the system a perspex box (C) measuring $10 \times 8 \times 14\,\text{cm}$ is placed on a digital balance. The box is separated into two unequal compartments by a wall with an opening 1 cm below the top. A small persistaltic pump circulates a nutrient solution without NO_3^- and K^+ from the left to the right compartment, resulting in a constant level in the right compartment. With a syphon this constant level is transmitted to vessel W, so that in the measuring system the total volume of the solution is kept constant. The transpiration of the plant can be determined continuously by weighing the solution lost from box C. The nitrate concentration is measured using an Orion liquid ion-exchange membrane electrode 92-07, potassium using a Philips glass electrode G15-K and pH using an Electrofact 7G112 glass electrode. The reference is an Electrofact gel-filled electrode, type SR20. After amplification the electrode signals are permanently presented to a HP85 computer with a 3421A data acquisition/control unit. When the K^+ and NO_3^- concentrations measured in the system are below a present level, known amounts of KNO_3 and $Ca(NO_3)_2$ are titrated to compensate for the uptake

by the plant. Because the nitrate uptake usually exceeds the potassium uptake the potassium uptake is compensated with KNO_3 and the NO_3^- uptake with an additional amount of $Ca(NO_3)_2$.

Because nitrate is the only nitrogen source, the pH usually raises due to OH^- or HCO_3^- production of the root system. This bicarbonate production is titrated with 0.25 mol sulfuric acid to achieve a constant pH of 4.5. At this pH all HCO_3^- produced is liberated as CO_2. The O_2 electrodes are Electrofact Clark-type electrodes type Z152. One is positioned in the circulating nutrient solution, the second in the aeration vessel. Root respiration is measured and regulated by pumping a known amount of oxygen-rich nutrient solution from the aeration vessel to the root vessel using a CFF dosing pump type E 603. At the same time a same amount of the circulated nutrient solution flows back passively to the aeration vessel, keeping the water level constant. From the difference in oxygen concentration and the flow rate the net amount of oxygen pumped into the circulated system can be calculated. In a steady state situation the oxygen concentration in the root vessel is constant and this net amount of oxygen equals the amount taken up by the plant roots. In case of a non-stable oxygen concentration root respiration is corrected for the change in concentration. Because only a limited amount of aerated nutrient solution is pumped into the root vessel the oxygen concentration in the circulated nutrient solution is always lower than in the aeration vessel.

The amount of oxygen supplied to the roots can be manipulated by changing the ratio of nitrogen and oxygen gas used for aeration. Nitrogen and oxygen was obtained from cylinders with compressed gas and mixed in the correct ratio using two adjustable mass flow controllers (Precision Flow Devices type 112). The reduction in root respiration however depends not only on the reduction in oxygen content of the gas mixture but also on the root respiration under optimal conditions, which is a function of the root size. In practice it is impossible to reduce the root respiration to a preset value. Therefore the reduction of root respiration and the effects on root activity of a number of comparable plants are averaged.

Every 15 minutes a series of 19 relevant data about the state of the system and the amounts titrated is registered on the HP85 tape cartridge. It is possible to register 15 successive days on one tape. The cumulative uptake of NO_3^-, K^+, H^+ and water were first plotted versus time and from periods of constant activity the uptake rates were calculated using regression analysis. In order to demonstrate the influence of oxygen stress on root activity independent of the root size, this influence is expressed as a percentage of the root activity under optimal conditions. Except for water uptake no significant differences in root activity have been found between the day and night period. Therefore the root activities under oxygen stress are calculated as the percentage of the activity averaged over the control period of about 24 h. Only the influence of oxygen stress on water uptake is calculated as the percentage of the water uptake of the corresponding period during the control day.

Root growth was measured as increase in root volume using a method described before (Veen, 1977).

Nitrate reductase activity (NRA) was measured using the intact tissue assay of Jones and Sheard (1977). Samples of 2-g fresh-root tissue were taken in duplicate from root systems of stressed and control plants and vacuum infiltrated under N_2 in 25 ml of a medium containing 50 mM KNO_3, 1%

n-propanol and 100 mM phosphate buffer at 27°C. Nitrite released into the medium after 30 and 60 minutes was determined with sulfanilamide and 0.02% N-1-naphtylethylene diamine. Optical density was measured at 540 nm.

Growing conditions

Maize (*Zea mays*, var. LC11), tomato (*Lycopersicum esculentum* L. var. Counter) and cucumber (*Cucumis sativus* L. var. Farbio) seeds were sown in moist sand in a climate room at 20°C, 80% relative humidity and a daylength of 16 hours. The light intensity was about 50 μE m^{-2} sec^{-1}. After 7 days a number of uniform plants of each species were selected and grown on half strength Hoagland nutrient solution containing (mM): KNO_3, 2.5; $Ca(NO_3)_2$, 2.5; KH_2PO_4, 0.5; $MgSO_4$, 1.0 and trace element (mg l^{-1}) B, 0.1; Mn, 0.1; Zn, 0.01; Cu, 0.004; Fe, 7.5; Mo 0.01. Light intensity was about 150 μE m^{-2} sec^{-1} with a daylength of 16 h. The temperature was 22°C and the relative humidity 70%. After 3 to 4 weeks a plant was fixed in the measuring system and grown at the same environmental conditions. After an adaption period of about 3 days the respiration rate and other root activities were measured under optimal supply of oxygen and nutrients for one day. The day after this control day the respiration rate was reduced by choosing an appropriate ratio of N_2 and O_2 gas that was obtained from gas cylinders, and the effect on root growth, water and ion uptake and bicarbonate production was measured.

Results

To compare the three plant species the root activities at optimal conditions on basis of unit root weight or unit leaf area are given in Table 1. Maize showed a relatively high root respiration rate, the potassium uptake per unit root weight of cucumber was significantly lower than that of tomato and maize. There was no significant difference in nitrate uptake between the three species, but the bicarbonate production of cucumber was significantly lower than that of tomato and maize. Tomato had the highest water uptake rate per unit leaf area and maize, being a C_4 plant, the lowest. The effect of a

Table 1. Oxygen uptake (O_2), potassium uptake (K^+), nitrate uptake (NO_3^-), bicarbonate production (Z) and water uptake rate (H_2O) of three plant species under climate room conditions at optimal oxygen supply. Values followed by the same letter are not significantly different ($P > 0.05$)

Species	O_2 $mg\,g^{-1}h^{-1}$	K^+ $\mu mol\,g^{-1}h^{-1}$	NO_3^- $\mu mol\,g^{-1}h^{-1}$	Z $\mu mol\,g^{-1}h^{-1}$	H_2O $mg\,cm^{-2}h^{-1}$
Cucumber	0.13[a]	2.3[c]	5.3[e]	1.5[f]	8.0[h]
Tomato	0.15[ab]	3.8[d]	7.1[e]	2.8[g]	10.9[i]
Maize	0.20[b]	3.9[d]	6.8[e]	2.9[g]	6.2[h]

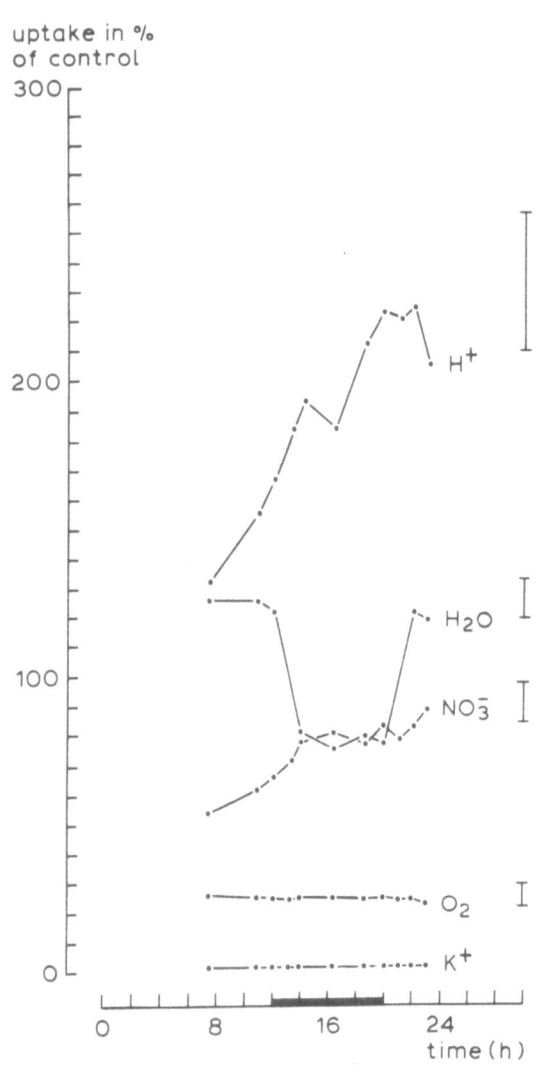

Fig. 2. Time course of bicarbonate production (H^+), uptake of water (H_2O), nitrate (NO_3^-), potassium (K^+) and oxygen (O_2) of a maize root system after reduction of oxygen supply. *Vertical bars* indicate standard deviation, the horizontal bar the dark period.

reduction in root respiration to 20% of optimal on a number of root activities of maize was measured (Fig. 2). K^+ uptake stopped immediately after reduction of oxygen supply, K^+ concentration in het root environment even raised due to leakage of K^+ out of the roots. Nitrate uptake was reduced to about 40% of the rate at optimal oxygen supply, but increased in time to about 90% of the level at optimal oxygen supply. The reduction of water uptake about 16 h after onset of oxygen stress occurred during the dark period. There was a 20% reduction of water uptake in the dark whereas there was no influence of oxygen stress in the light.

The most striking effect of reduction of oxygen supply to the roots was the increase in alkalinization of the nutrient solution by increased HCO_3^- production of the roots. This is even more striking for cucumber (Fig. 3) where the rate of alkalinization of the nutrient solution after 24 h of oxygen stress was about 4× the control rate. A reduction in oxygen supply to 20% of the maximal uptake rate induced an immediate reduction of the nitrate uptake to 40% of the control rate, but during the experimental period the nitrate uptake increased and after 24 h the uptake rate had recovered to 80%. Potassium uptake completely stopped at low-oxygen conditions and as with maize, K^+ concentration in the root medium increased due to leakage from the root system. Reduction of the oxygen supply to 20% of the maximal respiration rate caused death of the root tips of tomato, reduction of the water uptake rate and hence wilting of the plant. To study adaption of tomato plants to oxygen stress, oxygen supply was reduced to 40% of the maximal respiration rate. In spite of this lower oxygen stress, nitrate uptake was relatively more reduced. The recovery after 24 h treatment was only 50% of the control value (Fig. 4). Also water uptake showed a decrease towards the end of the stress period, and although the rate of bicar-

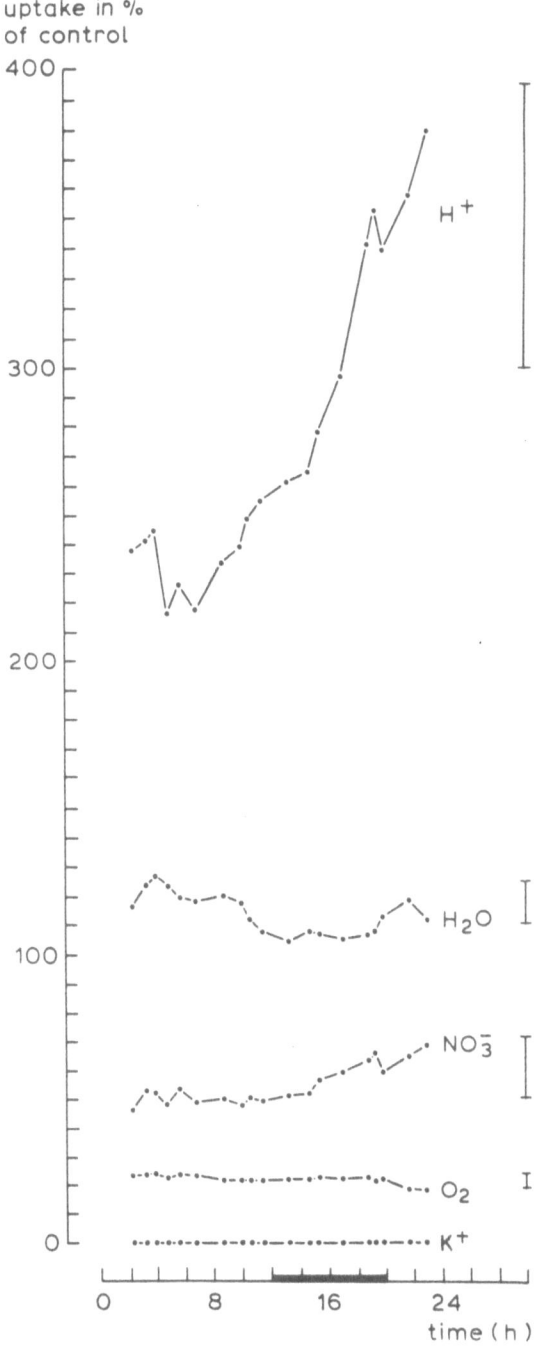

Fig. 3. Time course of bicarbonate production (H$^+$), uptake of water (H$_2$O), nitrate (NO$_3^-$), potassium (K$^+$) and oxygen (O$_2$) of a cucumber root system after reduction of oxygen supply. *Vertical bars* indicate standard deviation, the *horizontal bar* the dark period.

bonate production increased, this increase was clearly less than for maize and cucumber. As for cucumber and maize potassium uptake stopped immediately after reduction of oxygen supply and K$^+$ leaked into the root medium.

Bicarbonate production of the root system is supposed to be a reflection of the nitrate reduction rate of the root system. Therefore, nitrate reductase activity (NRA) of roots of the three species under control conditions and after 24 h of oxygen stress was measured (Table 2).

Cucumber had a significantly lower NRA than tomato and maize. The relative increase in NRA due to oxygen stress was also highest in cucumber, but in absolute sense the NRA in tomato roots increased more. The NRA in maize roots showed no significant increase at oxygen stress and the activity was between that of cucumber and tomato.

Figure 5 shows the effect of a reduced respiration rate on nitrate uptake a few hours after start of the

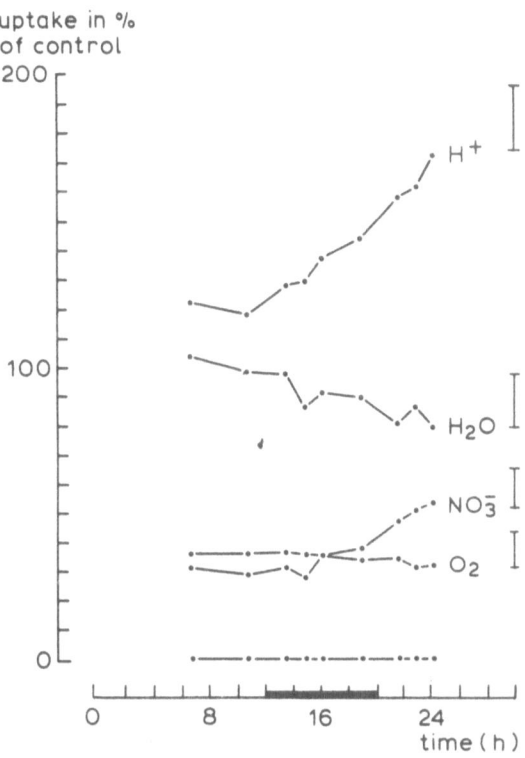

Fig. 4. Time course of bicarbonate production (H$^+$), uptake of water (H$_2$O), nitrate (NO$_3^-$), potassium (K$^+$) and oxygen (O$_2$) of a tomato root system after reduction of oxygen supply. *Vertical bars* indicate standard deviation, the *horizontal bar* the dark period.

Table 2. *In vivo* nitrate reductase activity (μmol.h^{-1} on a fresh weight basis) in the roots of three plant species, grown at optimal oxygen supply and after 24 h of oxygen stress. Values followed by the same letter are not significantly different ($P > 0.05$)

Species	Treatment	
	Control	O$_2$-stress
Cucumber	0.64[a] $\pm$ 0.16	1.22[b] $\pm$ 0.29
Tomato	2.50[b] $\pm$ 0.73	3.86[c] $\pm$ 0.36
Maize	1.74[b] $\pm$ 0.37	1.88[b] $\pm$ 0.53

treatment. The oxygen uptake of maize roots could be reduced to 50% of the control rate, without influencing nitrate uptake. Below that point nitrate uptake decreased nearly linearly with oxygen uptake. In cucumber roots a reduction of oxygen uptake to 80% of the control rate had no influence on nitrate uptake; below that point nitrate uptake decreased linearly with oxygen consumption. Tomato was the most susceptible to oxygen stress and although measurements in the range between 70 and 100% oxygen uptake are missing, the data are situated around the 45° line, suggesting an equivalent reduction of respiration and nitrate uptake.

Discussion

Root respiration is measured as the rate of oxygen uptake by roots from a circulating nutrient solution. It is assumed that the root medium is the only oxygen source for root respiration and that no oxygen is supplied to the roots by internal gas transport. A functional shoot to root internal gas transport was demonstrated in anaerobically grown pea roots (Armstrong, 1979). Several plant species, when grown under anaerobic condition, develop aerenchyma in the root cortex (Trought and Drew, 1980; Yu *et al.*, 1969), large gaseous spaces which provide low resistance channels for gas diffusion. Roots grown under aerated conditions contain also intercellular gas spaces. Considerable differences in root porosity exist between species which can be expressed as differences in specific weight of their roots. In our case the specific weights of the roots of maize, tomato and cucumber were 0.87, 0.96 and 0.97 g cm^{-3} respectively. The roots of our plants are grown under aerobic conditions and only one day of oxygen stress is a

too short period to induce aerenchyma. Not only the percentage porosity but also the continuity of the pores is an important factor determining the influence of gas filled pores on longitudinal oxygen transport. The functional importance of root porosity on oxygen transport from shoot to root under our experimental conditions was examined by cutting the shoot at the end of the experimental period and sealing the cut stem surface with a 1:4 mixture of bees wax and vaseline. No influence on the rate of oxygen uptake by the root system for more than one hour after this interference could be observed. So under our experimental conditions no oxygen transport from shoot to root occurred and the rate of oxygen uptake from the nutrient solution can be considered as equal to the root respiration rate. This conclusion is in accordance with results of Prioul and Guyot (1985) who found no influence of the oxygen concentration in the shoot environment on the rate of oxygen uptake from a nutrient solution of aerobic wheat roots. Wheat roots, grown for one week under anaerobic conditions, however, showed considerable internal oxygen transport, depending on the oxygen diffusion gradient. With all our plants the respiration could be reduced to a certain extent without influence on water uptake. Even reduction of the root respiration to 20% of the control rate had no influence on water uptake of cucumber and maize. Also Trought and Drew (1980) found no negative influence of anoxia on water uptake of young

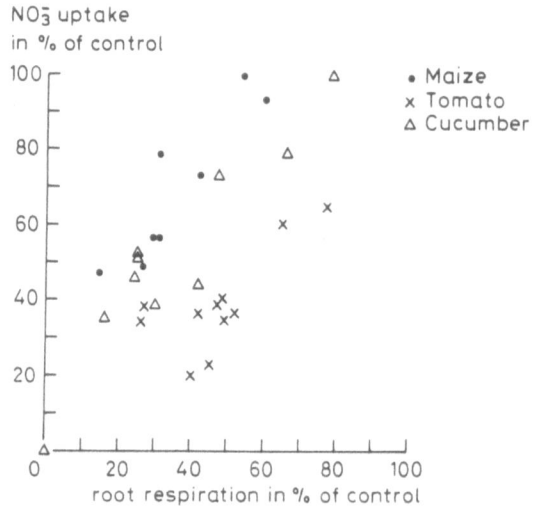

Fig. 5. Influence of root respiration on nitrate uptake of maize (●), tomato (×) and cucumber (△), about two hours after induction of oxygen stress.

wheat plants. Although only a limited number of data is available the results suggest that for maize the water uptake in the dark period is relatively reduced in case of oxygen stress (Fig. 2). This was not observed for cucumber and tomato. Tomato was much more sensitive to oxygen stress and even reduction of respiration to 40% caused a decrease in water uptake at the end of the experimental period.

Also Brouwer (1964) reduced respiration of bean roots with KCN and found no influence of the cyanide concentration on water uptake up to $10^{-5} M$, at higher concentrations the water uptake rate decreased. Brouwer (1964) also reports the ion uptake to be more sensitive to KCN. In our experiments the nitrate and potassium uptake are influenced by oxygen stress although not to the same extent. Nitrate uptake 2 h after start of the treatment was for maize not influenced by reduction of the respiration to 50% of maximal (Fig. 5). At further decrease of respiratory activity also nitrate uptake rate decreased. For cucumber the nitrate uptake decreased when the respiration rate was reduced more than 20%, whereas for tomato there was no reduction of respiration without immediate effect on nitrate uptake.

In the control period the uptake of nitrate and potassium and the efflux of bicarbonate of maize and tomato was in electrical equilibrium: nitrate uptake nearly equalled the sum of potassium uptake and bicarbonate efflux (Table 1). In the stress period, however, nitrate uptake decreased and the bicarbonate efflux increased (Figs. 2, 3 and 4). When bicarbonate efflux exceeded nitrate uptake, potassium leaked out of the roots, giving the impression of a passive movement in a direction that leads to maintenance of electrical neutrality. This is in agreement with data of Jackson *et al.* (1976) who consider the nitrate uptake as the driving force for net potassium uptake. The rate of potassium efflux could not be determined because the relative change in concentration was too small in a volume of 5 l nutrient solution.

The nitrate uptake a few hours after decrease of the respiration rate for cucumber and maize decreased less than the respiration rate, but tomato in general showed a stronger decrease in nitrate uptake than in respiration rate (Fig. 5). The recovery of nitrate uptake during the 24 h stress

period was also least for tomato. This recovery of nitrate uptake can be explained by the increase in nitrate reduction measured as HCO_3^- efflux which reduces the internal nitrate level and stimulates the uptake mechanism (Deane-Drummond and Glass, 1983). This stimulation of NRA is least in tomato with consequently a minor stimulation of NO_3^- uptake.

The nitrate reductase activity in the root system was determined in two different ways: as bicarbonate production of the root system and by an *in vivo* measurement using a representative sample of the root system and measuring the NO_2^- efflux in the dark under anaerobic conditions. Although both methods indicate that the NRA is increased after a period of oxygen stress, the extent of the increment depends on the method used. The second method is more a reflection of the potential NRA, because the procedure includes an optimalization of the conditions. Therefore the bicarbonate output is used as a measurement of the NRA under our experimental conditions. Immediately after reduction of root respiration the bicarbonate efflux increased during the 24 h experimental period (Figs. 2, 3 and 4) and also nitrate reductase activity increased (Table 2). This increase in NRA is a common feature of plants under oxygen stress (Garcia-Novo and Crawford, 1973, Lambers *et al.*, 1978) but the extent of this increase differs between the species. The increase in bicarbonate production of oxygen-stressed cucumber plants, up to 4 times the control rate, is regarded as a mechanism of plant roots to eliminate excess NADH by insufficient oxygen availability and to overcome a period of oxygen stress. Maize, however, with only a limited increase in HCO_3^- production was even more insensitive to oxygen stress than cucumber (Fig. 5). Its root porosity was, however, much greater than cucumber and tomato which for maize is regarded as the main mechanism to overcome a period of oxygen stress. In my opinion there is no internal transport of oxygen from shoot to root, but the porosity of the root system prevents radial oxygen gradients in the root by facilitating diffusive oxygen transport in the gas phase. In more compact roots of cucumber and tomato supply of oxygen to the central core and apical parts of the roots is more difficult, leading to a greater sensitivity at low oxygen conditions.

References

Armstrong W 1979 Aeration in higher plants. Adv. Bot. Res. 7, 226–332.

Brouwer R 1964 Water movement across the root. Symp. Soc. Exp. Biol. 19, 131–149.

Deane-Drummond C E and Glass A D M 1983 Short-term studies of nitrate uptake into barley plants using ion-specific electrodes and $^{36}ClO_3^-$. Plant Physiol. 73, 100–104.

Garcia-Novo F and Crawford R M M 1973 Soil aeration, nitrate reduction and flooding tolerance in higher plants. New Phytol. 72, 1031–1039.

Guyot C and Prioul J L 1985 Correction par la fertilisation minérale des effets de l'ennoyage sur le blé d'hiver. Experimentation sur sol. Agronomie 5, 743–750.

Jackson A W, Kwik K D, Volk R J and Butz R G 1976 Nitrate influx and efflux by intact wheat seedlings: Effects of prior nitrate nutrition. Planta (Berl.) 132, 149–156.

Jones R W and Sheard R W 1977 Conditions affecting *in vivo* nitrate reductase activity in chlorophyllous tissues. Can. J. Bot. 55, 896–901.

Lambers H, Steingröver E and Smakman G 1978 The significance of oxygen transport and metabolic adaptations in flood-tolerance in *Senecio* species. Physiol. Plant. 43, 277–281.

Prioul J L and Guyot C 1985 Role of oxygen transport and nitrate metabolism in the adaption of wheat plants to root anaerobiosis. Physiol. Vég. 23, 175–185.

Trought M C T and Drew M C 1980 The development of waterlogging damage in young wheat plants in anaerobic solution cultures. J. Exp. Bot. 31, 1573–1585.

Trought M C T and Drew M C 1981 Alleviation of injury to young wheat plants in anaerobic solution cultures in relation to the supply of nitrate and other inorganic nutrients. J. Exp. Bot. 32, 509–522.

Veen B W 1977 The uptake of potassium, nitrate, water and oxygen by a maize root system in relation to its size. J. Exp. Bot. 28, 1389–1398.

Yu P T, Stolzy L H and Letey J 1969 Survival of plants under prolonged flooded conditions. Agron. J. 61, 844–847.

B. C. Loughman et al. (Eds.), Structural and functional aspects of transport in roots, 231–233.
© 1989 by Kluwer Academic Publishers.

Effects of soil temperature and water on maize root growth

S.A. BARBER, A.D. MACKAY, R.O. KUCHENBUCH and P.B. BARRACLOUGH
Agronomy Department, Purdue University, West Lafayette, IN 47907, USA

Key words: nutrient uptake, phosphorus, root distribution, *Zea mays*

Maize (*Zea mays* L.) growth and the rate of nutrient uptake are influenced by the length and radius of roots in the surface 20-cm of soil. In many soils a large share of the available P and K is in the upper 20-cm because of cycling from greater depths by the plant roots and application of P and K fertilizers, hence most P and K uptake comes from this layer.

We have developed a mechanistic mathematical model that predicts nutrient uptake by roots growing in soil (Barber, 1984). This model combined the rate of increase in root surface area, the kinetics of nutrient uptake as affected by nutrient concentration in the soil solution, and the rate of supply of nutrients from the soil to the root by mass flow and diffusion. Information on 11 parameters is needed for operation of the model. The size and geometry of the absorbing root surface and its rate of increase with time is described by L_0, initial root length; k, rate of root growth, and r_0, average root radius. The kinetics of nutrient absorption are assumed to follow Michaelis-Menten kinetics and the relation is described using parameters I_{max}, the maximal influx; K_m, the solution concentration at the root surface where net influx, In, is $\frac{1}{2}I_{max}$; and C_{min}, the concentration in soil solution at the root where In = 0. The supply of nutrients to the root by the soil is described by C_{li}, the concentration in the soil solution; b, the buffer power of the nutrients on the solid phase for nutrients in solution; D_e, the effective diffusion coefficient for nutrient diffusion through the soil, v_0, the rate of water flux into the root, and r_1 the half-distance between root axes.

Predicted uptake using the model agrees closely with observed uptake when the uptake process follows the model assumptions (Barber, 1984). The significant assumptions are that the root absorbs

nutrients from solution and exudates do not influence the value of C_{li} by changing soil pH, *etc.*, and that the effect of root hairs and mycorrhizae are insignificant. Where these effects do arise, the model can be modified to include them. The model was used in a sensitivity analysis where the value of each parameter was changed to 0.5, 1.5 and 2.0 of the initial value while keeping the values of the remaining parameters constant and uptake calculated. The result of a sensitivity analysis for P uptake is shown in Fig. 1. For the same degree of change, root growth rate, k, and root radius, r_0, had the largest effect on uptake. Because of this we chose to investigate factors that may change k and r_0 for the root growth occurring in the upper 20-cm of soil.

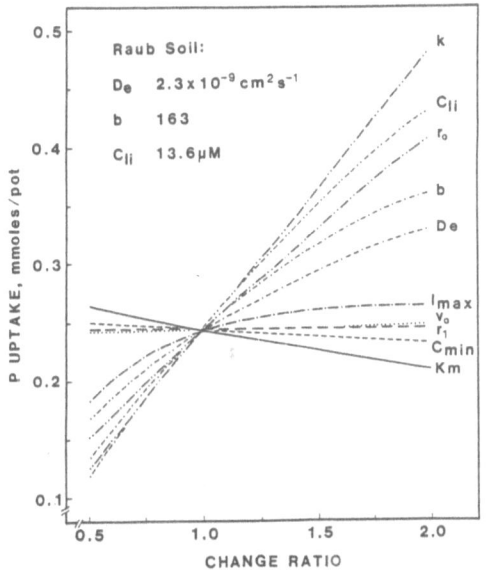

Fig. 1. Sensitivity analysis of parameters used in a mechanistic model describing P uptake by soybeans (*Glycine max* L.). Each parameter was varied while holding the remaining parameters at their initial level.

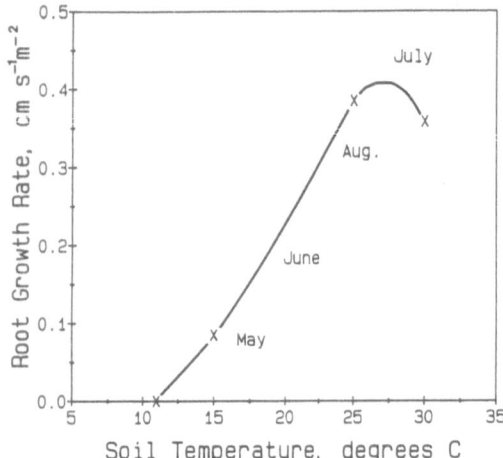

Fig. 2. Relation between maize root growth rate and soil temperature. Values are shown for mean soil T at 10-cm under bare soil at Lafayette, IN, USA, for each month.

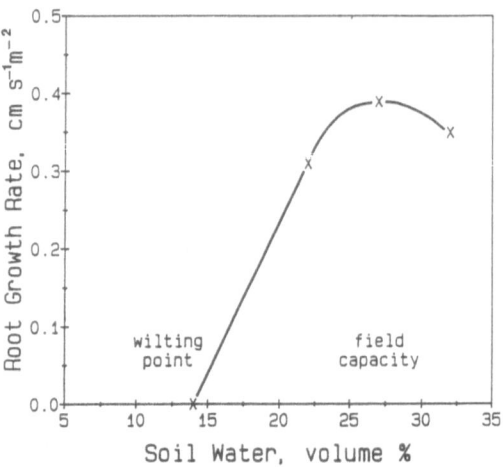

Fig. 3. Relation between volumetric soil water level and rate of root growth in pot culture.

While there may be differences among cultivars, maize root growth generally starts at about 9°C and reaches a maximum growth rate at 28°C (Fig. 2). This relation plus a knowledge of soil temperature variation with time can be used to calculate the effect of temperature on root growth both in terms of variation with soil depth and with time. Figure 2 was developed from experimental data of Mackay and Barber (1984) and Walker (1969). At Lafayette, Indiana, USA, the mean temperature under 10-cm of soil is 16°C for May, 22°C for June, 26°C for July, and 25°C for August. Actual temperatures in a field where maize is growing may be less due to shading. Using this temperature-root growth relation, root growth as a fraction of maximum would be 0.28 for May, 0.65 for June, 0.97 for July and 0.96 for August. Root growth in the 0- to 20-cm layer would not be limited significantly by temperature during July and August.

In addition to the general effect on root growth due to changes in temperature with time, in spring and summer the upper layers will also be warmer than deeper layers so that growth rate, if only affected by temperature, will be greater in the upper soil layers.

The second soil property that influences root growth rate is soil water. Data of Mackay and Barber (1985) show that rate of root growth on Raub silt loam (Aquic Argiudoll) is a maximum at a volumetric water content, θ, of field capacity (0.33 kPa) and it ceases to grow at wilting point (1500 kPa). Available water is the soil water between 0.33 and 1500 kPa. The relation of volumetric soil water to rate of root growth is shown in Fig. 3. When available water was 60% of field capacity, root growth was 70% of the maximum rate. Most soils wet and dry from the surface downward and as the top layers dry, root growth rate increases in deeper, wetter layers. However since much of the available P and K are frequently in the 0- to 20-cm layer, the root growth in layers below 20-cm may not contribute greatly to the P and K supply to the plant.

The 0- to 20-cm layer will have the widest fluctuations in both soil water and soil temperature so that root growth in this layer will be greatly dependent on weather conditions. Using the climate conditions at Lafayette, Indiana, USA, we find that because the soil profile is usually at field capacity at planting and because water use by transpiration is usually less than precipitation for the first 6 weeks of maize growth, the water level from 5-cm to 20-cm will stay near field capacity during this growth period. Temperatures are below optimum for root growth at this time of year so soil temperature rather than soil water will have the greatest effect on root growth. The 0- to 5-cm layer may dry rapidly in bare (little plant cover) soil so that root growth in this layer may be influenced by rapid soil water changes due to evaporation losses after each addition from precipitation.

The data for our location indicate that the primary factor affecting root distribution during the first four to six weeks of growth is soil tem-

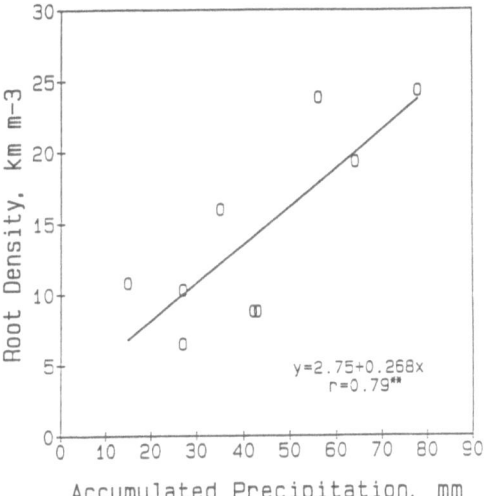

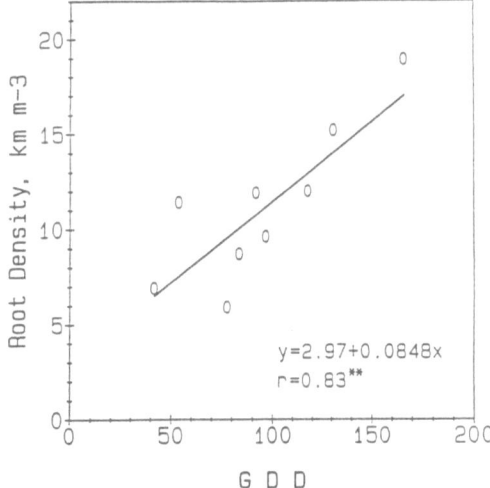

Fig. 4. Relation between precipitation for 3 weeks before anthesis and root density in the 0- to 15-cm layer of Raub silt loam in the field. Data are for 9 years.

Fig. 5. Relation between accumulated growing degree days for 2 weeks after planting and maize root density at 60- to 75-cm on Raub silt loam. Data are for 9 years.

perature. Soil water then becomes the principal factor regulating root growth and distribution since soil temperature in the 0- to 20-cm layer is high enough for near maximum root growth rate. At our location, precipitation during the four weeks prior to anthesis is usually less than evapotranspiration so that the soil water content decreases. Since reduction starts from the surface, the primary reduction is in the upper layers which contain the most of the available P and K.

Root distribution in the soil at anthesis was measured in nine separate years. Root density in the 0- to 15-cm layer was highly correlated with precipitation occurring during the three weeks prior to anthesis (Fig. 4). This indicates that soil water level was the primary factor affecting root growth rate.

Soil temperature affects root growth more in the deeper soil layers because they are cooler at our location. The major difference among the nine years was in the temperature during the first few weeks after planting. Root growth is more sensitive to the range of soil temperatures occurring at this time. With increased heat applied to the soil as the season progressed, roots grew deeper since the deeper soil layers became warm enough for more rapid root growth. The relation between accumu-

lated heat units expressed as growing degree days (a growing degree day is the mean °C above 10°C per day) and root density at 60- to 75-cm is shown in Fig. 5. There was no relation between temperature and root density in the 0- to 15-cm layer because soil water level had the greater impact. Temperature even with no soil water variable had the greatest effect at 60- to 75-cm because there was little variation in the water content of this layer with year at this stage of growth. In other soils with differing climate differences from our observations could occur. The objective of this paper is to show that both temperature and precipitation affect root distribution within the soil profile.

References

Barber S A 1984 Soil Nutrient Bioavailability: A mechanistic approach. John Wiley and Sons, New York, NY.

Mackay A D and Barber S A 1984 Soil temperature effects on root growth and phosphorus uptake by corn. Soil Sci. Soc. Am. J. 48, 818–823.

Mackay A D and Barber S A 1985 Soil moisture effects on root growth and phosphorus uptake by corn. Agron. J. 77, 519–523.

Walker J M 1969 One-degree increments in soil temperatures affect maize seedling behavior. Soil Sci. Soc. Am. Proc. 33, 729–736.

B. C. Loughman et al. (Eds.), Structural and functional aspects of transport in roots, 235–239.
© 1989 by Kluwer Academic Publishers.

Changes in the osmotic potential of the root as a factor in the decrease in the root-shoot ratio of *Zea mays* plants under water stress

PETER M. SCHILDWACHT
Department of Plant Ecology, University of Utrecht, Lange Nieuwstraat 106, 3512 PN Utrecht, The Netherlands

Key words: osmoregulation, osmotic value, root, shoot, water stress, *Zea mays*

Abbreviations. P, turgor pressure; RWC, relative water content

Introduction

Mature plants cells regulate turgor pressure by osmoregulation. After the onset of water stress the osmotic potential decreases due to loss of water, in addition there is a net influx of osmotica into mature cells or synthesis of osmotically active compounds (Turner and Jones, 1980). Full restoration of turgor pressure has been observed within a period of 48 h for leaves (Michelena and Boyer, 1982 with Zea; Takami *et al.*, 1982 with Helianthus). It has frequently been found that osmoregulation was not capable of full turgor pressure restoration (Hellebust, 1976; Turner and Jones, 1980).

For enlarging cells the influx of osmotica is also needed to sustain the volume growth by water uptake. The newly formed cell volume in the process of cell enlargement requires an additional influx of osmotic materials. Theoretically, a limited supply of osmotica to these cells can lead to either a full turgor pressure maintenance at a low enlargement rate or to a partial turgor pressure maintenance at a somewhat higher enlargement rate.

So far few studies have been done on roots. For roots of cotton (Bernstein, 1961) and maize root tips (Sharp and Davies, 1979) full restoration of turgor pressure (P) was found, in the same experiments only partial turgor restoration of maize leaves was observed.

Greacen and Oh (1972) suggested that roots have a greater capacity for osmotic adjustment than leaves. If this is true, the increase in the root to shoot ratio after the onset of water stress may be explained by the growth rate (Lockhart, 1965):

$$r = [(m \cdot L)/(m + L)][\Delta\psi + P - Y] \qquad (1)$$

in which r is the relative rate of change in cell volume (s^{-1}), m a constant of proportionality $(s^{-1} Pa^{-1})$, L the hydraulic conductance $(m\,s^{-1} Pa^{-1})$, $\Delta\psi$ the water potential difference between extending cells and the cell wall (Pa), and Y the turgor pressure threshold value (Pa).

In cell extension osmotica are needed to sustain the water potential difference and the turgor pressure. It is obvious that at a low external water potential a high influx rate (or rate of synthesis) of osmotica in enlarging root cells may result in a high cell enlargement rate.

After a lowering of the plant water potential a difference in influx capabilities of extending root and leaf cells will result in a shift in the amount of newly formed roots and shoot parts, and hence in a shift in the root/shoot ratio.

The aim of this work is to compare the capacity for osmotic adjustment of roots and leaves under conditions of water stress and to discuss its effect on root/shoot ratios.

Materials and methods

Plant material

Zea mays L. (cv. Caldeira 535) seeds were germinated in moist sand. After 5 or 6 days, when the length of the primary root was 100–150 mm, plants were transferred into an aerated full strength Hoagland solution at 25 ± 2°C. The pH was between 4.0 and 4.5. The photon flux density was 230 μmol m^{-2}s^{-1}, 16 h a day. Twenty-four hours after transfer to nutrient solution plants were treated with different amounts of polyethylene glycol-400 (PEG-400, BDH, Poole, England). After each addition of PEG-400 the pH and temperature were adjusted.

Three root segments were sampled, the extension zone of the root from 0 to 10 mm behind the root tip, and 0–10 and 60–70 mm from the root base where the root was fully expanded. Location of the extension zone was studied by placing markers at 1 mm intervals up to 20 mm behind the root tip.

Osmotic potentials

Ten root segments were pooled. Adhering water was removed by centrifugation with a tabletop-centrifuge at 1500 rpm for 1 minute. Segments were frozen at −20°C and, after thawing, cell sap was pressed out of the segments. Osmotic potential was measured with a dewpoint psychrometer (Wescor). Osmotic potential of the first leaf was measured on the first 3 cm of the leaf tip.

Water potential and turgor pressure

Water potential of the leaves was measured with a pressure chamber. Turgor pressure of the primary leaf was calculated as the difference between the osmotic potential and the pressure chamber reading the leaf. Turgor pressure of roots was calculated as the difference in osmotic potential of cells and nutrient solution.

Potassium measurement

Measurement of K$^+$ concentration in the expressed sap of root segments was done with a flame-spectrophotometer.

Amino acids

Frozen root segments were treated according to Borstlap (1972) and amino acids were analyzed by gas chromatography.

Results

Osmotic potential

The osmotic potential of the root cells differed from −1.03 to ±0.16 MPa for the root tips to

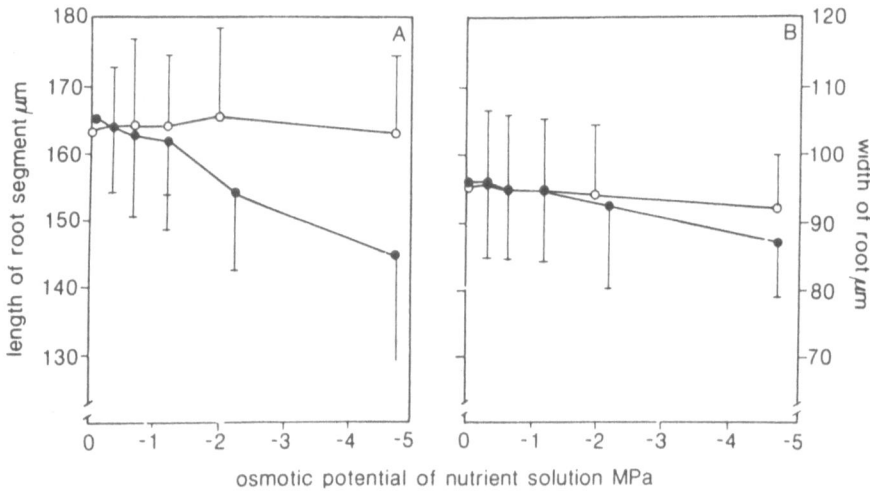

Fig. 1. Length (**A**) (values divided by 10), and width (**B**) of 1640 μm long root segments cut out of the root zone with elongating cells, 3 mm behind the root tip. Measurements were made 1 h after replacement of the nutrient solution with nutrient solutions with different amounts of PEG (—●—) or with a new nutrient solution (—○—). *Bars* indicate SE, n = 66.

-0.94 ± 0.12 and -0.88 ± 0.14 MPa for the middle and basal part of the root. A decrease of the osmotic potential of the nutrient solution lowered the osmotic potential of root cells, depending on the level of osmotic stress and the location along the root axis. The root base always had a higher osmotic potential than that of the nutrient solution. The osmotic potential of the middle part of the root was intermediate between the other 2 segments (data not shown).

One hour after addition of PEG to the nutrient solution, the increase in length of $1640 \mu m$ long root segments cut out the region with elongating cells (0 to 10 mm from the root tip) decreased with

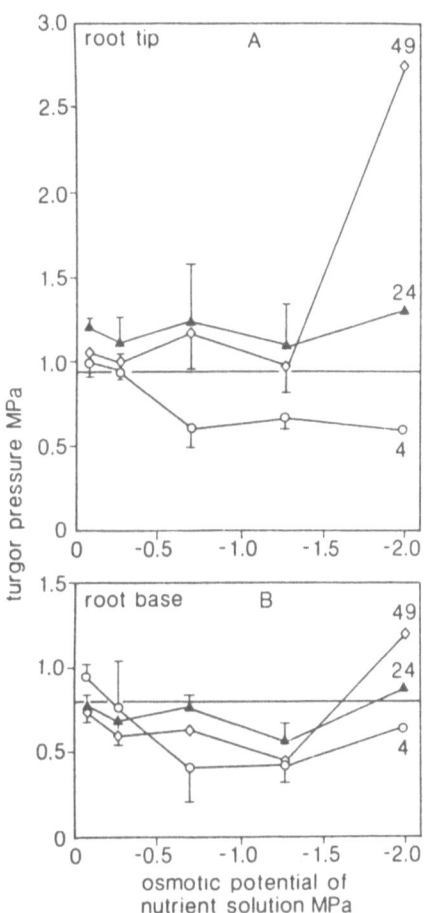

Fig. 2. Turgor pressure of root tips (**A**) and turgor pressure of the cells of the root basis (**B**) 4 (—○—), 24 (—▲—) or 49 (—◇—) after the start of the water stress. Water stress was applied by decreasing the osmotic potential of the nutrient solution with PEG-400. *Bars* indicate SE, n = 3–4. One experiment was conducted at an osmotic potential of 2.0 MPa. The pressure at the start of the experiment is indicated by the *horizontal bar* in the figures.

decreasing osmotic potential of the nutrient solution, whereas the width of these segments decreased less (Fig. 1). Maximal shrinkage was measured within 9 minutes of stress. Using the data of Fig. 1, the volume reductions were calculated at decreased osmotic potentials of the nutrient solution, and -2.0 MPa. Measured after 4 h, the decrease in osmotic potential of root tips could maximally be attributed for 3 or 6% to cell volume decrease itself at osmotic potentials of the nutrient solution of -1.0 and -2.0 MPa, respectively.

Decrease of the osmotic potential of the nutrient solution from -0.07 to -2.0 MPa, decreased the RWC of the leaf and root tip from 95.8 to 88.4%, and from 95.5 to 89.6% respectively. The observed decrease in RWC corresponds with the degree of observed shrinkage.

Cell turgor pressure

Cells of the apical cm of the root had a cell turgor pressure of 0.95 MPa (Fig. 2A). The value for the root basal cm and the first leaf were 0.80 and 1.05 MPa, respectively (Figs. 2B and 3). In the root tip, the (calculated) turgor pressure declined with increased osmotic stress, measured after 6 hour stress. After 24 hour, the turgor pressure had recovered to a higher level especially at severe stress (Fig. 2A). In the root base, full turgor restoration was completed within 24 hour to the value at time zero (Fig. 2B). The first fully elongated leaf partly restored the turgor pressure, 4 h after the start of the osmotic stress (Fig. 3). The rate of turgor restoration was much lower than for root cells. At the highest stress applied P remained lower than under unstressed conditions.

Composition of the osmotic potential

The concentration of potassium in 12 middle parts of roots was $83.0 \pm 27.2 \mu mol\, g\, FW^{-1}$ at the start of the experiment. After 4 h osmotic stress no increase in K^+ concentration was found, and after 27 h the concentration was somewhat lower (Fig. 4).

The concentration of amino acids hardly increased. The concentration of γ-aminobutyric acid and valine was increased after 5 h. Thirty hours

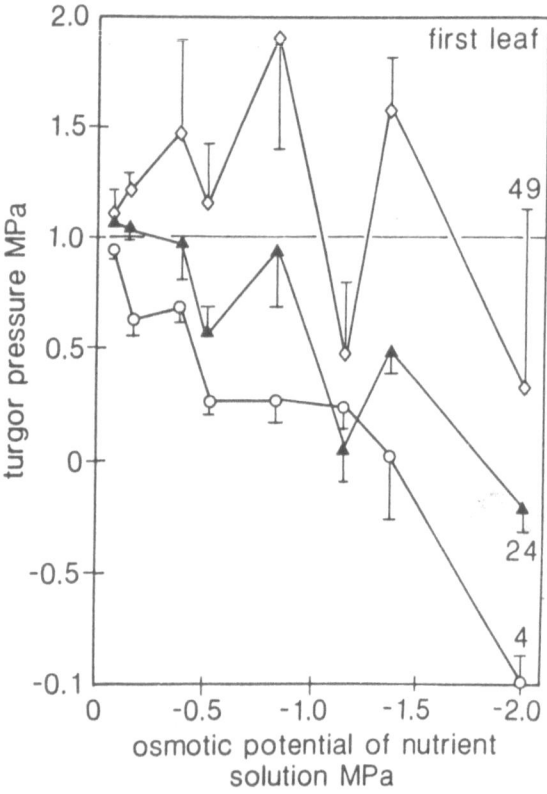

Fig. 3. Turgor pressure of the first leaf 4 (—○—), 24 (—▲—) or 49 (—◇—) h after the start of the water stress. Water stress was applied by decreasing the osmotic potential of the nutrient solution with PEG-400. *Bars* indicate SE, n = 6.

after the onset of osmotic stress the concentration of γ-aminobutyric acid, valine, threonine and iso-leucine was increased (Table 1). There was no

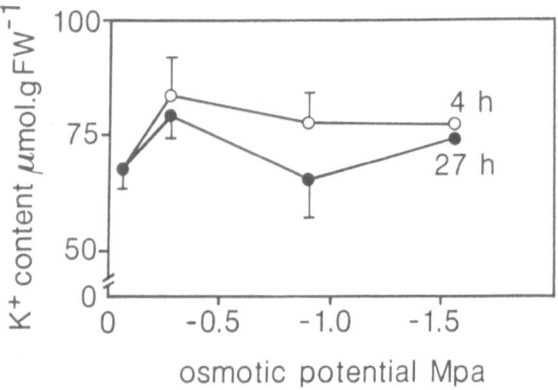

Fig. 4. Potassium concentration in the middle part of the root 4 and 27 h after the start of the decrease in the osmotic potential of the nutrient solution with PEG-400 to the indicated values. *Bars* indicate mean ± SE, n = 8–18.

change in the concentration of proline. Evidently, K^+ and amino acids did not contribute to osmo-regulation.

Discussion

Turgor

After application of osmotic stress mature parts of roots were capable of full turgor restoration over a wide range of applied water potentials. The mature portion of the primary leaf was also capable of full P restoration. The rate of P restoration in mature tissues was slower for leaves than for roots. P restoration was accomplished by a combination of cell shrinkage and transfer of osmotica to the vacuoles. The osmotica in the vacuole became more concentrated in the process of shrinking. A high elasticity of the cell wall contributes to maintenance of turgor under water stress. Root cells and leaf cells showed equal elasticity: a change in water potential to − 1.2 MPa caused extending root cells to shrink 11% and elongating leaf cells 13% (data not shown). Differences in elasticity of leaf and root cells then could not explain differences in turgor restoration. It can be argued that P of the extending root cells was not known exactly. The water potential of those cells could differ from that of the surrounding cell walls because of the influx of water, giving a $\Delta\psi \neq 0$ in eqn. (1) as was found for extending leaf cells (Westgate and Boyer 1984). So the true value of P may be somewhat lower.

Root/shoot ratio

An increase in root/shoot ratio has frequently been observed in response to water stress (Brouwer, 1983). This increase is caused by a shift in the partitioning of dry matter between roots and shoots but the mechanisms underlying this shift are unclear (Lambers, 1983). Carbohydrates have been thought to limit growth rate (Brouwer, 1983; Thornley, 1977), but not all authors agree with this concept (Lambers, 1983). The observed difference in P maintenance of roots and leaves could cause differences in the enlargement rate between root and shoot as a result of water stress. This in turn could lead to the observed shift in dry matter shoot/

Table 1. Amino acids in *Zea mays* roots that significantly changed in concentration after 30 h in a nutrient solution with −0.9 MPa PEG-400, or in a standard Hoagland solution (−0.07 MPa). Concentrations are expressed as μmol amino acids per g FW ± SD, n = 4. Difference are significant ($p < 0.05$)

Amino acid	Water potential (MPa)		Change (%)
	−0.07	−0.9	
γ-Aminobutyric acid	0.45 ± 0.24	1.20 ± 0.82	167
Valine	0.41 ± 0.08	0.60 ± 0.13	46
Iso-leucine	0.21 ± 0.04	0.30 ± 0.07	43
Serine	1.29 ± 0.24	0.96 ± 0.33	−26
Total amino acids	15.25	16.37	

root ratio. The turgor of the elongation region of these young plants was not measured. Elongation zones of the 5th to 9th leaf were, at least under our conditions, not capable of full P maintenance (Schildwacht, unpublished) as opposed to the root tips (Fig. 2A). A difference in P maintenance between leaf and root meristem would affect the enlargement rate according to eqn. (1).

Osmoregulatory substances in the roots

Turgor restoration was accomplished by the uptake of osmotica in the vacuoles. Most of the osmotica in maize roots were carbohydrates (Acevedo *et al.*, 1971). However Cram (1976) demonstrated that K^+ and amino acids also play a rôle and that their significance depended on environmental conditions. It was concluded that K^+ did not contribute to osmoregulation of these roots. The incorporation of all amino acids into protein was probably inhibited by water stress (Steward and Hansson, 1980) and this allows the possibility of amino acids being used osmotically. However, no significant increase in the total amount of amino acids occurred (Table 1). It is clear that roots adjusted osmotically faster than leaves and this may provide a mechanism for the change in root/shoot ratio observed after the start of water stress. Potassium and amino-acids were not used for osmotic adjustment.

Acknowledgement

The author thanks J van Alphen, J Hoppener and W van Kleunen and for skilful technical assistance.

References

Acevedo E, Hsiao T C and Henderson D W 1971 Immediate and subsequent growth response of maize leaves to changes in water status. Plant Physiol. 48, 631–636.

Bernstein L 1961 Osmotic adjustment of plants to saline media. I. Steady state. Am. J. Bot. 48, 909–918.

Borstlap A C 1972 Changes in the free amino acids of *Spirodela polyrhiza* (L) Schleiden during growth inhibition by L-valine, L-isoleucine or L-leucine. Acta Bot. Neerl. 21, 404–416.

Brouwer R 1983 Functional equilibrium: Sense or nonsense. Neth. J. Agric. Sci. 31, 335–348.

Cram W J 1976 Negative feedback regulation of transport in cells: The maintenance of turgor, volume and nutrient supply. *In* Transport in Plants. II. Part B Tissues and Organs. Eds. U Luettge and M G Pitman. pp 284–316. Springer-Verlag, Berlin. ISBN 3-540-07453-8.

Greacen E L and Oh J S 1972 Physics of root growth. Nature 235: 24.

Hellebust J A 1976 Osmoregulation. Annu. Rev. Plant Physiol. 27, 485–505.

Lambers H 1983 The functional equilibrium: Nibbling on the edges of a paradigm. Neth. J. Agric. Sci. 31, 305–311.

Lockhart J A 1965 An analysis of irreversible plant cell elongation. J. Theor. Biol. 8, 264–275.

Michelena V A and Boyer J S 1982 Complete turgor maintenance at low water potentials in the elongation region of maize leaves. Plant Physiol. 69, 1145–1149.

Sharp R E and Davies R E 1979 Solute regulation and growth by roots and shoots of water stressed maize plants. Planta 147, 43–49.

Steward C R and Hanson A D 1980 Proline accumulation as a metabolic response to water stress. *In* Adaptation of Plants to Water and High Temperature Stress. Eds. N C Turner and P J Kramer. pp 173–189. Wiley and Sons, New York. ISBN 0-471-05372-4.

Takami S, Rawson H M and Turner N C 1982 Leaf expansion of four sunflower (*Helianthus annuus* L.) cultivars in relation to water deficits. II. diurnal patterns during stress and recovery. Plant Cell Environ. 5, 279–286.

Thornley J H M 1977 Root-shoot interactions. Sym. Soc. Exp. Biol. 31, 367–389.

Turner N C and Jones M M 1980 Turgor maintenance by osmotic adjustment: A review and evaluation. *In* Adaptation to Water Stress and High Temperature Stress. Eds. N C Turner and P J Kramer. pp 87–103. Wiley and Sons, New York. ISBN 0-471-05372-4.

Westgate M E and Boyer J S 1984 Transpiration- and growth-induced water potentials in maize. Plant Physiol. 74, 882–889.

B. C. Loughman et al. (Eds.), Structural and functional aspects of transport in roots, 241–246.
© 1989 by Kluwer Academic Publishers.

Induction of abscisic acid in excised maize roots by osmotic and salt stress

D.A. BAKER[1] and D.R. LACHNO[2]
[1]*Department of Biochemistry and Biological Sciences, Wye College, University of London, Ashford, Kent, TN25 5AH, UK and* [2]*School of Biological Sciences, University of Sussex, Sussex, BN1 9QG, UK*

Key words: abscisic acid, root, stress, water balance, xylem exudate, *Zea mays*

Moderate water stresses in the range 0 to 0.6 MPa applied with polyethylene glycol (PEG) 6000 to excised roots of *Zea mays* L. var. LG 11, induced increases of up to four-fold in the amount of free and bound abscisic acid (ABA) detected in the root after a 12 h period of xylem exudation. The ABA concentration in xylem exudate collected after a 2 h PEG-induced water stress to excised roots also increased by up to four-fold. Salt stresses, induced with NaCl solutions, resulted in similar increases in the free and bound ABA concentrations in root tissue and in the ABA concentration of the xylem exudate. The ABA concentrations in both root tissue and xylem exudate were highest 4 h after removal of the stress and then declined over the next 8 h. These results are interpreted as support for the concept that root-produced ABA may have a role in the fine control of the plant's water balance.

Introduction

The hormonal content of plant tissues often changes in response to stress conditions, enabling the plant to continue growth in spite of unfavourable conditions. The role of abscisic acid (ABA) in the responses of plants to water stress has now become well established following the original observation by Wright and Hiron (1969) of a dramatic increase in the ABA content of wheat leaves after wilting and the subsequent demonstration by Jones and Mansfield (1970) that ABA application causes stomatal closure.

Although there is abundant information concerning the responses of shoots when stresses are exerted *via* the root medium (*e.g.* Bengtson *et al.*, 1977; Mizrahi *et al.*, 1970), little is known of the part played by the roots in terms of their capacity for the synthesis and export of hormones to the shoot in response to stress conditions.

Milborrow and Robinson (1973) reported a modest rise in ABA concentrations when excised avocado and sunflower roots were allowed to dry in air. Ten-fold increases in the ABA content of red kidney bean roots were observed after one hour in response to a water stress of -0.4 MPa, induced by immersion in PEG 6000 and a sixteen-fold increase after two hours of stress. Similar results were obtained with pea and sunflower roots (Walton *et al.*, 1976). Biosynthesis of ABA from [^{14}C] mevalonic acid has been demonstrated for intact roots (Barr, 1973), and thus there is some evidence that an initial rapid response of the shoot to stress conditions may be root mediated. The ability of excised maize roots to synthesize ABA and export it in the xylem sap in response to both water and salt stresses has been investigated in the present study.

Materials and methods

Seeds of *Zea mays* L., var. LG 11 soaked in running tap water for 12 h were planted out on 13-l plastic tanks containing $0.1 \, mol \, m^{-3} \, CaCl_2$. The seeds were supported 10 mm above the liquid surface on a nylon grid of 6-mm mesh and kept moist with a covering of wet tissue. The solution was vigorously aerated which also served to create spray and hence keep the seeds moist. Germination and seedling growth were always in darkness at $25 \pm 1°C$.

Collection of xylem exudates

Seminal roots from 7-day-old seedlings were excised just below the region of lateral root primordia, giving lengths of 70–150 mm. The cut end of each root was sealed into one end of a 100 μl glass micro-capillary pipette. Batches of 80–100 micropipettes with attached roots were held vertically around 50 mm diameter supporting drums and immersed in 1 litre of osmoticum so that the tissue micropipette junctions were just above the liquid surface. The control solution was 0.1 mol m^{-3} CaCl$_2$. Different levels of water stress, between −0.1 and −0.6 MPa, were obtained by dissolving the appropriate amounts of polyethylene glycol 6000 (PEG) in 0.1 mol m^{-3} CaCl$_2$ (Michel and Kaufmann, 1973). Two levels of salt stress were obtained from solutions with potentials of −0.2 and −0.4 MPa, prepared by dissolving NaCl in 0.1 mol m^{-3} CaCl$_2$ (Lang, 1967).

Roots were subjected to a 2 h stress and transferred to 1 mol m^{-3} KCl + 0.1 mol m^{-3} CaCl$_2$ for 4, 8 or 12 h in order to promote xylem exudate production. Application of stress treatment and collection of exudate were carried out at 25 ± 0.5°C in darkness. All solutions were aerated throughout. At the end of the exudation period, the roots were detached, the exudate expelled into 20 ml screw-top vials and weighed. Typically, 5–8 g of exudate were obtained per 100 roots, the greatest quantities coming from the control treatments and the least from the stressed tissue. Roots from which ABA was to be extracted were rapidly blotted, weighed, and frozen in liquid nitrogen. Others were weighed fresh and then oven dried overnight at 60°C before recording dry weights.

Extraction and purification of ABA from root tissue

The methods used for extraction and purification of ABA from root tissue were designed to remove the free acid without simultaneously releasing any from conjugated (bound) forms. A reliable estimate of conjugated ABA could then be made after further extraction. Details of the extraction and purification procedure have been presented elsewhere (Lachno and Baker, 1986).

Results and discussion

An initial experiment without any internal standard added had confirmed the presence of ABA in maize root tissue and xylem exudate under control conditions. Very small quantities of the 2-*trans*, 4-*trans*-ABA isomer were also detected in most experiments, but it never amounted to more than 5% of the total free ABA. Recoveries of the internal standard in these experiments varied between 40 and 80%, largely depending upon the level of endogenous ABA present. There was a steady rise in both free and bound ABA concentration with increasing severity of stress (Fig. 1). The concentration of ABA in the xylem exudate under control conditions ranged between 0.38 and 0.59 ng ml^{-1} (Fig. 2). Increasing water stress resulted in higher concentrations of ABA in the exudate such that a stress of −0.6 MPa induced a 4-fold increase.

An interesting aspect of the response of maize roots to very mild stresses (−0.1 to −0.2 MPa) was the consistent fall in ABA content in the xylem exudate relative to the control. Thus a water stress of at least −0.2 MPa was required before positive stimulation occurred. This is not readily explained

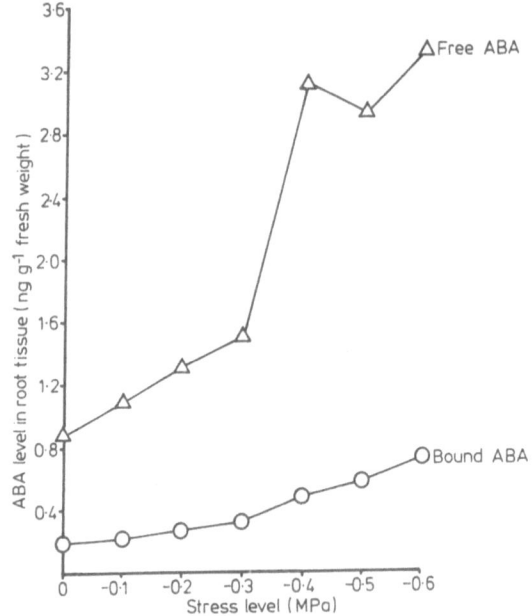

Fig. 1. The change in free and bound ABA levels in maize root tissue after 2 h of water stress in PEG 6000 and a subsequent 12 h period of xylem exudation (in 1 mol m^{-3} KCl + 0.1 mol m^{-3} CaCl$_2$) at 25°C.

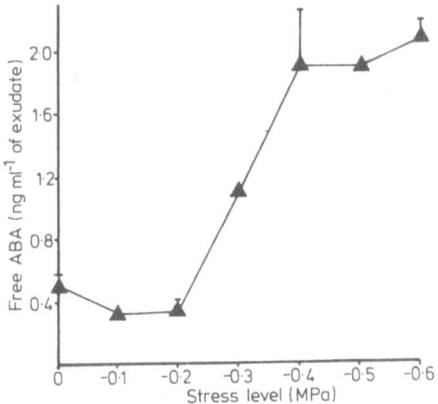

Fig. 2. The change in free ABA concentration of xylem exudate from maize roots in response to water stress. After 2 h of water stress in PEG 6000 at 25°C exudates were collected over a 12 h period of incubation in 1 mol m^{-3} CaCl$_2$. *Points with vertical bars are means of 2 or more separate experiments. Bars are one standard deviation.* The fall in ABA concentration when the control is compared with the -0.2 MPa stress treatment was significant at the 10% level.

in terms of the known reactions of plants to water stress, and may be an experimental artifact. Nevertheless, it does seem as though a mechanism exists in plant roots to raise the ABA concentration of the sap exported to the shoots in response to a moderate water stress.

When total free ABA in the root is considered (*i.e.* amount in tissue plus xylem exudate) a sigmoid relationship is found between percentage stimulation and increasing water stress (Fig. 3). Levels of free and bound ABA in the tissue, and free ABA in xylem exudate, were at their highest 4 h after removal of the stress, but then fell rapidly over the next 4 h (Fig. 4). The decline in ABA concentrations then became less marked in the period from 8 to 12 h after stress removal.

In most respects the effects of NaCl-induced salt stress closely resembled those reported above for water stress with PEG 6000 (Fig. 5). Free and bound ABA levels in root tissue increased with increasing salt stress. The pattern of response was the same as for water stress, but the effects were more pronounced. Thus a salt stress of -0.4 MPa induced a 5-fold increase in free ABA in the tissue (Fig. 5), whereas a 3-fold increase was obtained from the same level of water stress (Fig. 1). The free ABA concentrations in the xylem exudates after salt stress were almost identical to levels produced by the corresponding water stresses. The reduction

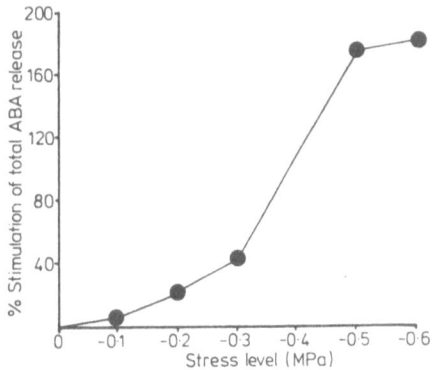

Fig. 3. The percentage stimulation of total free ABA release by maize roots (in tissue + xylem exudate) over a 12 h exudation period (in 1 mol m^{-3} KCl + 0.1 mol m^{-3} CaCl$_2$) after a 2 h period of water stress in PEG 6000 at 25°C. Calculated as the difference between stressed and control levels as a percentage of control.

in ABA concentration in response to a water stress of -0.2 MPa noted above was also evident in the salt stress data. Again the higher level of salt stress (-0.4 MPa) induced a significant increase in xylem sap ABA concentration compared with the control.

Tietz (1975) found that levels of ABA in pea roots of plants grown in water culture were much reduced through excretion of the hormone into the

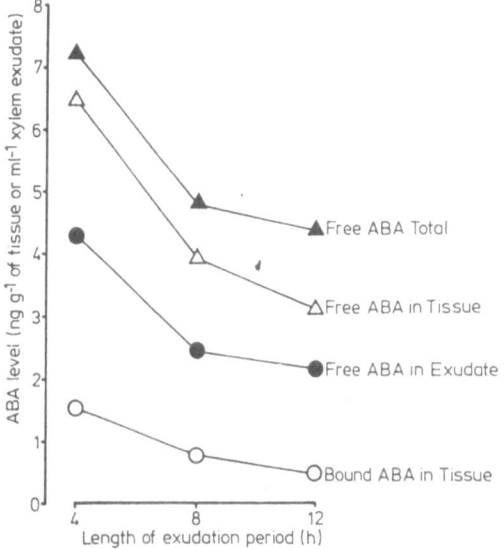

Fig. 4. Time dependence of ABA levels in maize root tissue and xylem exudate following removal of water stress, in the form of PEG 6000 (-0.4 MPa) at 25°C, applied for 2 h. Exudates were collected after different periods of incubation in 1 mol m^{-3} KCl + 0.1 mol m^{-3} CaCl$_2$. (Total free ABA production in tissue + xylem exudate, expressed per g fresh weight of tissue.)

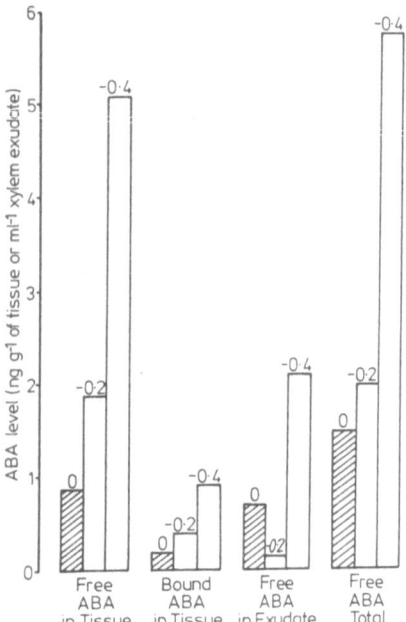

Fig. 5. ABA levels in maize root tissue and xylem exudate following a 12 h period of exudation (in 1 mol m⁻³ KCl + 0.1 mol m⁻³ CaCl₂) preceded by a 2 h period of NaCl-induced salt stress at 25)C. Stress levels −0.2 and −0.4 MPa: Free ABA in exudate was calculated per g of exudate. Total free ABA was the sum of quantities in the tissue and xylem exudate expressed per g of tissue.

medium. This phenomenon has since been investigated in maize roots (LG 11) by Rivier *et al.* (1983) under conditions of osmotic stress. They found that ABA levels in root tissue fell to 25% of initial values after 30 min incubation in liquid medium and that osmotic stresses between − 0.3 and − 0.8 MPa induced higher levels of ABA in both the tissue and osmoticum. Similarly, in an experiment using mannitol as the osmoticum (Lachno and Baker, 1986) maize roots were also found to excrete ABA into solution. The mannitol treatment was equivalent to a stress of − 0.5 MPa and induced a 4-fold increase in ABA excretion into the medium over an 18 h period. However, of the total ABA released during the experiment, the greater portion was located in the tissue in both control and stress treatments.

The amounts of free ABA detected here (Fig. 1) are comparable to those recorded by Rivier *et al.* (1983) using a similar experimental system. Their roots, however, were not excised prior to stress application. The 16-fold increase of ABA in bean root tissue reported by Walton *et al.* (1976) follow-

ing an osmotic stress of − 0.4 MPa in the root medium is at first sight less easily interpreted when compared with the present results. However, their conclusion that it originated from the roots themselves is not entirely consistent with their data.

The effects of water stress on the ABA concentration in the xylem sap of sunflower roots were investigated by Hoad (1975). A water stress of − 1.2 MPa applied to whole plants through the root medium induced a 40-fold increase in xylem sap ABA concentration. The presence of shoots would, therefore, seem to be the key to the vast difference in response between his plants and the excised maize roots used in the present study.

Hoad (1973, 1975) concluded that the principal site of ABA synthesis in stressed plants is the leaves, and that the appearance of high ABA concentrations in other organs such as the roots is dependent upon redistribution in the vascular system. Later work has confirmed the existence of very high ABA concentrations in phloem sap (Hoad, 1978; Zeevaart, 1977). The data presented here show that some independent synthesis of ABA also occurs in the roots.

Wright (1977) found a sigmoid relationship between increasing ABA content and decreasing water potential in wheat leaves; the hormone began to accumulate before wilting symptoms appeared. Similarly, the ABA increase in maize roots observed here (Fig. 3) could have been initiated as soon as tissue water potential was slightly reduced by a mild osmoticum. The trigger for the more rapid increase in ABA synthesis in response to a stress of − 0.3 to − 0.4 MPa may have been through a reduction in tissue turgor (Pierce and Raschke, 1980). Whilst the root tissue, stressed at -- 0.6 MPa, could not have been fully turgid, it nevertheless retained a positive turgor as no signs of flaccidity were apparent on removing roots from the osmoticum.

Recovery patterns similar to those presented in Fig. 4 have been found for free ABA in leaves by Hiron and Wright (1973) and Zeevaart (1980), and in phloem exudate by Hoad (1978). The close similarity between the fall in ABA concentration in the xylem exudate and the fall in the ABA content of the root tissue suggests a concentration-dependent transport of the hormone into the xylem.

Mizrahi *et al.* (1970, 1972) reported that a salt

stress in the root medium produced a dramatic increase in the ABA content of tobacco leaves. The data presented here suggest that a small part of the increase in leaf ABA that they observed may well have resulted from transport of ABA in the xylem after synthesis in the roots, although the bulk must have been produced in the leaves themselves, probably through water stress effects (Mizrahi *et al.*, 1971).

The close similarity between the responses of maize roots to moderate salt and water stresses (Figs. 3 and 5) indicates that the main action of salt in stimulating ABA synthesis is through the water stress that it imposes. The capacity for ABA synthesis in roots seems to be very much lower than that of leaves. Whether ABA is synthesised in chloroplasts (Milborrow, 1974; Milborrow and Robinson, 1973; Railton *et al.*, 1974) or in cytoplasm (Hartung *et al.*, 1981) is open to debate. What is certain is that most of the ABA in non-stressed leaves is located in the chloroplasts and that a mild water stress induces a redistribution, such that cytoplasmic ABA increases rapidly with respect to chloroplastic ABA (Hartung *et al.*, 1981; Loveys, 1977). It is conceivable that root plastids (Milborrow, 1974, 1978) are responsible for the release of ABA that occurs in response to moderate water stresses of the type reported here. It is, therefore, suggested that the vast differences in the capacity for ABA synthesis that exists between roots and leaves may be due to differences in plastid densities. The other possibility is that roots lack the necessary supply of mevalonate precursor for ABA synthesis when links with the shoot are severed.

Stresses between 0 and -0.6 MPa are within the range to which plants are frequently exposed from day to day. Evidence was found in the present study for a hormonal response mechanism operating in maize roots over this mild to moderate stress range. ABA levels in root tissue were raised in response to stress and increased quantities of the hormone were translocated out of the roots into the xylem sap. These observations give some support to the hypothesis of Blake and Ferrell (1977) that roots may be drought-sensing organs and also suggest two possible ways in which fine control of water balance could be mediated through changes in ABA synthesis in roots.

Davies *et al.* (1980) proposed a mechanism for control of stomata in water stressed plants. At a critical level of water stress, ABA was thought to act as a 'distress signal' moving from chloroplasts to guard cells, where it induced stomatal closure and hence promoted maintenance of leaf turgor, which is essential to growth. It is proposed here that ABA synthesised by roots in response to moderate soil water deficits may be translocated in the transpiration stream directly to the guard cells where it could contribute to the control of stomatal aperture. Weyers and Hillman (1980) have shown that guard cells rapidly accumulate ^{14}C ABA and any high boiling point material moving in the apoplast would be concentrated in the vicinity of sites of evaporation such as the stomata. This proposal would fit in with the classical concept of hormone action in intact plants.

The second method by which root produced ABA could help to maintain plant turgor would be through an increase in the hydraulic conductivity of the root. Glinka and Reinhold (1971) and Glinka (1973) reported an increase in permeability to water following ABA treatment and a stimulation of root exudation from tomato plants within 30 min. Results from Collins and Kerrigan (1974) and Collins and Morgan (1980) support this observation and, additionally, Glinka (1980) has demonstrated a promotion of exudation from detopped sunflowers 6 min after addition of ABA to the roots. An increase in hydraulic conductivity lowers root resistance to water uptake and enables the plant to maintain a positive water flux in spite of reduced soil water potential. In addition ABA could stimulate ion release into the xylem, thus facilitating osmotic water transport.

Systems such as the ones suggested would enable the plant to respond rapidly to environmental fluctuations. ABA produced in the root in response to changes in soil water potential could control water fluxes through the plant by acting directly on plasma membranes and guard cells without the need to invoke the release of large quantities of ABA from chloroplasts with the inevitable consequence of long term closure of stomata.

References

Barr M L 1973 Biosynthesis of abscisic acid: Incorporation of radioactivity from ^{14}C mevalonic acid by intact roots. Plant Physiol. 51, Suppl. 47.

Bengtson C, Falk S O and Larsson S 1977 The after-effect of water stress on transpiration rate and changes in abscisic acid content of young wheat plants. Physiol. Plant. 41, 149–154.

Blake J and Ferrell W K 1977 The association between soil and xylem water potential, leaf resistance, and abscisic acid content in droughted seedlings of Douglas fir (*Pseudotsuga menziesii*). Physiol. Plant. 39, 106–109.

Collins J C and Kerrigan A P 1974 The effect of kinetin and abscisic acid on water and ion transport in isolated maize roots. New Phytol. 73, 309–314.

Collins J C and Morgan M 1980 The influence of temperature on the abscisic acid stimulated water flow from excised maize roots. New Phytol. 84, 19–26.

Davies W J, Mansfield T A and Wellburn A R 1980 A role for abscisic acid in drought endurance and drought avoidance. *In* Plant Growth Substances 1979. Ed. F Skoog. pp 242–253. Springer-Verlag, Berlin.

Glinka Z 1973 Abscisic acid effect on root exudation related to increased permeability to water. Plant Physiol. 51, 217–219.

Glinka Z 1980 Abscisic acid promotes both volume flow and ion release to the xylem in sunflower roots. Plant Physiol. 65, 537–540.

Glinka Z and Reinhold L 1971 Abscisic acid raises the permeability of plant cells to water. Plant Physiol. 48, 103–105.

Hartung W, Heilmann B and Gimmler H 1981 Do chloroplasts play a role in abscisic acid synthesis? Plant Sci. Lett. 22, 235–242.

Hiron R W P and Wright S T C 1973 The role of endogenous abscisic acid in the response of plants to stress. J. Exp. Bot. 24, 769–781.

Hoad G V 1973 Effect of moisture stress on abscisic acid levels in *Ricinus communis* L. with particular reference to phloem exudate. Planta 113, 367–372.

Hoad G V 1975 Effect of osmotic stress on abscisic acid levels in xylem sap of sunflower (*Helianthus annuus* L.). Planta 124, 25–29.

Hoad G V 1978 Effect of water stress on abscisic acid levels in white lupin (*Lupinus albus* L.) fruit, leaves and phloem exudate. Planta 142, 287–290.

Jones R J and Mansfield T A 1970 Suppression of stomatal opening in leaves treated with abscisic acid. J. Exp. Bot. 21, 714–719.

Lachno D R and Baker D A 1986 Stress Induction of abscisic acid in maize roots. Physiol. Plant. 68, 215–221.

Lachno D R, Harrison-Murray R S and Audus L J 1982 The effects of mechanical impedance to growth on the levels of ABA and IAA in root tips of *Zea mays* L. J. Exp. Bot. 33, 943–951.

Lang A R G 1967 Osmotic coefficients and water potentials of sodium chloride solutions from 0 to 40°C. Aust. J. Chem. 20, 2017–2023.

Loveys B R 1977 The intracellular location of abscisic acid in stressed and non-stressed leaf tissue. Physiol. Plant. 40, 6–10.

Michel B E and Kauffmann M R 1973 The osmotic potential of polyethylene glycol 6000. Plant Physiol. 51, 914–916.

Milborrow B V 1974 Biosynthesis of abscisic acid by a cell-free system. Phytochemistry 13, 131–136.

Milborrow B V 1978 Abscisic acid. — *In* Phytohormones and Related Compounds: A Comprehensive Treatise. Vol. 1. Eds. D S Letham *et al*. pp 295–347. Elsevier/North Holland Biomedical Press, Amsterdam.

Milborrow B V and Robinson D R 1973 Factors affecting the biosynthesis of abscisic acid. J. Exp. Bot. 24, 537–548.

Mizrahi Y, Blumenfeld A and Richmond A E 1970 Abscisic acid and transpiration in leaves in relation to osmotic root stress. Plant Physiol. 46, 169–171.

Mizrahi Y, Blumenfeld A, Bittner S and Richmond A E 1971 Abscisic acid and cytokinin contents of leaves in relation to salinity and relative humidity. Plant Physiol. 48, 752–755.

Mizrahi Y, Blumenfeld A and Richmond A E 1972 The role of abscisic acid and salination in the adaptive response of plants to reduced aeration. Plant Cell Physiol. 13, 15–21.

Pierce M and Raschke K 1980 Correlation between loss of turgor and accumulation of abscisic acid in detached leaves. Planta 148, 174–182.

Railton I D, Reid D M, Gaskin P and MacMillan J 1974 Characterisation of abscisic acid in chloroplasts of *Pisum sativum* cv. Alaska by combined gas chromatography-mass spectrometry. Planta 117, 179–182.

Rivier L, Leonard J-F and Cottier J-P 1983 Rapid effect of osmotic stress on the content and exodiffusion of abscisic acid in *Zea mays* roots. Plant Sci. Lett. 31, 133–137.

Tietz A 1975 Einfluß der Kulturmethode auf die Ausscheidung von Phytohormonen durch Erbsenwurzen. (Influence of the culture method on the phytohormone excretion of pea roots) Biochem. Physiol. Pflanzen 167, 371–378.

Walton D C, Harrison M A and Cote P 1976 The effects of water stress on abscisic-acid levels and metabolism in roots of *Phaseolus vulgaris* L. and other plants. Planta 131, 141–144.

Weyers J D B and Hillman J R 1980 Effects of abscisic acid on $^{86}Rb^+$ fluxes in *Commelina communis* L. leaf epidermis. J. Exp. Bot. 31, 711–720.

Wright S T C 1977. The relationship between leaf water potential (leaf) and the levels of abscisic acid and ethylene in excised wheat leaves. Planta 134, 183–189.

Wright S T C and Hiron R W P 1969 (+)-Abscisic acid, the growth inhibitor induced in detached wheat leaves by a period of wilting. Nature 224, 719–720.

Zeevaart J A D 1977 Sites of abscisic acid synthesis and metabolism in *Ricinus communis* L. Plant Physiol. 59, 788–791.

Zeevaart J A D 1980 Changes in the levels of abscisic acid and its metabolites in excised leaf blades of *Xanthium strumarium* during and after water stress. Plant Physiol. 66, 672–678.

B. C. Loughman et al. (Eds.), Structural and functional aspects of transport in roots, 247–249.
© 1989 by Kluwer Academic Publishers.

Ionic composition and metabolic changes in salt stressed roots

J.M. STASSART

Laboratorium Plantenfysiologie, Vrije Universiteit Brussel, 65 Paardenstraat, B-1640 Sint-Genesius-Rode, Belgium

Key words: barley, calcium, potassium, sodium, transport, xylem

Introduction

The primary effect of salt stress is experienced by the roots. Discrimination between different ions is of vital importance for the survival of a plant and the present investigation was designed to evaluate root behaviour during ion uptake and xylem transport under salt stress conditions. It is well known that in low salt conditions potassium is much favoured over sodium (Neirinckx and Stassart, 1975). Our aim was to find out whether the root could still, in an acceptable way, discriminate between the two ions at high sodium levels. We used barley roots, since barley, being a glycophyte, nevertheless is rather tolerant to salt, compared to other monocot crop plants (Stassart and Bogemans, 1987). The effect of the presence of calcium was also studied, since it is known (Stassart and Neirinckx, 1980; Stassart *et al.*, 1985) that calcium interferes with sodium uptake at different steps of the sodium pathway through the rootsystem: at the absorption sites on the cell wall (Stassart *et al.*, 1981), the primary transmembrane uptake, the vacuolar loading (Neirinckx and Stassart, 1979) and on the sodium transport through the xylem (Stassart and Neirinckx, 1981).

Material and methods

Barley (*Hordeum vulgare* c.v. Menuet) was used in all experiments. After germination overnight at 20°C, in wet sand, barley seedlings were placed on gauze over a $0.25\,mM$ $CaSO_4$ filled tank at 20°C. Seven days after germination roots were excised and brought into plexi-glass transport-boxes.

The different uptake solutions used were:

$100\,mM$ NaCl, $100\,mM$ Nacl + $10\,mM$ KCl, $100\,mM$ NaCl + $10\,mM$ $CaCl_2$ and $100\,mM$ NaCl + $10\,mM$ KCl + $10\,mM$ $CaCl_2$. These solutions were alternatively labelled with ^{22}Na; ^{36}Cl; ^{86}Rb and ^{45}Ca as tracers for transport studies.

All individual experiments were done using Low-Salt-Roots (LSR), coming straight from the $0.25\,mM$ $CaSO_4$ culture solution into the uptake solution, or using High-Salt-Roots (HSR) which were first equilibrated for 24 hours in an equivalent non-labelled solution of identical composition.

Experiments were carried out at 1°C and 20°C. By subtracting fluxes measured at 1°C (passive flux) from those measured at 25°C (passive plus active fluxes) the active component of ion transport in the roots was obtained.

Radioactive tracers were measured by GM-counting, after rinsing the root sample and determining the fresh weight. The sample was evaporated in a planchette using a detergent solution, so that the extracted radioactivity could spread evenly over the surface of the planchette. The specific activity of labelling was between 2000 and 5000 cpm per micro-equivalent of the ion concerned. Uptake and transport were lasting between 18 and 24 hours, and fluxes were expressed in $\mu eq\,g^{-1}\,fr\,wt\,h^{-1}$ (micro-equivalent per gram fresh weight per hour).

For measurements where the radioactivity was close to the background activity, results show N.M. (non-measurable) and indicate the flux was non-existent or at a very low rate for the part concerned.

Transport-boxes were divided in three parts, with a loading compartment containing the labelled solution and the apical part of the root. A guard-space, to prevent eventual leakage, and a recipient space, where the xylem exudate was collected, both these compartments containing an equivalent non-labelled solution where the basal cut end of the root was protruding.

We could measure three ion flux components:

A. Label found in the apical root segment, in direct contact with the labelled solution, could be defined as root accumulation (Acc.).

B. Label found in the basal root segment, protruding in the guard-space and recipient-space, and therefore not in direct contact with the labelled solution, could be described as ion redistribution in the root (Red.).

C. Label found in the solution of the recipient-space was defined as xylem exudation (Xyl.).

Since all compartments of the transport-box contained an equivalent solution no osmotic drive force was applied on the roots, the ion pumping into the xylem was enough to drive the long distance transport.

Data given represent the average for 8 to 16 replicates. For reasons of clarity we limited data of our results to the active fluxes of the different ions.

Results and discussion

Sodium in LSR (Table 1a)

The addition of potassium to the uptake solution interfered with the primary accumulation of sodium, but it did not have any effect on the sodium xylem transport. If calcium was added, its addition did not interfere with accumulation of sodium, but it reduced very much the sodium mobility in the root. No active redistribution could be found and a severe reduction of the xylem transport occurred.

In high salt roots the xylem exudation of sodium became predominant (if compared with the accumulation). Roots were, due to the equilibrating pretreatment, already full of sodium. So the strongest effects of calcium could be seen on the xylem flux of sodium. Calcium completely blocked the xylem loading of sodium but sodium accumulated in the apical part of the root. Adding potassium in addition to calcium reduced the sodium accumulation very strongly, and potassium apparently carried some sodium into the xylem flux.

Potassium (Table 1b)

Potassium was provided in the uptake solution with sodium at an outside ratio of K/Na = 1/100. A K/Na ratio of 1/2 in accumulation and of 1/2 in xylem exudation, and a ratio of 2/1 in redistribution. When calcium was added to the outside solution the ratio's for LSR became 1/5 in accumulation and 2/3 for xylem exudation. In HSR the effect was even more pronounced, the accumulation ratio was 2/1 and the redistribution ratio was 4/1. Addition of calcium increases the accumulation ratio to 10/1.

Chloride (Table 1c)

Chloride moved very much in the same way as sodium (and as the sum of sodium and potassium when the latter was added to the solution). Moreover, it was remarkable that in LSR the passive component of chloride movement was rather large compared with its active component. Chloride presumably was dragged by sodium movement into the root, by H^+ efflux or by cotransport.

Combining the results found for the different ions, one can arrive at a certain conclusion for the ion movements into roots exposed to a relatively severe salt stress. For an integrated picture, our results were expressed as percentages, the total influx being the sum of accumulation, redistribution and xylem exudation, equalling 100% for the highest sum measured (treatment sodium without addi-

Table 1. Fluxes for: a, sodium; b, potassium; c, chloride, all expressed in μeq g^{-1} fr wt h

Solution			LSR			HSR		
Na	K	Ca	Acc	Red	Xyl	Acc	Red	Xyl
a. Fluxes for sodium								
100	0	0	2.62	0.18	0.63	0.54	N.M.	1.23
100	10	0	0.91	0.12	0.69	0.19	0.05	0.33
100	0	10	1.42	N.M.	0.32	0.35	N.M.	N.M.
100	10	10	0.97	0.17	0.37	0.01	N.M.	0.16
b. Fluxes for potassium								
100	10	0	0.39	0.24	0.34	0.36	0.21	0.27
100	10	10	0.22	0.22	0.22	0.14	0.22	0.27
c. Fluxes for chloride								
100	0	0	2.20	0.14	0.46	0.58	0.02	1.23
100	10	0	0.80	0.17	0.92	0.01	0.07	0.65
100	0	10	0.87	0.08	0.36	N.M.	N.M.	0.04
100	10	10	1.23	0.44	0.85	0.34	N.M.	0.51

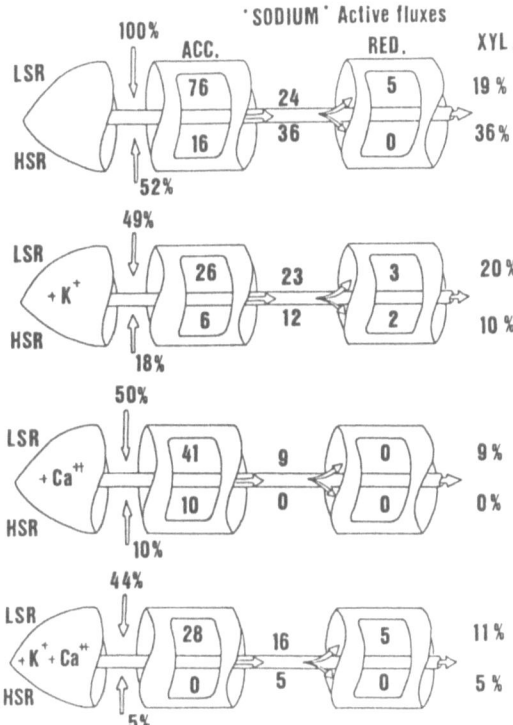

Fig. 1. Integrated results for active sodium fluxes, expressed in percentages.

assium indicates that the cell membranes were functioning properly. To explain the effect of calcium on the sodium movements was not easy, but the fact that sodium accumulation was much less affected than xylem transport indicates that in the presence of calcium sodium was effectively sequestered into the vacuole, as was found in LSR. Due to a high vacuolar sodium accumulation, less sodium was present in the symplast and therefore less sodium reached the stele and was loaded into the xylem. In HSR the effect was even more evident, because xylem transport was by far the most important flux through the root. Calcium reduced xylem transport of sodium and the redistribution to non-measurable quantities (Table 1a). Further research is needed with roots of a less tolerant barley variety to verify the above explanation. Such experiments could show whether the selective mechanisms for absorption would collapse under stress conditions. Possibly, collapse of control of Na transport could occur at the site of the xylem loading. When this last step would break down, heavy loads of sodium and chloride would reach the above ground parts of the plant and cause lethal damage to the shoot.

tion of any other ion). The changes became more evident as shown in Fig. 1, for sodium fluxes. Figure 2 shows the integrated data for sodium, potassium and chloride fluxes in LSR roots. Even though the barley roots were put under considerable Na stress, they still managed to cope with the abundance of sodium and chloride in the medium. The fact that the K/Na ratios of the different fluxes inside the root were relatively in favour of pot-

References

Neirinckx L and Stassart J M 1975 Reconsideration of the Viets effect of calcium by excised barley roots. XII Int. Bot. Congr., Leningrad.

Neirinckx L and Stassart J M 1979 The effect of calcium on the uptake and distribution of sodium in excised barley roots. Physiol. Plant. 47, 235–238.

Stassart J M and Bogemand J 1987 Intervarietal ionic composition changes under salt stress. Gen. Asp. of Plant Min. Nutr. Eds. W H Gabelman and B C Loughman. pp 127–137. Nijhoff Publishers, Dordrecht, The Netherlands.

Stassart J M and Neirinckx L 1980 The interaction of calcium on sodium uptake and transport in low salt barley roots. Plant Membrane Transport. Elsevier/North Holland, 401–402.

Stassart J M, Bogemand J and Neirinckx L 1985 Mineral composition changes in barley under salt stress. Arch. Int. Physiol. Biochem. 93, 19.

Stassart J M, Neirinckx L and Dejaegere R 1981 The interaction between cations and calcium during their adsorption on cell walls of barley roots compared with adsorption and absorption in intact barley roots. Ann. Bot. 47, 647–652.

Stassart J M and Neirinckx L 1981 Transport and distribution of sodium in barley roots. XIII. Int. Bot. Congr. Sydney, p. 89.

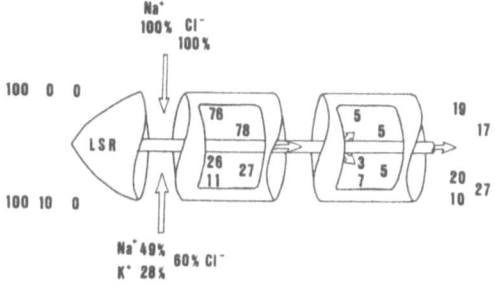

Fig. 2. Integrated results for Na, K and Cl fluxes in LSR, expressed in percentages.

B. C. Loughman et al. (Eds.), Structural and functional aspects of transport in roots, 251–255.
© 1989 by Kluwer Academic Publishers.

Physiological responses of lupin roots to soil compaction

B.J. ATWELL

CSIRO, Dryland Crops and Soils Research Program, Wembley, W.A. 6014, Australia

Key words: compaction, lupin, roots, soil

The compaction of soils and natural high impedance of sub-soils often constitute a major barrier to root growth, especially at low moisture content. The consequences for plant growth and development are dramatic (Barley and Greacen, 1967; Scott Russell and Goss, 1974; Taylor and Ratcliff, 1969).

Morphology and anatomy of compacted roots

An increase in root diameter in response to high soil strength is well documented in a number of annual crop species, including barley (Wilson *et al.*, 1977), pea (Eavis, 1967), maize (Barley, 1962) and cotton and peanuts (Taylor and Ratcliff, 1969). More information on the occurrence of radial swelling of roots of woody species is required before universality of the phenomenon is accepted. In *Lupinus* spp., which are notoriously tolerant to compaction in coarse-textured soils (Jarvis, pers. comm.), radial swelling of the root is pronounced in compacted zones of the soil (Fig. 1).

The explanation for the paradoxical swelling response, whereby root diameter behind the apex increases as the path for root growth is restricted, has as yet not been adequately explained. In an elegant analysis of the phenomenon, Abdalla *et al.* (1969) and Barley and co-workers (*e.g.* Barley, 1968) made a strong case that soil strength is diminished in front of the growing apex as radial swelling of the main axis occurs resulting in accelerated growth of the apex into this zone of reduced compaction. The expression of radial swelling in lupins, which are tolerant to compaction, is consistent with the claim that radial swelling helps to overcome high soil strength; a more conclusive survey of genera would be useful.

Figure 2 shows that the increase in root diameter in response to compaction is due to an increase in

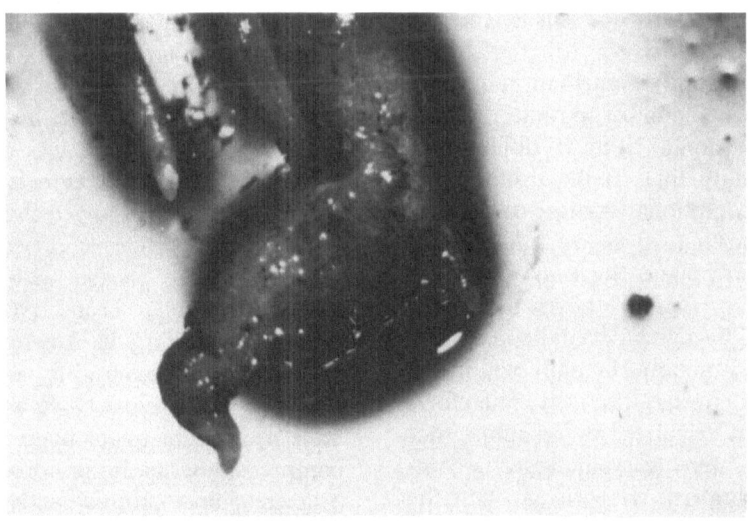

Fig. 1. Tap root of *L. angustifolius* growing in a compacted sand in field conditions.

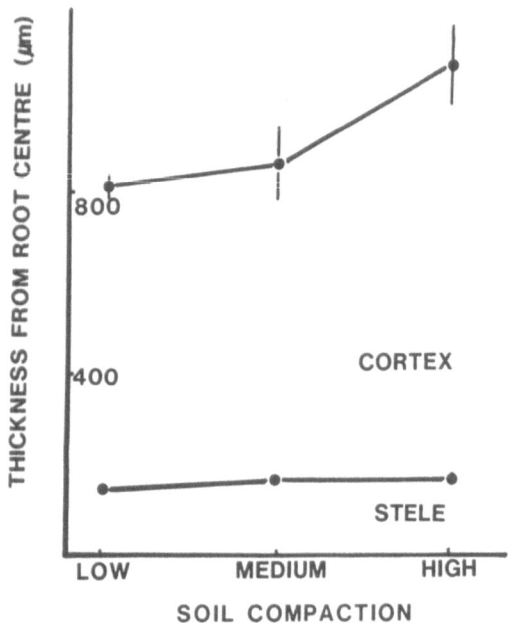

Fig. 2. Radius of the stele and thickness of the cortex in roots of *L. angustifolius* grown at three levels of soil compaction in a sandy clay loam. Bulk densities were as follows: low (1.23 Mg.m^{-3}), medium (1.42 Mg.m^{-3}) and high (1.64 Mg.m^{-3}).

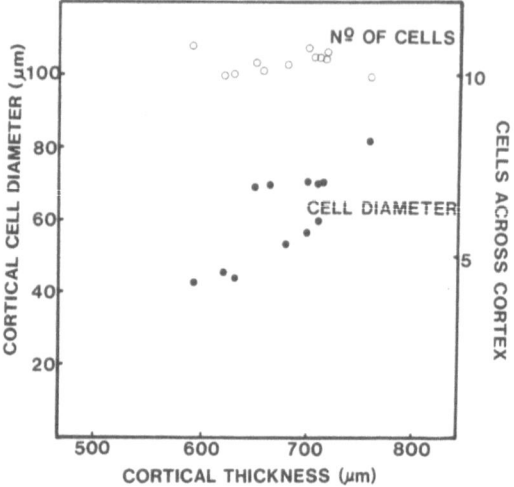

Fig. 3. Number of cells across the cortex and mean cortical cell diameter in roots of *L. angustifolius* grown at three levels of compaction in a sandy clay loam. Increasing cortical thickness on the abcissa is the result of soil compaction. For bulk densities see Fig. 2.

thickness of the cortex, while the stele remains constant in diameter. This substantiates the earlier observations of Wilson *et al.* (1977) and Scholefield and Hall (1985) who demonstrated the unchangeable dimension of stele when roots grow through restricted pores. The means by which root dimensions are modified by environment are now the subject of renewed interest (Jackson and Stead, 1983); the role of ethylene in altering rheological and cytological properties of root cells is a central theme in these studies.

It is salutory to look at the contribution of cortical cell diameter and cell number to radial swelling of the root. This should help to differentiate between the possibility that (i) the proliferating meristem generates additional cortical cell files in compacted roots and that (ii) the existing cells expand in the transverse plane. Such an analysis in *Lupinus angustifolius* (Fig. 3) shows the radial swelling is explained by a greater cell diameter; cell *lengths* were reduced by only 5% and 24% in the inner cortex and epidermis, respectively. Therefore, cortical cell volumes, calculated by assuming that the cell is a cylinder, were generally larger in compacted roots of lupin. This is consistent with the observations of Wilson *et al.* (1977) on barley; they

also found that cells in the epidermis were more affected by external pressure than those of the inner cortex. Cell lengths always decreased in response to compaction but in cortical and epidermal layers this was more than compensated by an increase in cell diameter; cell volumes were up to 50% higher in impeded roots.

Lateral root proliferation

The inhibition of growth of main axes (tap roots or seminal roots) leads to a diminished assimilate demand in the terminal apex and exploitation of a smaller soil volume by roots, unless lateral root formation provides compensation. Crossett *et al.* (1975) showed that the loss of apical dominance causes lateral roots of cereals to proliferate. Similarly, Goss (1977) showed that barley roots growing in ballotini beads, produced laterals freely when the beads were packed with a pore spacing of 160 μm. In finer beads (70 μm pores), which mimicked packing in fine-textured soils, lateral root length was reduced by 50%. The same roots, however, had almost twice as many primary laterals per centimetre of main axis when grown in compressed beads. In lupin roots growing through a layer of fine-textured sandy clay loam, by contrast, there was little or no proliferation of laterals

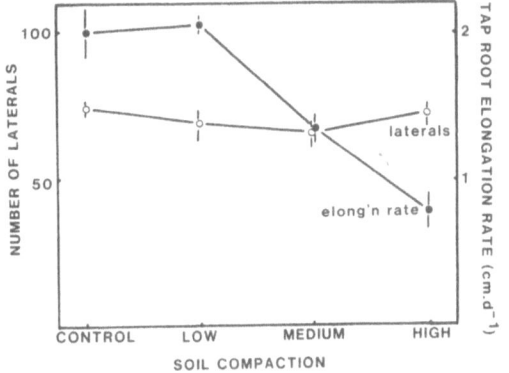

Fig. 4. Numbers of primary lateral roots and elongation rate of the tap root of *L. angustifolius* grown in three levels of compaction in a sandy clay loam (low, medium, high) or in uncompacted white sand (control). For bulk densities see Fig. 2.

due to compaction, possibly reflecting a differential response between monocots and dicots (Fig. 4). Dawkins *et al.* (1983) point out the need to distinguish between increased lateral root numbers *per unit length* of main axis and initiation of new lateral root primordia as a direct consequence of compaction.

There are other consequences of impaired growth of the main root axis by mechanical impedance. The zone of lateral root growth can advance to a point much nearer to the apex of the main axis when roots are compacted (Goss, 1977; Fig. 5). This will have the effect of maximizing the volume of soil explored by these roots per unit length of the main axis. It should be remembered

that the tissue from which the laterals emanate may not be *younger* than the more distal tissue from which laterals emerge in freely-growing roots.

Substrate levels in root apices

The changes in anatomy and morphology of compacted roots are paralleled by metabolic perturbations; assimilate distribution and ion levels in particular are affected. A major dilemma arising in roots growing through compacted soil is that radial swelling and lateral root proliferation, immediately below the apex compete with the apex itself for imported substrates. The need for additional turgor pressure in the apex may exacerbate this problem.

Are root apices growing in compacted media substrate-limited? Greacen and Oh (1972) showed that 5 mm root apices of pea were able to 'osmoregulate' in compacted soil with an efficiency of 70%. That is, for each additional pressure unit of mechanical impedance to root growth there was an extra 0.7 unit of osmotic pressure generated to counter-balance. As a result growth rates decline as impedance increases until the root stops growing. The osmotic agents in these pea roots were not identified but the external supply of inorganic ions (K^+, Na^+, Cl^-) was probably inadequate to provide the additional root osmotica; soluble carbohydrates presumably also accumulated.

By comparison, lupins growing in sandy clay loam showed a small increase in osmotic pressure

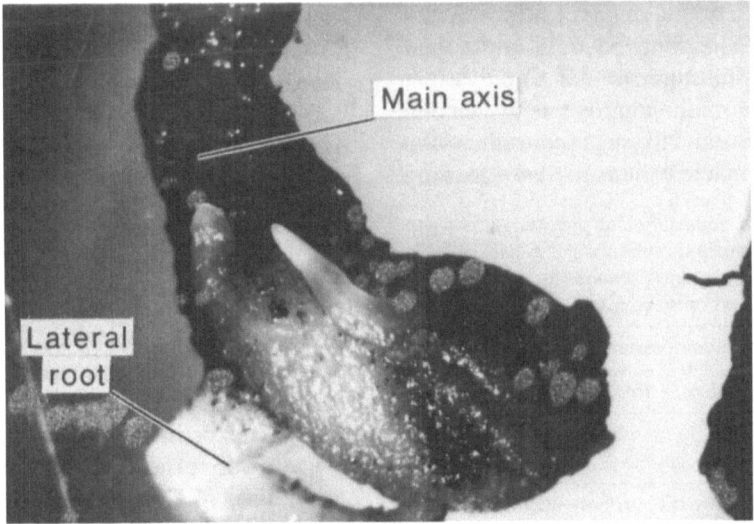

Fig. 5. Tap root of *L. angustifolius* growing in compacted sandy clay loam in field conditions.

Table 1. Osmotic pressures (OP), elongation rate, root diameters, and root volume increase in roots of *L. angustifolius* grown at three levels of compaction in a sandy clay loam. The 'increase in OP due to less growth' is calculated from the formula (Volume (Low)/Volume (Med, High)) × 1.16 MPa and the 'actual increase in OP' is the amount by which the OP exceeds 1.16 MPa. Osmotic pressure is measured in 0–5 mm root apices (expanding tissue) by freezing, thawing, expressing sap and measuring freezing point depression (Roebling, W Germany). Based on 8 replicates (means $\pm$ SEM)

Soil compaction	OP in 0–5 mm zone (MPa)	Elong. rate (mm.d^{-1})	Root diam. (mm)	Volume increase (mm^3 d^{-1})	Increase in OP due to less growth (Mpa)	Actual increase in OP (MPa)
Low	1.16 $\pm$ 0.04	45.5	1.56	87	–	–
Medium	1.30 $\pm$ 0.05	33.2	1.68	74	0.20	0.14
High	1.47 $\pm$ 0.04	28.1	1.81	72	0.24	0.31

when the soil was compacted to a bulk density of 1.64 Mg.m^{-3} (*cf.* 1.23 Mg.m^{-3} in control pots). The increase in mechanical impedance over the same range was difficult to assess because soils containing clay generate high skin friction against penetrometer probes. Osmotic pressures, which approximately equalled turgor pressures in this system (Ψ close to zero), were 1.16 and 1.47 MPa, at low and high compaction respectively (Table 1). In spite of the higher turgor pressure, there was a 38% lower root elongation rate in the compacted soil, indicating a substantial increase in resistance to root growth. This system is qualitatively similar to that used by Greacen and Oh (1972). The implications for osmotic adjustment will be discussed below.

The nature of the osmotic agents in lupin root apices was analysed (Table 2). Major osmotica in roots such as potassium, sucrose, fructose and glucose were measured in the expressed sap and they were found (together with a divalent anion, to balance potassium) to contribute 65–67% of the total osmotic pressure. This suggests that (i) potassium and sugars were the main components of osmotic pressure in lupin roots and (ii) sugar concentrations were higher in compacted than in freely-growing

root apices, in spite of the carbohydrate demand for growth of thicker root axes and the increased number of lateral roots adjacent to the apex. Apparently root apices were not substrate limited for growth. In fact, there seems to be a buildup of substrates due to slower expansion of apical tissue in compacted soil. A 'passive' buildup such as this would account for the relatively constant ratios of potassium to sugars in freely-growing and compacted roots (1.35 and 1.12 mmol K^+.g^{-1} sucrose equivalents, respectively). These ratios are in the same order as the 0.37 to 0.54 mmol K^+.g^{-1} sucrose observed by Hocking *et al.* (1978) in the phloem sap of lupin suggesting that osmotica were being delivered to the apex via the phloem in a constant ratio. Sugars subsequently were respired thereby increasing the K^+:sucrose ratio. This argues against the more complex phenomenon of 'osmoregulaton' whereby specific regulatory mechanisms (*e.g.* ion transport processes, respiration) are activated in response to osmotic requirements (see Reed, 1984).

Table 1 shows an analysis of the rate of expansion of apices and the increase in osmotic pressure when roots are compacted. The latter is calculated, relative to the 'low compaction' treatment, and denoted as 'Actual increase in OP'. This accords closely with the 'passive' increase in osmotic pressure predicted from the lower rates of tissue expansion ('Increase in OP due to less growth'). Therefore, there is again evidence that growth impairment due to mechanical impedance may quantitatively explain the high osmotic pressure in compacted root apices. This interpretation also applies to the data of Greacen and Oh (1972) on peas and explains the incomplete 'osmoregulation' when root growth was impeded. Presumably, solutes *e.g.*

Table 2. Contribution of potassium and sucrose, glucose and fructose to osmotic pressure in expressed sap of *L. angustifolius*. For details see Table 1. Potassium is assumed to be balanced by divalent anion (A^{-2}). Based on 4 replicates

Soil compaction	Contribution to osmotic pressure:		
	sucr. + gluc. + fruc.[a]	K^+ + A$^{-2\Psi}$	Total
Low	47%	18%	65%
High	48%	15%	63%

[a] Sugars are in the following proportions; sucrose ($\sim$10%), glucose ($\sim$37%), fructose ($\sim$53%).
Ψ K^+ is accompanied by an anion (A^{-2}).

ions in the peas were exported from apices or sugars were metabolized when they accumulated, thus creating the impression of 'partial osmoregulation'.

Concluding remarks

The impedance of roots by compacted soils is a major agricultural problem. The structure and function of roots is still however poorly understood. Radial swelling of roots is generally observed and is probably an adaptive trait in response to soil compaction. Radial swelling is primarily achieved by the yielding of tangential and radial cell walls of the outer cortex with little or no proliferation of new files of cells. In lupins, the reduction in cell *length* is commensurate with the impaired rate of root elongation suggesting a direct response of the meristematic zone to external pressure. Lateral roots proliferate nearer to the apex of compacted root axes.

Osmotically active substances accumulate in the apices of lupin roots when compacted. Potassium (plus the counter-anion) and sugars provide most of the osmotica. The ratio of potassium to sugars is preserved in impeded roots suggesting a simple buildup of solutes due to impaired growth.

Further evidence for passive buildup of solutes is provided by the analysis of growth; volume expansion is impaired in compacted roots to about the same extent as solutes accumulate. This militates against the idea that 'osmoregulation' is an adaptive trait for soil compaction. Work on radial swelling and root dimensions may yield positive results for screening lines for tolerance to compaction.

Acknowledgements

The help provided by Ms Shellie Carter and Ms Georgette Elliott with microscopy and Ms Elaine Mathews with assays is gratefully acknowledged. Travel funds were supplied by the Wheat Industry Research Committee of Western Australia.

References

Abdalla A M, Hettiaratchi D R P and Reece A R 1969 The mechanics of root growth in granular media. J. Agric. Eng. Res. 14, 236–248.

Barley K P 1962 The effects of mechanical stress on the growth of roots. J. Exp. Bot. 13, 95–110.

Barley K P 1968 Deformation of the soil by the growth of plants. Trans. Ninth Intern, Cong. Soil Sci. 1, 759–768.

Barley K P and Greacen E L 1967 Mechanical resistance as a soil factor influencing the growth of roots and underground shoots. Adv. Agron. 19, 1–43.

Crossett R N, Campbell D J and Stewart H E 1975 Compensatory growth in cereal root systems. Plant and Soil 42, 673–683.

Dawkins T C K, Roberts J A and Brereton J C 1983 Mechanical impedance and root growth — the role of endogenous ethylene. *In* Growth Regulators in Root Development. Eds. M B Jackson and A D Stead. pp 55–71. British Plant Growth Regulator Group, Monograph No. 10, Oxford, UK.

Eavis B W 1967 Mechanical impedance and root growth. Inst. Agric. Eng. Symp. Silsoe, Paper No. 4/F/39.

Goss M J 1977 Effects of mechanical impedance on root growth in barley (*Hordeum vulgare* L.). I. Effects on the elongation and branching of seminal root axes. J. Exp. Bot. 28, 96–111.

Greacen E L and Oh J S 1972 Physics of root growth. Nature 235, 24–25.

Hocking P J, Pate J S, Atkins C A and Sharkey P J 1978 Diurnal patterns of transport and accumulation of minerals in fruiting plants of *Lupinus angustifolius* L. Ann. Bot. 42, 1277–1290.

Jackson M B and Stead A D 1983 (Eds.) Growth Regulators in Root Development. British Plant Growth Regulator Group. Monograph No. 10, Oxford, UK.

Reed R H 1984 Use and abuse of osmo-terminology. Plant Cell Environ. 7, 165–170.

Scholefield D and Hall D M 1985 Constricted growth of grass roots through rigid pores. Plant and Soil 85, 153–162.

Scott Russell R and Goss M J 1974 Physical aspects of soil fertility — the response of roots to mechanical impedance. Neth. J. Agric. Sci. 22, 305–318.

Taylor H M and Ratcliff L F 1969 Root elongation rates of cotton and peanuts as a function of soil strength and soil water content. Soil Sci. 108, 113–119.

Wilson A J, Robards A W and Goss M J 1977 Effects of mechanical impedance on root growth in barley, *Hordeum vulgare* L. II. Effects on the development in seminal roots. J. Exp. Bot. 28, 1216–1227.

B. C. Loughman et al. (Eds.), Structural and functional aspects of transport in roots, 257–261.
© 1989 by Kluwer Academic Publishers.

Physiological and structural changes under flooding of whole corn plants

GALINA M. GRINIEVA and T.V. BRAGINA
K.A. Timiriazev Institute of Plant Physiology, USSR Academy of Sciences, SU-127296 Moscow, USSR

Key words: corn, exudation, flooding, oxygen, respiration

It is well known that flooding the soil decreases root growth, transport of ions, accumulation of dry matter and crop yield (Bradford and Yang, 1981; Jackson *et al.*, 1978; Grinieva, 1975; 1981; Grinieva *et al.*, 1986; Trought and Drew, 1980).

Corn plants are relatively tolerant of wetland conditions and well adapted to flooding (Grinieva, 1975). However, the effect of flooding on whole plants has not been firmly established. A specific feature of flooding is that increasing hypoxia is associated with permanent watering (Wample and Reid, 1979).

The aim of the present work is to study physiological and structural changes in shoots and roots of various ages after exposure of whole plants to flooding. Ten seedlings of 10-day-old *Zea mays* L. cv. Odesskaja were germinated in gravel culture and flooded in 5 l vessels so that the plants were completely under stagnant water. Illumination was at $70\,W.m^{-2}$, to 23°C and the treatment period lasted 24, 48 and 72 h. Plants were retained in normal humidity conditions after the end of the treatment period, the time of restoration lasting 6–8 days. The control plants grew on wet gravel (Grinieva *et al.*, 1986).

The oxygen content of the aqueous medium surrounding the roots was determined on a Radiometer ABC-1 gas analyzer (Denmark). Data shown (Fig. 1) indicate that pO_2 declined in four hours from the start of flooding (30 mm Hg) but oxygen content declined more slowly in the middle and upper parts of the vessel while the pH of the former was 7.3 and the latter 6.1.

Table 1 shows that the growth of shoots and roots was retarded under flooding conditions. Leaf length and thickness decreased and dry matter accumulation was reduced (Atwell and Greenway,

1987). The ratio of dry mass of the shoots to that of the root system was identical with the control value. It follows that flooding of whole plants inhibited not only root growth but that of the shoots also. Formation of adventitious roots took place, but their growth was also retarded (Grinieva and Bragina, 1987; Wiedenroth and Erdmann, 1985). Typical changes appeared in adventitious roots when compared with control plants. Flooding appeared to promote vessel expansion and weaken-

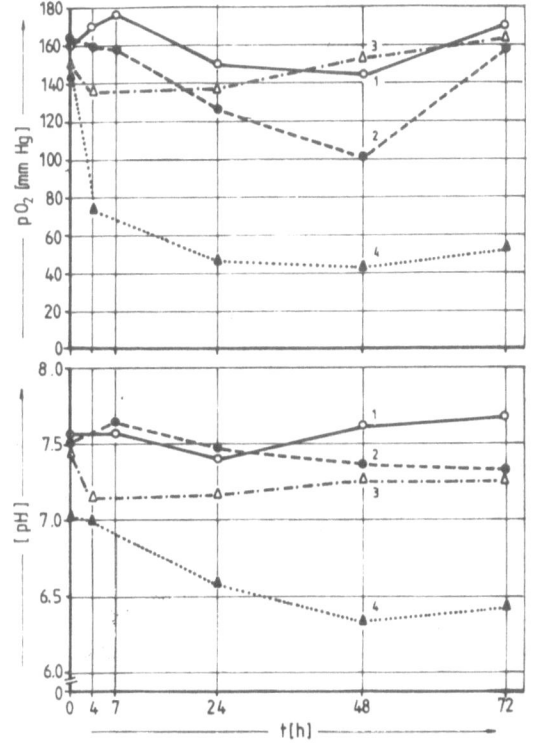

Fig. 1. Oxygen content (pO_2 mm Hg) and pH in the aqueous medium. **1, 2**—zone of leaves, **3, 4**—zone of roots. 1, 3—control plants, 2, 4—flooded whole plants.

258   *Grinieva and Bragina*

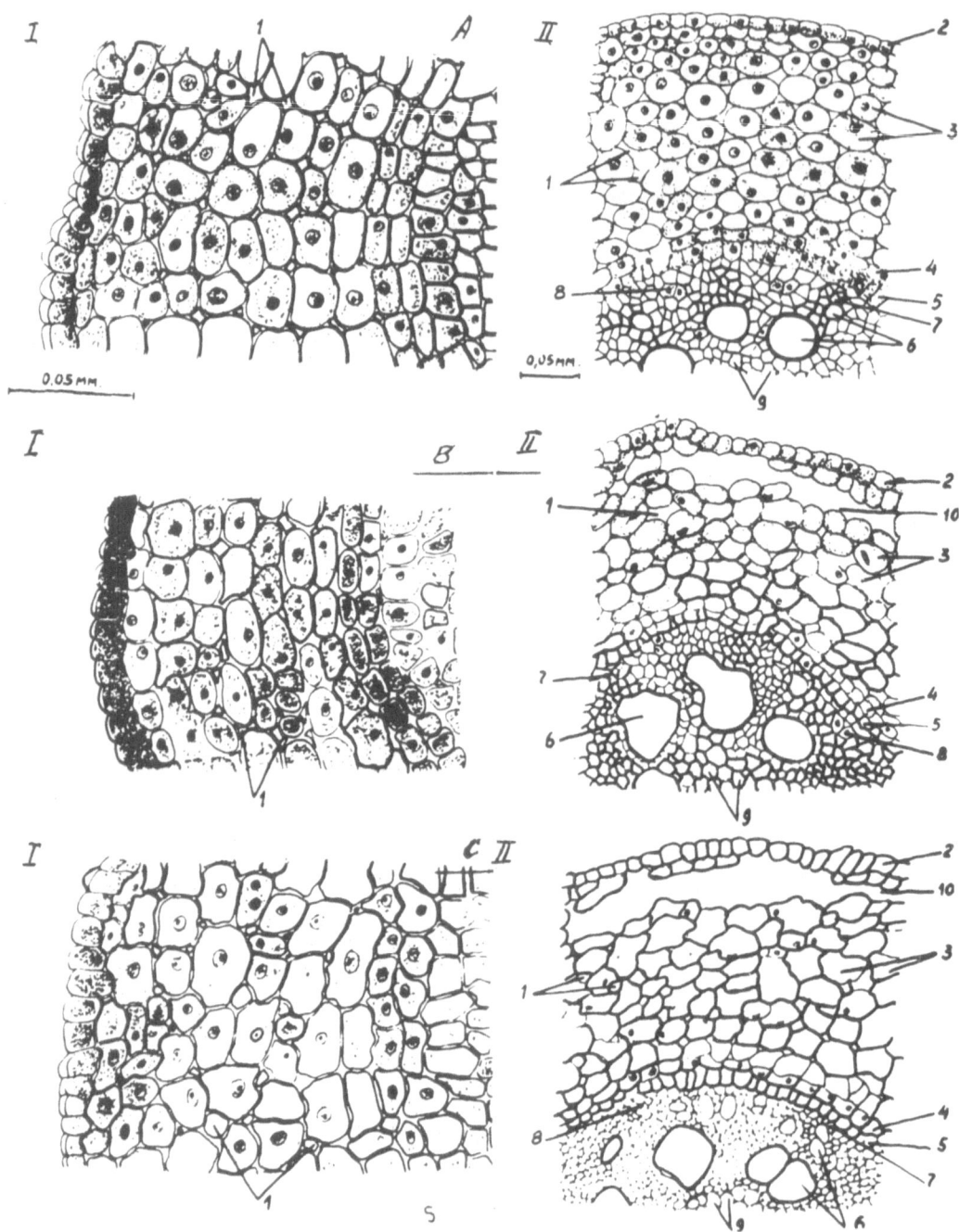

Fig. 2. Anatomical characteristics of adventitious roots after whole plant flooding. **I** — meristem; **II** — zone of elongation. **A** — 24 h; **B** — 48 h; **C** — 72 h periods of flooding. **1** — Intercellular spaces; **2** — rhizodermis; **3** — parenchyma of primary cortex; **4** — endodermis; **5** — pericycle; **6** — metaxylem; **7** — protoxylem; **8** — phloem; **9** — pith; **10** — aerenchyma.

ing of the schlerenchyma. Aerenchyma developed in adventitious roots and deformation of the endodermis occurred. The area of primary cortex was increased and the zone of elongation showed more damage than the meristematic region.* The leaves of flooded plants retained characteristic changes

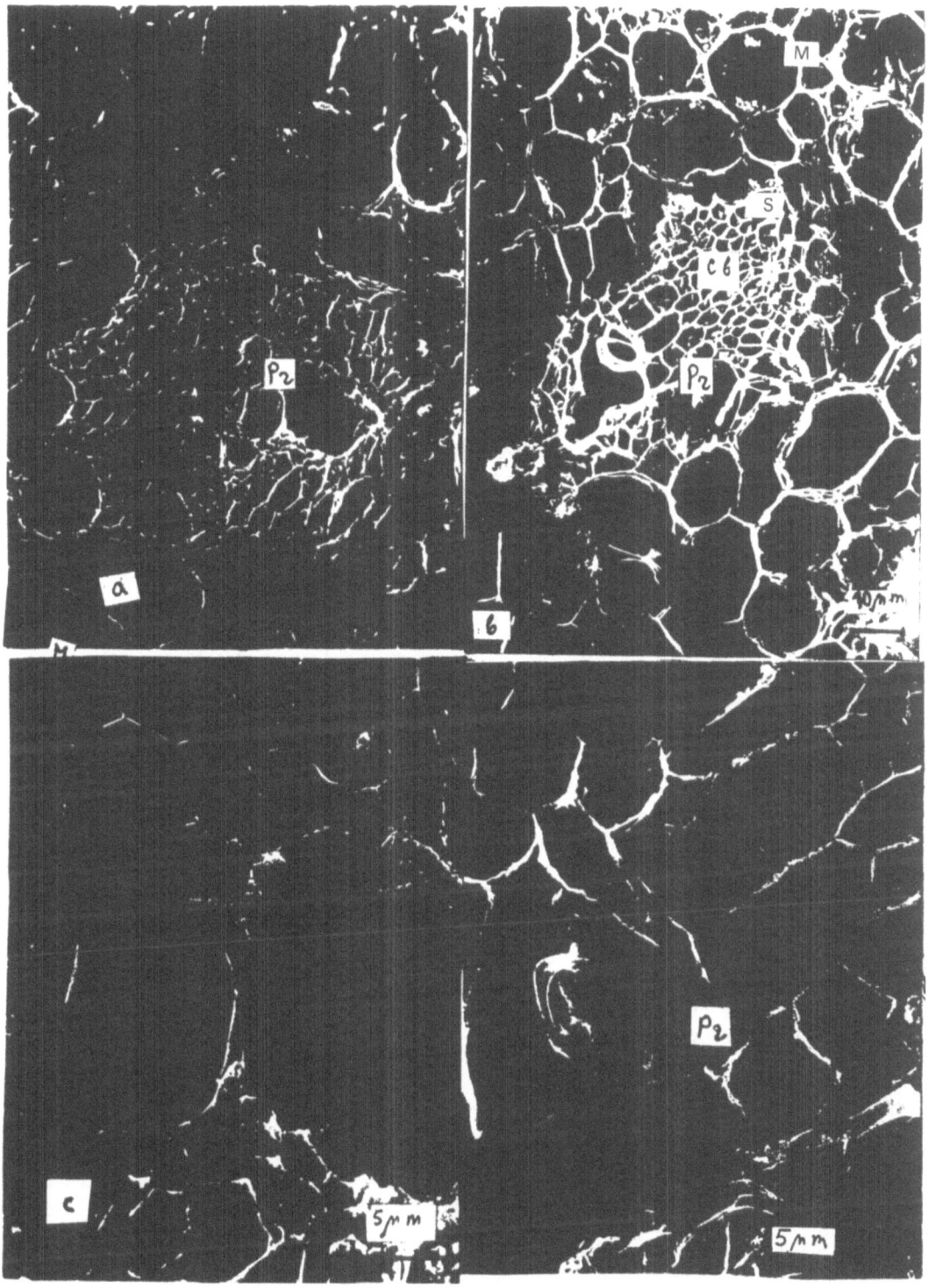

Fig. 3. Scanning electron micrographs of the stem base after 72 h of whole plant flooding. **a, b** — medular conducting bundles × 1200. **c, d** — medular conducting bundles × 4000. **a, c** — control; **b, d** — flooding. **cb** — conducting bundles; **Pr** — protoxylem. M — metaxylem; S — sclerenchyma.

Table 1. Morphological characteristics of corn plants after 48 h of flooding and 7 days recovery ($P < 0.05$)

Trait	Control	After flooding
Height of plant, cm	35.3 ± 1.1	26.5 ± 0.6
Length of leaf blade (2-d leaf), cm	26.7 ± 1.0	18.7 ± 1.1
Thickness of leaf, μm	154.0 ± 5.1	141.4 ± 3.1
Dry mass of shoot, mg	109 ± 0.01	75 ± 0.02
Length of adventitious roots, cm	21.2 ± 1.25	7.2 ± 0.22
Length of main root, cm	29.0 ± 1.2	damaged
Number of adventitious roots	4.55 ± 0.47	6.4 ± 0.4
Dry mass of root, mg	84 ± 0.005	60 ± 0.005
The ratio: dry mass of shoots to roots	1.30	1.25

after restoration. The bundles had larger clearances, the cell walls were thinner and the process of lignification was retarded. It was observed previously (Grinieva, 1975) that flooding of roots promoted expansion of the stem base and that this was a characteristic trait of adaptation to soil flooding, but this effect was not observed in the experiments reported here.

For observation in a scanning microscope (Hitachi HSM 22D, Japan) sections were taken from the base of the stem on the level of adventitious roots after 72 h of flooding (Fig. 3). In the conducting bundles flooding promoted expansion of vessels and weakening of schlerenchyma.

The water-conducting elements of large bundles were represented by protoxylem vessels with spiral thickenings. The diameter of protoxylem vessels significantly exceeds that of protoxylem vessels of such bundles in the control plants and large air-bearing cavities developed in the bundle. Structural

Table 2. Rate of exudation (Jv) in adventitious roots after different periods of flooding ($P < 0.05$) of whole plants

Variant	Jv = μl.h^{-1}.cm^{-2}	
	Control	Flooding
48 h of flooding	0.77 ± 0.07	0.68 ± 0.06
48 h of flooding and 6 days of restoration	0.43 ± 0.04	0.82 ± 0.08
72 h of flooding	0.73 ± 0.08	0.33 ± 0.04

characteristics of the conducting system at the base of the stem determined the nature of the water regime in the root system, especially that of adventitious roots arising from the stem base.

It can be seen from the data presented in Table 2 that the rate of exudation declines in adventitious roots after 48 h of flooding. However, after 6 days recovery the rate of exudation increased about twofold. Flooding for 72 h decreased the rate of exudation significantly. Thus, flooding of whole plants affected the rate of exudation in adventitious roots unequally depending on different periods of flooding.

In addition, differences in the rate of oxygen consumption were observed in root systems and overground organs and the data presented in Fig. 4 indicate that oxygen uptake decreased after flooding in all organs except the leaves.

However, after 8 days of recovery the oxygen consumption was higher in all organs in comparison with the control. The responses to flooding conditions for whole plants have similarities with those found under total anoxia in N_2. The progressive reduction of oxidative activity took place

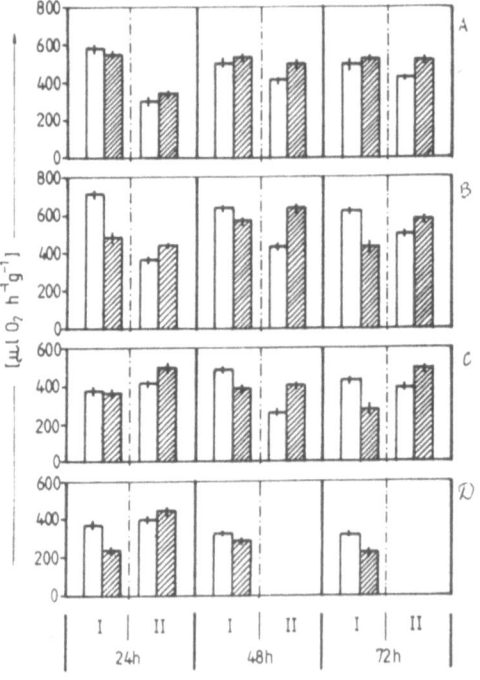

Fig. 4. Oxygen uptake (μl of O_2 h^{-1}.g^{-1} of row mass) by different organs after whole plant flooding (**I**) and 8 days of recovery (**II**). **A**—leaves; **B**—stem base; **C**—adventitious roots; **D**—main root. Unhatched columns-control; hatched columns-flooding.

under anoxia over the period 24–72 h. Short-term treatment increased the rate of respiration. Prolonging anoxia up to 48–72 h resulted in irreversible destruction of the ultrastructure of all cell organells which then led to a total disruption of the energetics of the cell (Grinieva, 1975). It was also found that in flooded plants the adaptive responses appeared only if the period of flooding lasted about 24–36 h. The data obtained confirm our assumption that the stem base and adventitious roots play an important regulatory role in the development of adaptive responses of corn plants under flooding conditions. In higher plants the adaptive reactions to oxygen deficiency depend primarily on whether cessation of oxygen supply applied to the whole plant or to the root system alone. In conditions of general anoxia the regulatory system is switched off and consequently the adaptive mechanisms may reveal their effect only during limited periods. Under conditions of root hypoxia when the oxygen supply to shoots is maintained the plants are able to survive.

References

Atwell B J and Greenway H 1987 The relationship between growth and oxygen uptake in hypoxic rice seedlings. J. Exp. Bot. 38, 454–465.

Bradford K J and Fa Yang S 1981 Physiological responses of plants to waterlogging. Hort. Science. 16, 25–30.

Jackson M B, Gales K and Campbell D J 1978 Effect of waterlogging soil conditions on the production of ethylene and on water relations in tomato plants. J. Exp. Bot. 29, 183–193.

Grinieva G M 1975 Control of plants metabolism under oxygen deficiency. Moscow: Nauka Publishers (in Russian).

Grinieva G M 1981 The effects of flooding on metabolism and structure of maize roots. *In* Structure and Function of Plant Roots. Eds. R Brouwer *et al.* Martinus Nijhoff Publishers, The Hague-Boston-London, 323–326.

Grinieva G M, Borisova T A and Garkavenkova A F *et al.* 1986 Effect of flooding time on exudation, respiration, and anatomical structure of corn roots. Fisiol. rast. 33, 987–996 (*In Russian*). Sov. Plant Physiol. 760–767.

Grinieva G M and Bragina T V 1987 Physiological and structural changes under flooding of whole corn plants, 3rd Internat. Symposium, Abstract, Nitra, Czechoslovakia, p. 48.

Trought M C T and Drew M C 1980 The development of waterlogging damage in wheat seedlings (*Triticum aestivum* L.). I. Shoot and root growth relation to changes in the concentrations of dissolved gases and solutes in the soil solution. Plant and Soil 54, 77–94.

Wample R L and Reid D M 1979 The role of endogenous auxins and ethylene in the formation of adventitious roots and hypocotyl hypertrophy in flooded sunflower plants (*Heliantus annuus*) Physiol. Plant. 45, 219–226.

Wiedenroth E M and Erdmann B 1985 Morphological changes in wheat seedlings (*Triticum aestivum* L.) following root anaerobiosis and partial pruning of the root system. Ann. Bot. 56, 307–316.

B. C. Loughman et al. (Eds.), Structural and functional aspects of transport in roots, 263–267.
© 1989 by Kluwer Academic Publishers.

Recovery of ultrastructure in water stressed root epidermal cells of *Zea mays*

MILADA ČIAMPOROVÁ

Institute of Experimental Biology and Ecology, CBES, Slovak Academy of Sciences, Dúbravská cesta 14, CS-81434 Bratislava, Czechoslovakia

Key words: rehydration, root epidermis, ultrastructure, *Zea mays* L.

Abbreviations. PEG 4000, polyethylene glycol, molecular weight 4000; ER, endoplasmic reticulum; ψ_w, water potential

Introduction

Ultrastructural changes within maize root cells caused by non-lethal water stress recover within 45 min to 4 hrs (Nir *et al.*, 1969). During rehydration of water-stressed embryos of wheat (Marinos and Fife, 1972) and maize (Crèvecoeur *et al.*, 1976) all the cell components recovered their structure. In *Pleurozium schreberi*, the organelles recover their structure in the following order: vacuoles and mitochondria, in 10 min; dictyosomes in 1 h; ER, ribosomes and chloroplasts, in 1 to 2 h; and the nucleus, in 3 to 4 h (Noailles, 1978). Repair of cell components within 24 h of rehydration of the palisade cells in drought tolerant species *Selaginella leipdophylla* was shown to coincide with the recovery of respiratory and photosynthetic activities (Bergtrom *et al.*, 1982).

PEG 4000-induced water stress inhibits growth of maize roots and affects almost all the cell components in the root epidermis and these effects may be reversible (Čiamporová, 1976; 1977). Water potential in maize root tips increased rapidly in the first 30 min of rehydration (from -3.86 to -2.90 MPa) and continues to increase more slowly during the next 48 h of rehydration. Root growth recovers after 2 h of rehydration (Čiamporová and Eliáš, 1987). The ultrastructure recovery of root epidermal cells within these two hours is documented in this paper.

Materials and methods

Seedlings of *Zea mays* L., inbred line A15, were cultivated in half strength Hoagland solution ($\psi_w = -0.037$ MPa) and stressed in PEG 4000 solution ($\psi_w = -1.52$ MPa) for 24 h as described earlier (Čiamporová and Eliáš, 1987). The stressed seedlings were transferred back to $\frac{1}{2}$ Hoagland solution and 1-mm segments of the 3-mm apical part of the primary roots were fixed after 5 min to 2 h of rehydration. For electron microscopy, routine techniques of fixation in glutaraldehyde and OsO_4, dehydration in ethanol and embedding in Epon were used. Ultrathin sections cut at 0.5, 1, and 2 mm from the root cap junction were examined in a Tesla BS 500 electron microscope.

Results and discussion

Ultrastructure of maize root epidermal cells was completely restored following 24 h of rehydration (Čiamporová 1976; 1977). The study of the process of structural repair required the investigation of shorter rehydration periods. Our experiments have shown that the structural damage to stressed epidermal cells was significantly repaired within several minutes. In the mitochondria which were devoid of cristae due to the previous water stress, the development of internal membranes could be

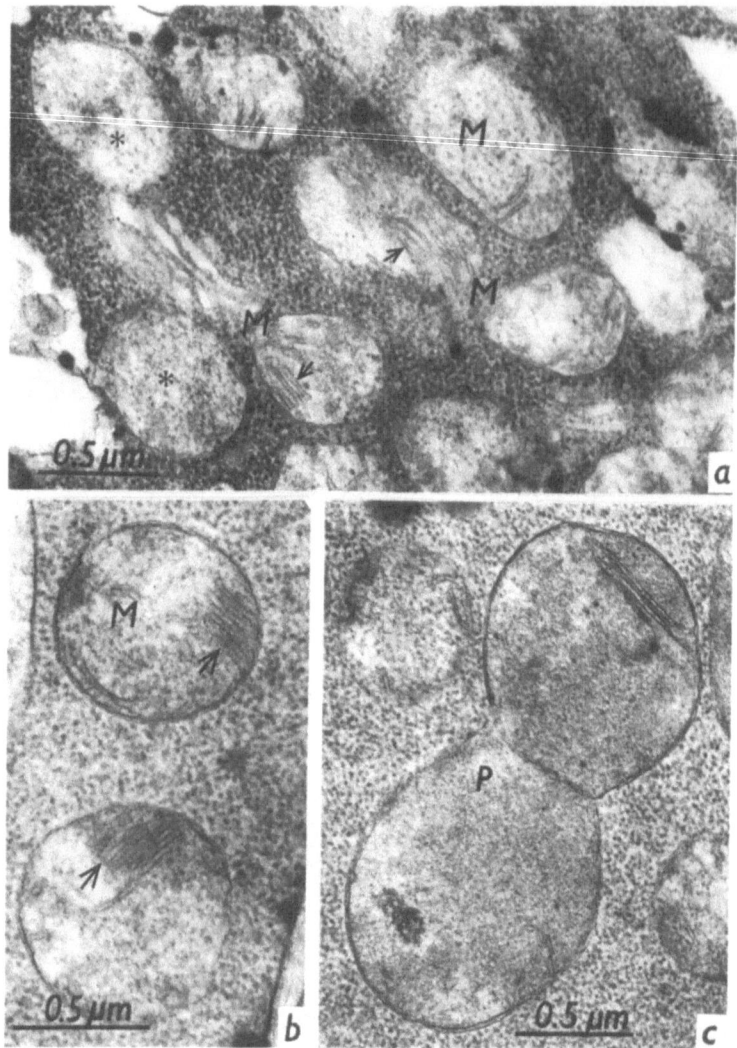

Fig. 1. Meristematic epidermal cells rehydrating for 5 (**a**) and 10 (**b, c**) min. contain mitochondria (**M**) still devoid of cristae (*asterisks*) and mitochondria where the formation of cristae often in parallel arrangement (*arrows*) has already begun. Plastids (**P**) often undergo division.

observed after only 5 min of rehydration (Figs. 1a and b). The newly formed cristae often revealed parallel orientation (Figs. 1b and 3a). Elongated mitochondria with constrictions were frequently observed indicating that the increase in mitochondrial number might occur by means of organelle division during the early stages of rehydration. Similarly, the number of plastids could increase by the division of organelles since these were frequently observed after 10 min of rehydration (Fig. 1c). Parallel complexes of ER induced by water stress (Čiamporová, 1976; 1977; Nir *et al.*, 1969) also disintegrate rapidly. Fragmentation of the long ER elements and dilation of the fragments (Figs. 2 and

3a) were visible after 5 to 10 min of rehydration. Dictyosomes with short cisternae still loosely arranged were present near to these ER complexes (Fig. 3a). Vesicles, probably of ER origin, appeared between both organelles (Fig. 3a). This structural appearance possibly indicates that the regenerating ER system contributes to the structural repair and later also to recovery of the functional activity of dictyosomes. Fully reversed dictyosomes with numerous large vesicles in their vicinity could not be observed until after 30 min of rehydration (Fig. 3b). In approximately the same period, numerous polyribosomes appeared in the cytoplasm of epidermal cells. In most cells, the nuclear chroma-

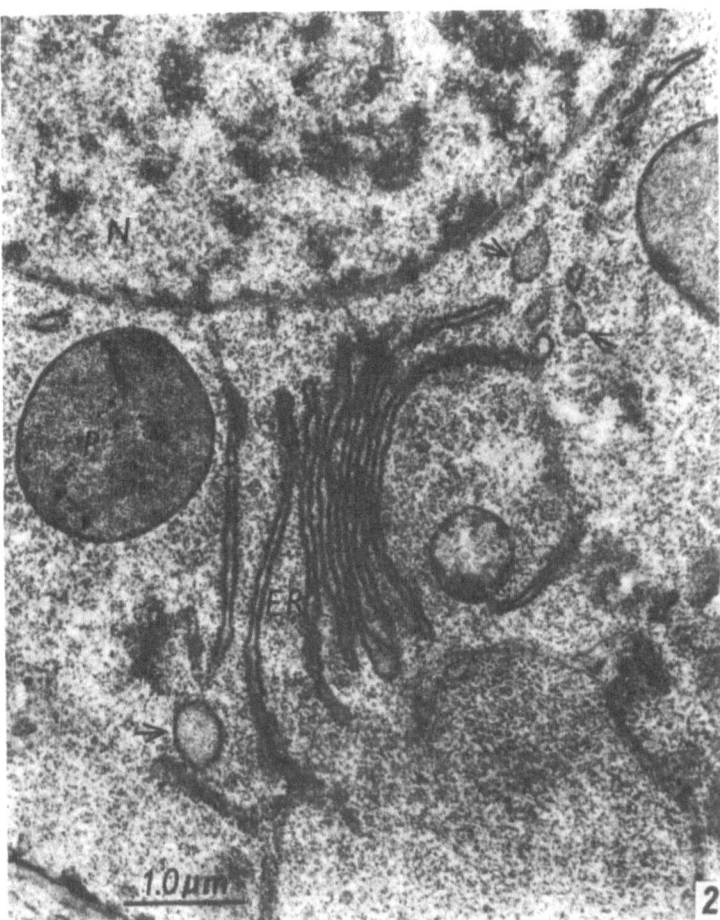

Fig. 2. Meristematic cell rehydrating for 5 min. Fragmentation and dilation of membrane (*arrows*) of the parallel ER complex and a nucleus (N) with dispersed chromatin.

tin aggregations were dispersed after 10 min of rehydration (Fig. 2) and there were no nuclei with condensed chromatin in the cells rehydrated for 2 h.

Electron microscopic investigation has shown that ultrastructure damage brought about by mild water stress is reversible (Bergtrom, 1982; Crève-coeur *et al.*, 1976; Marinos and Fife, 1972; Nir *et al.*, 1969; Noailles, 1978). The maintenance of cell membrane continuity of the individual cell compartments has been considered to be crucial for the capacity to repair ultrastructure (Fellows and Boyer, 1978). Water stress induced by PEG 4000 did not cause membrane fragmentation and there were still some mitochondria with a quite well preserved internal structure in epidermal cells after 24 h (Čiamporová, 1976; 1977). These circum-stances might represent an important prerequisite for the recovery of the cell.

The pattern of structural recovery of the in-dividual cell components was similar in epidermal cells at different distances form the root cap junc-tion. There was a variability in the rapidity of ultrastructure recovery within the cells, indepen-dent of their age. A normal ultrastructure of both meristematic and more vacuolated cells was largely recovered by 2 h of rehydration.

According to the literature, the rate of ultrastruc-tural recovery seems to be quite high, *i.e.* within several minutes to 4 or 24 h. It would be difficult to discuss the sequence of stages of recovery because there are few detailed studies examining different cells and plant species. Nevertheless, the rapid re-pair of mitochondria reported in *Pleurozium* (Noailles, 1978) and in root epidermal cells reveals that the regeneration of an energy supply is essen-tial. A comparison of the development of structural

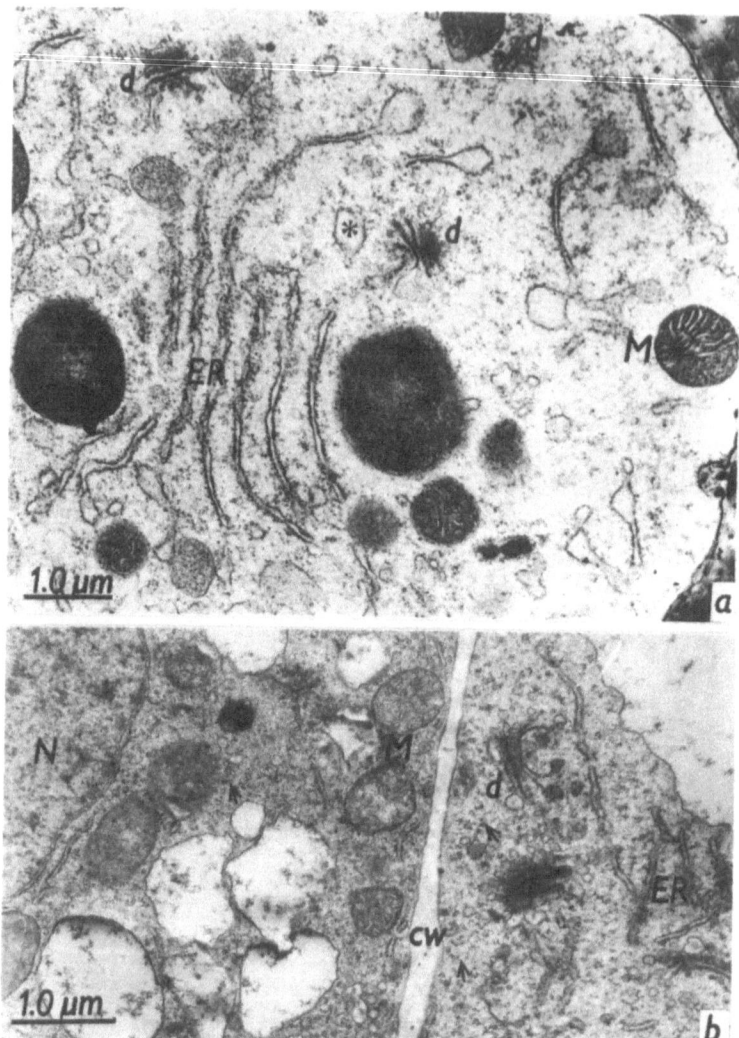

Fig. 3a. Epidermal cell at 2 mm from the root cap junction rehydrating for 10 min ER-derived vesicles (*asterisk*) between ER and dictyosomes (**d**). Mitochondria with parallel cristae (**M**).

Fig. 3b. Part of two epidermal cells divided by the cell wall (**cw**) after 30 min of rehydration with well recovered structures of dictyosomes (**d**), mitochondria (**M**), nucleus (**N**), ER, and polyribosomes (*arrows*).

responses of maize root epidermal cells to water stress (Čiamporová, 1987) with the structural recovery during rehydration reveals that the latter process proceeds more rapidly. This may be of importance from the point of view of plant adaptation to water shortage. The occurrence of maximum repair corresponds with the steep rise of water potential in the root tips within the first 30 min of rehydration (Čiamporová and Eliáš, 1987). This rise in water potential might create the conditions necessary for the rapid recovery of root growth and functions.

References

Bergtrom G, Schaller M and Eickmeier W G 1982 Ultrastructural and biochemical bases of resurrection in the drought-tolerant plant, *Selaginella lepidophylla*. J. Ultrastruct. Res. 78, 269–282.

Crèvecoeur M, Deltour R and Bronchart R 1976 Cytological study on water stress during germination of *Zea mays*. Planta 132, 145–149.

Čiamporová M 1976 Ultrastructure of meristematic epidermal cells of maize root under water deficit conditions. Protoplasma 87, 1–15.

Čiamporová M 1977 Ultrastructure of differentiating epidermal cells of maize root under water deficient conditions. Biologia Plant. 19, 107–112.

Čiamporová M 1987 The development of structural changes in epidermal cells of maize roots during water stress. Biologia Plant. 29, 290–294.

Čiamporová M and Eliáš P 1987 Growth and water potential of maize roots under rapidly induced water stress and rehydration. Biológia 42, 3–8.

Fellows R J and Boyer J S 1978 Altered ultrastructure of cells of sunflower leaves having law water potentials. Protoplasma 93, 381–395.

Marinos N G and Fife D N 1972 Ultrastructural changes in wheat embryos during a 'presowing drought hardening' treatment. Protoplasma 74, 381–396.

Nir I, Klein S and Poljakoff-Mayber A 1969 Effects of moisture stress on submicroscopic structure of maize roots. Austr. J. Biol. Sci. 22, 17–33.

Noailles M-C 1978 Étude ultrastructurale de la récupération hydrique après une période de sécheresse chez une hypnobryale: *Pleurozium schreberi* (Willd.) Mitt. Ann. Sci. Naturelles, Bot. Biol. Végétale 19, 249–265.

B. C. Loughman et al. (Eds.), Structural and functional aspects of transport in roots, 269–274.
© 1989 by Kluwer Academic Publishers.

Growth, root respiration and phosphorus utilization of normal and *Agrobacterium rhizogenes* transformed potato plants

S.C. VAN DE GEIJN[1,2], J. HELDER[1], H.G. VAN HOOREN[1] and CH.H. HÄNISCH TEN CATE[1]
[1]*Research Institute ITAL, P.O. Box 48, 6700 AA Wageningen, The Netherlands,* [2]*Present address: Centre for Agrobiological Research, P.O. Box 14, 6700 AA Wageningen, The Netherlands*

Key words: *Agrobacterium rhizogenes*, genetic transformation, phosphorus utilization, potato, relative growth rate, root respiration

Introduction

The use of *Agrobacterium* species as a vector for the transfer of genetic material, carrying desired properties to plants has become a well described and increasingly successful method (reviewed by Klee *et al.*, 1987). Most attention has been given to the use of strains of *A. tumefaciens*. The wild type of this species causes crown gall disease. Infection of wounded plant parts with *A. rhizogenes* induces a genetic transformation that results in the proliferation of rapidly growing roots. This phenomenon is also known as 'hairy root disease'. The hairy root traits are encoded on the bacterial root inducing (Ri) plasmid (Chilton *et al.*, 1982).

It has been observed that the genetic stability of the plant material regenerated after transformation with *A. tumefaciens* is often poor. More stable results can be obtained using *A. rhizogenes* by regeneration from transformed root cultures (Hänisch ten Cate *et al.*, 1987). This finding has prompted a study of the properties of *A. rhizogenes* transformed plants.

Ri-transformed plants (Ri-plants) of the potato cultivar Bintje grew more vigourously *in vitro* than normal plants (Hänisch ten Cate *et al.*, 1988). Moreover, it has been suggested that plants, totally or partially provided with a Ri-transformed root system, perform better than normal ones under stress conditions such as water deficiency (Moore *et al.*, 1979; Tepfer, 1983).

To test whether root systems, resulting from these transformations, differ from normal roots with respect to growth, respiration and mineral uptake we have compared different Ri-plant clones of potato cv Bintje with untransformed plants. To study the effect of the transformed root system *per se* grafting experiments were included in the project.

Material and methods

Plant material and culture

Four different phenotypes of *Agrobacterium rhizogenes* transformed *Solanum tuberosum* L. cv Bintje plants (Ri-b, Ri-c, Ri-d and Ri-g) and one of cv Désirée (Ri-h) were regenerated from five distinct hairy root clones and propagated *in vitro* (Hänisch ten Cate *et al.*, 1988).

Subsequently, the plants were transferred (t = 0) to vermiculite and drenched with half strength Hoagland solution. After 10 days the plants were transferred to a well aerated nutrient solution of the same composition (10 plants on 15 dm^3 solution, pH 6.8). The nutrient solution was replaced twice a week.

In some experiments normal Bintje and various Ri-plants were grown on normal (0.35 mM) and low (0.035 mM) phosphate concentration in otherwise unchanged conditions.

All experiments were performed in a climate room at 20°C, RH of 90% and 18 h light period (20 klux). Groups of at least five plants were used in each treatment. Growth was measured by determining the fresh weight of all plants twice a week.

270   *van de Geijn* et al.

Root respiration measurements

O$_2$ consumption of roots was measured according to Lambers *et al.* (1983). The non-cytochrome linked activity of the alternative pathway was determined using 5 mM salicylhydroxamic acid (SHAM). To eliminate the disturbing effect of peroxidase activity gentisic acid was added to a final concentration of 0.5 mM (Spreen Brouwer *et al.*, 1986). Residual respiration was determined after subsequent addition of KCN to a final concentration of 0.4 mM.

Mineral content

Shoots of 3-week-old tomato plants (*Lycopersicon esculentum* L. cv marathon) were grafted onto cv Bintje (control) and Ri-Bintje root systems. At harvest tops were separated from roots, and the material was dried (48 h at 70°C), homogenised and digested in a salicylic acid/sulfuric acid mixture. The digests were made clear with H$_2$O$_2$. Mineral content was determined by Spectrophotometry (N and P), Inductively Coupled Plasma (B) or Atomic Absorption Spectrophotometry (Mg, Zn, K, Mn and Fe).

Results and discussion

Grafting experiments

In order to determine the characteristics of the Ri-root system, shoots of control Bintje plants were grafted onto control and Ri-Bintje roots. When grafted onto Ri-Bintje roots, the shoots after two days showed callus-like outgrowths on the leaves, mainly close to the nervation (Fig. 1). This reaction was not unique for Ri-roots; Bintje shoots grafted onto tomato root systems also developed these structures. The outgrowths disappeared after a few days, but in general survival of graftings on Ri-roots was very poor.

To avoid this problem, tomato shoots were used in further grafting experiments. After four weeks on a complete nutrient solution the apical, newly formed tomato leaves of shoots grafted onto Ri-root systems became slightly chlorotic. The mineral content in roots and shoots did show significant differences between treatments for various elements (Table 1). The phosphorus-content in Ri-roots and in shoots grafted onto Ri-roots was always significantly higher than in the controls.

Not only tomato shoots grafted onto Ri-Bintje roots developed chlorosis (at normal P-levels in the

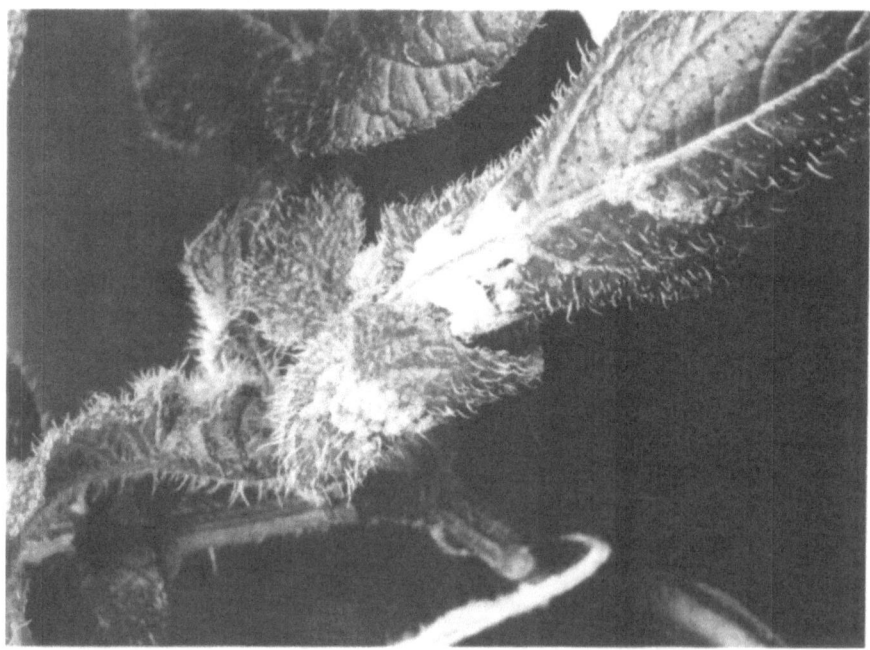

Fig. 1. Detail of a leaf of cv Bintje two days after grafting onto a Ri-root system. White callus-like outgrowths were formed close to the nervation.

Table 1. Comparative mineral content of grafted plants: tomato shoots onto control (Bi) and transformed Bintje (Ri-g) roots after 4 weeks. (NS = difference not significant at $P = 0.05$; nd = not determined)

Element	Tomato-shoot	Potato-roots
N	NS	NS
P	Ri-g > Bi	Ri-g > Bi
K	NS	Ri-g > Bi
Mg	Bi > Ri-g	NS
Mn	nd	NS
Zn	Ri-g > Bi	NS
Fe	NS	nd
B	Ri-g > Bi	NS

Table 2. Phosphorus concentration (mg P.g^{-1} DM) in the tops of control and Ri-plants of potato on a nutrient solution with normal (0.35 mM) and low (0.035 mM) phosphate concentration after 30 days growth

	Bintje	Ri-phenotype		
		Ri-c	Ri-d	Ri-h
Normal P	3.8a[a]	5.1c[b]	5.3c[b]	5.8c[b]
Low P	2.9a	2.6a	3.0a	4.6b[b]

[a] values followed by the same letter do not differ significantly at $P = 0.05$.
[b] plants showing chlorosis after 30 days.

nutrient solution), non-grafted Ri-potato plants also turned slightly yellow in the apical meristem and P-content was high (Table 2). Phosphorus content of the Ri-Bintje plants of the low P-treatment declined to the control level and chlorosis disappeared after 22 days. Only in the cv Désirée phenotype Ri-h the chlorosis in the top leaves persisted and again this chlorosis was accompanied by a high P-content (Table 2).

On the leaves of control Bintje plants in the low-P treatment callus-like outgrowths, similar to those observed after grafting onto Ri-roots, appeared after more than one month of culture in these conditions.

Growth rate

At the end of the *in vitro* propagation Ri-plants always had higher fresh weight than plantlets of the control Bintje. This difference was due to a larger root mass, and the lighter green leaves and stem were also larger and sturdier (Fig. 2).

The relative growth rate (RGR) of all potato

Fig. 2. Plantlets of cv Bintje (**A**), Ri-g (**B**) and Ri-b (**C**) cultured on vermiculite with half strength Hoagland solution for one week.

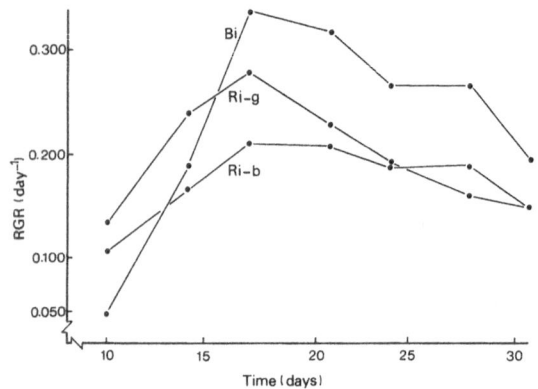

Fig. 3. Relative growth rate (RGR) of Bintje and phenotype Ri-b and Ri-g cultured on half strength Hoagland solution for one month.

Table 3. Relative growth rate (RGR) of control Bintje and Ri-plants on nutrient solution with normal (0.35 mM) and low (0.035 mM) phosphate concentration, after 3 weeks of growth.

	Bintje	Ri-phenotype		
		Ri-c	Ri-d	Ri-h
Normal P	0.148b[a]	0.132a	0.125a	0.130a
Low P	0.160b	0.135a[b]	0.111a	0.113a

[a] values followed by the same letter do not differ significantly at $P = 0.05$.

Root respiration

The O_2 consumption of control Bintje roots per unit of root mass gradually increased from 4 to 28 days (Fig. 4a, b). In contrast the rate of respiration of Ri-roots was almost constant. The phosphate concentration of the nutrient solution did not affect the O_2 consumption of roots.

Two weeks after transfer from the *in vitro* culture the alternative respiration in the control roots was slightly but significantly higher than in the Ri-roots. This difference was not found after three weeks, when this pathway in all cases was responsible for about 20–30% of the total oxygen consumption. The residual O_2 consumption always was 12 to 18%.

plants increased after transfer from vermiculite to the nutrient solution. However, this increase was greater and lasted longer for Bintje than for the Ri-plants (Fig. 3). The RGR of Bintje exceeded that of the Ri-plants after 3 weeks culture but at that time the weight of Bintje plants was still less or about equal to that of Ri-plants. Ri-plants had a lower shoot to root ratio.

Comparison of the growth of the plants at normal and low phosphate concentration showed that, after a lag phase of 2–4 days after transfer from vermiculite, the RGR did not depend upon the phosphate concentrations (Table 3). The RGR of Bintje was again significantly higher than that of the Ri-transformed clones.

General characterization

The Ri-transformed potato plantlets clearly differed from Bintje in phenotype at the start of the

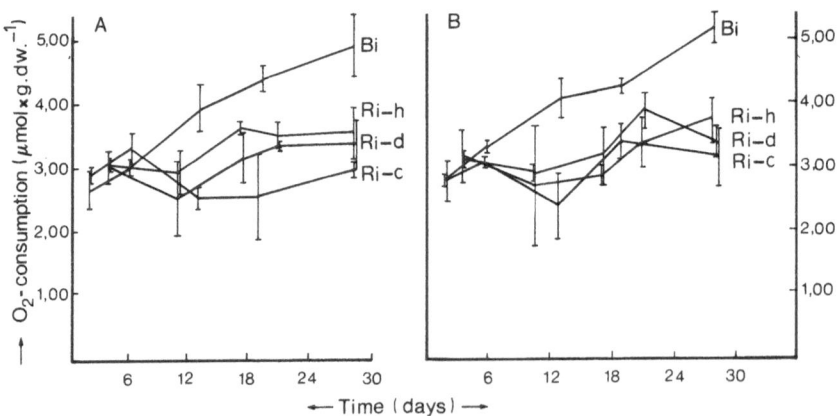

Fig. 4. Root respiration (O_2 consumption) of cv Bintje and Ri-plants cultured on nutrient solution with: **A.** normal (0.35 mM) and **B.** low (0.035 mM) phosphate during one month of growth.

culture on nutrient solution. The larger and more sturdy Ri-phenotypes did, however, lose their lead in size as Bintje eventually acquired a higher relative growth rate. This discrepancy in growth characteristics of Ri-plants and control Bintje plants *in vitro* as compared to a nutrient solution appears to be a general one in potato (Hänisch ten Cate *et al.*, 1988; Ooms *et al.*, 1985). It would be interesting to know whether this also holds for Ri-transformed plants of other species.

The most obvious difference between *in vitro* culture and nutrient solution culture is the presence or absence of sucrose in the medium. Consequently the growth on nutrient solution depends wholly on the energy supplied by photosynthesis. Apparently the lighter green Ri-plants with their lower increase of RGR and an almost constant rate of root respiration have a lower net rate of photosynthesis.

Among the minerals only the phosphorus content appeared to be affected in both roots and shoots. The Ri-plants were higher in P-content than Bintje. It is unclear whether this can be ascribed merely to the phenotypical characteristics (highly branched fine roots) or that differences in absorption kinetics are also involved.

The higher phosphate content may have caused an unfavourable Fe/P ratio in the developing tissues which may result in chlorosis (Mengel and Kirkby, 1978). The disappearance of the chlorosis in the tops of the Ri-Bintje when grown on low-P medium corroborates this hypothesis. The late response to the lowering of the P-supply may be due to the fact that solutions were refreshed twice weekly. In the course of the experiment it appeared that even at $0.035\,mM$ phosphate the nutrient medium was not depleted between refreshments until the plants were large enough, after about three weeks. The continuing uptake and the extensive recirculation of P inside the plant may have accounted for the lag in the decline of the plant P-content and the disappearance of the chlorosis.

Callus-like outgrowths were formed on the leaves of Bintje shoots when grafted onto Ri-root systems, and tomato roots. Similar malformations were observed when control Bintje plants were grown at low phosphate levels. This might be related to an interrupted internal P-circulation after grafting or a deficiency at a low phosphate supply.

The present results do not allow an extrapolation to field conditions. However, it would be interesting to investigate the performance of Ri-transformed plants in situations with a relatively low availability of phosphorus. A higher phosphate uptake could be achieved by phenotypical changes such as a finer root system. In soils this might be a major factor, as P is extremely immobile and a better exploration of the soil volume at equal root weight might be possible (Barber, 1984; Jungk and Claassen, 1986). The observation that in field conditions the phosphorus content of potato plants decreases from about 0.8% in spring to about 0.12% in autumn (J Vos and J Groenwold, personal communication) may corroborate this explanation.

Acknowledgements

The many suggestions of, and helpful discussions with Prof Dr H Lambers and the critical reading of the manuscript by Dr H Breteler and Dr L van Vloten-Doting are gratefully acknowledged.

References

Barber S A 1984 Soil Nutrient Bioavailability. John Wiley and Sons, New York.

Chilton M D, Tepfer D A, Petit A, David C, Casse-Delbart F and Tempe J 1982 *Agrobacterium rhizogenes* inserts T-DNA into the genomes of host plant root cells. Nature 295, 432–434.

Hänisch ten Cate Ch H, Sree Ramulu K, Dijkhuis P and De Groot B 1987 Genetic stability of cultured hairy roots induced by *Agrobacterium rhizogenes* on tuber discs of potato cv Bintje. Plant Sci. 49, 217–222.

Hänisch ten Cate Ch H, Ennik E, Roest S, Sree Ramulu K, Dijkhuis P and De Groot B 1988 Regeneration and characterization of plants from potato root lines transformed by *Agrobacterium rhizogenes*. Theor. Appl. Genet. 75, 452–459.

Jungk A and Claassen N 1986 Availability of phosphate and potassium as the result of interactions between root and soil in the rhizosphere. Z. Pflanzenernaehr. Bodenkd. 149, 411–427.

Klee H, Horsch R and Roger St 1987 Agrobacterium mediated plant transformation and its further application to plant biology. Annu. Rev. Plant Physiol. 38, 467–486.

Lambers H, Day D A and Azcon-Bieto J 1983 Cyanide resistant respiration in roots and leaves. Measurements with intact tissues and isolated mitochondria. Physiol. Plant. 58, 148–154.

Moore L, Warren G and Strobel G 1979 Involvement of a plasmid in the hairy root disease of plants caused by *Agrobacterium rhizogenes*. Plasmid 2, 617–626.

Ooms G, Karp A, Burrell M M, Twell D and Roberts J 1985 Genetic modification of potato development using Ri T-DNA. Theor. Appl. Genet. 70, 440–446.

Spreen Brouwer K, Van Valen T, Day D A and Lambers H 1986 Hydroxamate-stimulated O_2 uptake in roots of *Pisum sativum* and *Zea mays*, mediated by a peroxidase. Plant Physiol. 82, 236–240.

Tepfer D 1983 The biology of genetic transformation of higher plants by *Agrobacterium rhizogenes*. *In* Molecular Genetics of the Bacteria-Plant Interaction. Ed. A Puhler. pp 248–258, Springer-Verlag, Berlin, Heidelberg.

Developments in Plant and Soil Sciences

1. J. Monteith and C. Webb, Eds., Soil water and nitrogen in Mediterranean-type environments. 1981. ISBN 90-247-2406-6
2. J. C. Brogan, Ed., Nitrogen losses and surface run-off from landspreading of manures. 1981. ISBN 90-247-2471-6
3. J. D. Bewley, Ed., Nitrogen and carbon metabolism. 1981. ISBN 90-247-2472-4
4. R. Brouwer, I. Gašparíková, J. Kolek and B. C. Loughman, Eds., Structure and function of plant roots. 1981. ISBN 90-247-2510-0
5. Y. R. Dommergues and H. G. Diem, Eds., Microbiology of tropical soils and plant productivity. 1982. ISBN 90-247-2719-7
6. G. P. Robertson, R. Herrara and T. Rosswall, Eds., Nitrogen cycling in ecosystems of Latin America and the Caribbean. 1982. ISBN 90-247-2719-7
7. D. Atkinson et al., Eds., Tree root systems and their mycorrhizas. 1983. ISBN 90-247-2821-5
8. M. R. Sarić and B. C. Loughman, Eds., Genetic aspects of plant nutrition. 1983. ISBN 90-247-2822-3
9. J. R. Freney and J. R. Simpson, Eds., Gaseous loss of nitrogen from plant-soil systems. 1983. ISBN 90-247-2820-7
10. United Nations Economic Commission for Europe. Efficient use of fertilizers in agriculture. 1983. ISBN 90-247-2866-5
11. J. Tinsley and J. F. Darbyshire, Eds., Biological processes and soil fertility. 1984. ISBN 90-247-2902-5
12. A. D. L. Akkermans, D. Baker, K. Huss-Danell and J. D. Tjepkema, Eds., *Frankia* symbioses. 1984. ISBN 90-247-2967-X
13. W. S. Silver and E. C. Schröder, Eds., Practical application of *Azolla* for rice production. 1984. ISBN 90-247-3068-6
14. P. G. L. Vlek, Ed., Micronutrients in tropical food crop production. 1985. ISBN 90-247-3085-6
15. T. P. Hignett, Ed., Fertilizer manual. 1985. ISBN 90-247-3122-4
16. D. Vaughan and R. E. Malcolm, Eds., Soil organic matter and biological activity. 1985. ISBN 90-247-3154-2
17. D. Pasternak and A. San Pietro, Eds., Biosalinity in action: Bioproduction with saline water. 1985. ISBN 90-247-3159-3
18. M. Lalonde, C. Camiré and J. O. Dawson, Eds., *Frankia* and actinorhizal plants. 1985. ISBN 90-247-3214-X
19. H. Lambers, J. J. Neeteson and I. Stulen, Eds., Fundamental, ecological and agricultural aspects of nitrogen metabolism in higher plants. 1986. ISBN 90-247-3258-1
20. M. B. Jackson, Ed., New root formation in plants and cuttings. 1986. ISBN 90-247-3260-3
21. F. A. Skinner and P. Uomala, Eds., Nitrogen fixation with non-legumes. 1986. ISBN 90-247-3283-2
22. A. Alexander, Ed., Foliar fertilization. 1986. ISBN 90-247-3288-3
23. H. G. v.d. Meer, J. C. Ryden and G. C. Ennik, Eds., Nitrogen fluxes in intensive grassland systems. 1986. ISBN 90-247-3309-X
24. A. U. Mokwunye and P. L. G. Vlek, Eds., Management of nitrogen and phosphorus fertilizers in sub-Saharan Africa. 1986. ISBN 90-247-3312-X
25. Y. Chen and Y. Avnimelech, Eds., The role of organic matter in modern agriculture. 1986. ISBN 90-247-3360-X
26. S. K. De Datta and W. H. Patrick Jr., Eds., Nitrogen economy of flooded rice soils. 1986. ISBN 90-247-3361-8
27. W. H. Gabelman and B. C. Loughman, Eds., Genetic aspects of plant mineral nutrition. 1987. ISBN 90-247-3494-0
28. A. van Diest, Ed., Plant and Soil: Interfaces and interactions. 1987. ISBN 90-247-3535-1
29. United Nations, Ed., The utilization of secondary and trace elements in agriculture. 1987. ISBN 90-247-3546-7
30. H. G. v.d. Meer, R. J. Unwin, G. C. Ennik and T. A. van Dijk, Eds., Animal manure on grassland and fodder crops: Fertilizer or waste? 1987. ISBN 90-247-3568-8

31. N. J. Barrow, Reactions with variable-charge soils. 1987. ISBN 90-247-3589-0
32. D. P. Beck and L. A. Materon, Eds., Nitrogen fixation by legumes in Mediterranean agriculture. 1988. ISBN 90-247-3624-2
33. R. D. Graham, R. J. Hannam and N. C. Uren, Eds., Manganese in soils and plants. 1988. ISBN 90-247-3758-3
34. J. G. Torrey and J. L. Winship, Eds., Applications of continuous and steady-state methods to root biology. 1989. ISBN 0-7923-0024-6
35. F. A. Skinner, R. M. Boddey and I. Fendrik, Eds., Nitrogen fixation with non-legumes. 1989. ISBN 0–7923–0059–9
36. B. C. Loughman, O. Gašparíková and J. Kolek, Eds., Structural and functional aspects of transport in roots. 1989. ISBN 0–7923–0060–2